The Latin American Studies Book Series

The Latin American Studies Book Series promotes quality scientific research focusing on Latin American countries. The series accepts disciplinary and interdisciplinary titles related to geographical, environmental, cultural, economic, political and urban research dedicated to Latin America. The series publishes comprehensive monographs, edited volumes and textbooks refereed by a region or country expert specialized in Latin American studies.

The series aims to raise the profile of Latin American studies, showcasing important works developed focusing on the region. It is aimed at researchers, students, and everyone interested in Latin American topics.

Submit a proposal: Proposals for the series will be considered by the Series Advisory Board. A book proposal form can be obtained from the Publisher, Juliana Pitanguy (juliana.pitanguy@springer.com).

More information about this series at http://www.springer.com/series/15104

Omar Reyes

The Settlement of the Chonos Archipelago, Western Patagonia, Chile

Springer

Omar Reyes
Centro de Estudios del Hombre Austral
Instituto de la Patagonia Universidad de
Magallanes
Punta Arenas, Chile

ISSN 2366-3421 ISSN 2366-343X (electronic)
The Latin American Studies Book Series
ISBN 978-3-030-54325-9 ISBN 978-3-030-54326-6 (eBook)
https://doi.org/10.1007/978-3-030-54326-6

This Springer imprint is published by the registered company Springer Nature Switzerland AG
The registered company address is: Gewerbestrasse 11, 6330 Cham, Switzerland

Acknowledgements

This book brings together the results of 12 years of archaeological investigation in the Chonos Archipelago, in the Aysén Region of Chile. The work has been supported by funding from a number of competitive research projects and other grants awarded by the National Fund for Scientific and Technological Development (FONDECYT Regular 1050139, 1100643, 1130151 and 1170726), which is dependent on the National Committee for Scientific and Technological Investigation of Chile. The information, both published and unpublished, was systematised in the framework of the Doctorate in Archaeology Programme of the Faculty of Philosophy and Letters of the University of Buenos Aires, Argentina, thanks to its free public education system.

The archaeology of the Patagonian Archipelago has always attracted us, and true friends were always there to listen, advise and support. Thanks to Manuel San Román and Valentina Trejo for encouraging me when I needed it and for their undying advice ever since we were students together. To César Méndez and Amalia Nuevo for their generosity and help during this investigation.

My thanks and congratulations to all the crew members of "HMS" Catiao and Cai Cai, our homes from home at sea. Archaeologists and friends joined to lend support both on voyage and during fieldwork, contributing, in the different field trips over all these years, to the generation of a large part of the information presented here: Manuel San Román, Pablo González, José Díaz, Manny Gómez, Erik Lukoviek and Pipo Godoy, Francisco Cayla, Javier Cárcamo, César Méndez, Ismael Martínez, Pedro Cárdenas and Valentina Trejo.

The paleoenvironmental and geomorphological information is the fruit of the contributions and guidance of Jean Pierre Francois and Juan García. Valentina Trejo, Jacqueline Galimany, Constanza De la Fuente and Susan Kusminsky helped with the bio-anthropological analyses. Mauricio Moraga and Eugenio Aspillaga contributed by reconstructing scattered archaeological information researched decades ago. Jimena Torres, together with Manuel San Román, helped with archaeofaunal information. Flavia Morello and César Méndez told me about lithic technology. Javier Cárcamo provided the malacological material. Mauricio Moraga, as Principal Researcher of Project 1100643, provided me with resources to create registers, prospections and analyses in the Chonos Archipelago.

Some of the drawings and pictures included here were created by Paulina Chávez, Mauricio Álvarez (#Mauricio_alvarez_art), Erik Lukoviek, Manny Gómez, Ismael Martínez and Francisco Cayla. Many of the photographs were taken by our dear friend and professional archaeologist Pablo González, who always supported us whatever the circumstances.

The Huilliche Nahuelquín-Delgado indigenous community, of Traiguén Island and Puerto Aysén, opened their doors and offered us their friendship; their help was of vital importance in the prospection work and excavations carried out in the south of the Chonos Archipelago and the Taitao Peninsula. My gratitude to Don Benjamín Nahuelquín Levipani, Doña Isabel Delgado and their children Esaú, Eduardo, Manuel, José, Mónica and Marcia.

I owe the impulse to carry out this doctoral research to Dr. Guillermo Mengoni Goñalons—I thank him deeply for his guidance and support. Mary Jo Figuerero also deserves special mention for the encouragement she gave me on each of my visits to Buenos Aires. Thanks to Rafael Goñi for his insistence, guidance and motivation; to Guti Tessone for his vital advice on isotope studies and his broad collaboration in the work on diet and subsistence; to Fabiana Martín, Flavia Morello and Luis Borrero for their support during different stages of this work. Thanks to Dr. Florencia Borella, Dr. Francisco Zangrando and Dr. Juan Bautista Belardi, who, as members of the evaluation committee, offered valuable comments and suggestions. Thanks to Jorge Rabassa for believing that this book could be published.

The Chilean Navy supported us on the field trip to the Estero Sur site—thanks to the crew of LSG Caldera (2012). The logistical support of Pituso (Leonardo Zuñiga) in Coyhaique was also fundamental. Francisco Mena and Mauricio Osorio provided information every time a find of bones was reported in the archipelago. I also wish to thank Alexander Cherkinsky for the datings he provided for the Posa Las Conchillas site and Pat Curry for the manuscript information of her voyage to the archipelago.

The Research Centre in Patagonian Ecosystems (CIEP) allowed me to use their installations for safekeeping and bio-anthropological analysis. We are also grateful for access to samples from the Chonos Collection of the Social Sciences Faculty of the University of Chile. The Austral Man Study Centre (CEHA) of the Patagonia Institute, University of Magallanes, gave me institutional backing throughout the years of my research.

Special thanks to the coastal communities and peoples of the Aysén Region, seamen, fishermen, seafood gatherers, marine carpenters, mechanics and others, who took us around the islands and always showed their interest and support for our work.

To my family, to those who have always been there and those who set out and became part of my world.

Finally, to Carolina Belmar, for her unconditional support at the most difficult moments. Thank you.

Introduction

Abstract This chapter introduces the research problem. The general archaeological objectives of this investigation focused on describing the human occupation of the Chonos Archipelago within the regional context; we explored human mobility, by marine hunter-gatherer groups, over time and its relation with the occupation of adjacent areas, and evaluated the relation of the human groups with their environment, in other words with the exploitation of marine resources, the technology used and the areas occupied during the Holocene. To understand the dynamics of this process, we considered the paleoenvironmental and geomorphological factors which affected the archaeological record, evaluating the formation of the existing record and the particular conditions under which we can record the various cultural contexts today.

Keywords: Western Patagonia • Chonos Archipelago • Maritime hunter-gatherers • Archipelagic settlements • Coastal Archaeology • Coastal adaptation • Marine subsistence • Coastal sites

This book brings together the results of more than a decade of archaeological investigations in the Chonos Archipelago. The information presented here includes unpublished data as well as specific subjects addressed in earlier publications, which we develop here at greater length. We hope that a wide public, including university teachers, students and researchers, will be interested in this compendium of the archaeological work carried out in the islands and channels of this Patagonian Archipelago. The specialist reader may observe passages with which he or she is already familiar, expressed in more colloquial terms; however, it was necessary to summarise previous work and provide something of the history of this research for the general public.

The aim of this book is to help readers to appreciate the heritage value of a way of life that no longer exists, that of the marine hunter-gatherer-fishers. It is of particular interest for the residents and communities of the Aysén coastline, the Chiloé Archipelago and the whole coast of southern Chile. Discovering how a territory carved up by glaciation was occupied by canoe groups over a period of

6,000 years is a way of giving identity and value to a region which for many may appear remote, deserted and hostile.

The reader must also bear in mind that the work involved over all these years may sometimes be disproportionate to the results obtained. Archaeological fieldwork in the channels of Patagonia is often unproductive. Results can be built up only slowly due to bad weather, poor visibility, the logistics of boat travel and the problems of administering the resources available. Nevertheless, we hope that this book will be adequate recompense.

Framework Project and Origin of the Problem

The Pacific coast of the southernmost part of South America is broken up into a complex network of channels, fjords and hundreds of islands known as the Austral Archipelagos, Western Channels or western Patagonia (e.g. Bird 1938, 1946, Emperaire 1963, Steffen 1944, Fig. 1). Almost the entirety of this area, extending over more than 1,600 km from north to south, lies in Chilean territory, between Reloncaví Sound and Cape Horn (41°30'–55°60'S). It comprises a total shoreline of between 19,000 km (Bird 1988) and 50,000 km (Camus 2001), and an area of approximately 240,000 km^2. Approximately 240 km of the northern shore of the Beagle Channel (south coast of Tierra del Fuego) lies in Argentina. In this vast area of islands and channels, the first clearly visible human adaptations to the marine ecosystem date to around 6,400 years BP (or 7,200 years cal. BP) [1], according to the data obtained in the area of the Beagle Channel, just short of the southern tip of the continent (Piana and Orquera 2007, Piana et al. 2012, Orquera and Piana 2009, Orquera et al. 2011).

In this context, various issues of continental and regional importance are discussed (e.g. Bailey 2004, Dillehay et al. 2008, Erlandson 2001); one of these is the geographical origin and antiquity of the adaptation of hunter-gatherers to the marine environment—when and where human groups achieved complete and almost exclusive dependence on the sea coast (e.g. specific technologies for boat travel, specialisation in hunting marine fauna). The alternative origins proposed for this adaptation to coastal life lie at the two extremes of the Patagonian Archipelago: on the one hand in the south-west, around the Magellan Strait, the Otway Sound and the area of the Beagle Channel south of Tierra del Fuego; and on the other in the north, in Chiloé Island and Reloncaví Sound (Legoupil 2003, 2005, Legoupil and Fontugne 1997, Orquera and Piana 2005, 2006, Prieto et al. 2013, Rivas et al. 1999). Inherent to the discussion are questions of how fast this adaptive process spread (Bird 1938, Orquera et al. 2011) and the continuity or discontinuity of cultural traditions in such an extensive territory (Legoupil 1997, 2003, Morello et al. 2002, 2015, Ocampo and Rivas 2004, Orquera et al. 2011, Piana and Orquera 2009, San Román et al. 2016a). These regional issues, far from being exhausted, are constantly reviewed in the light of new finds and interpretations of the technological

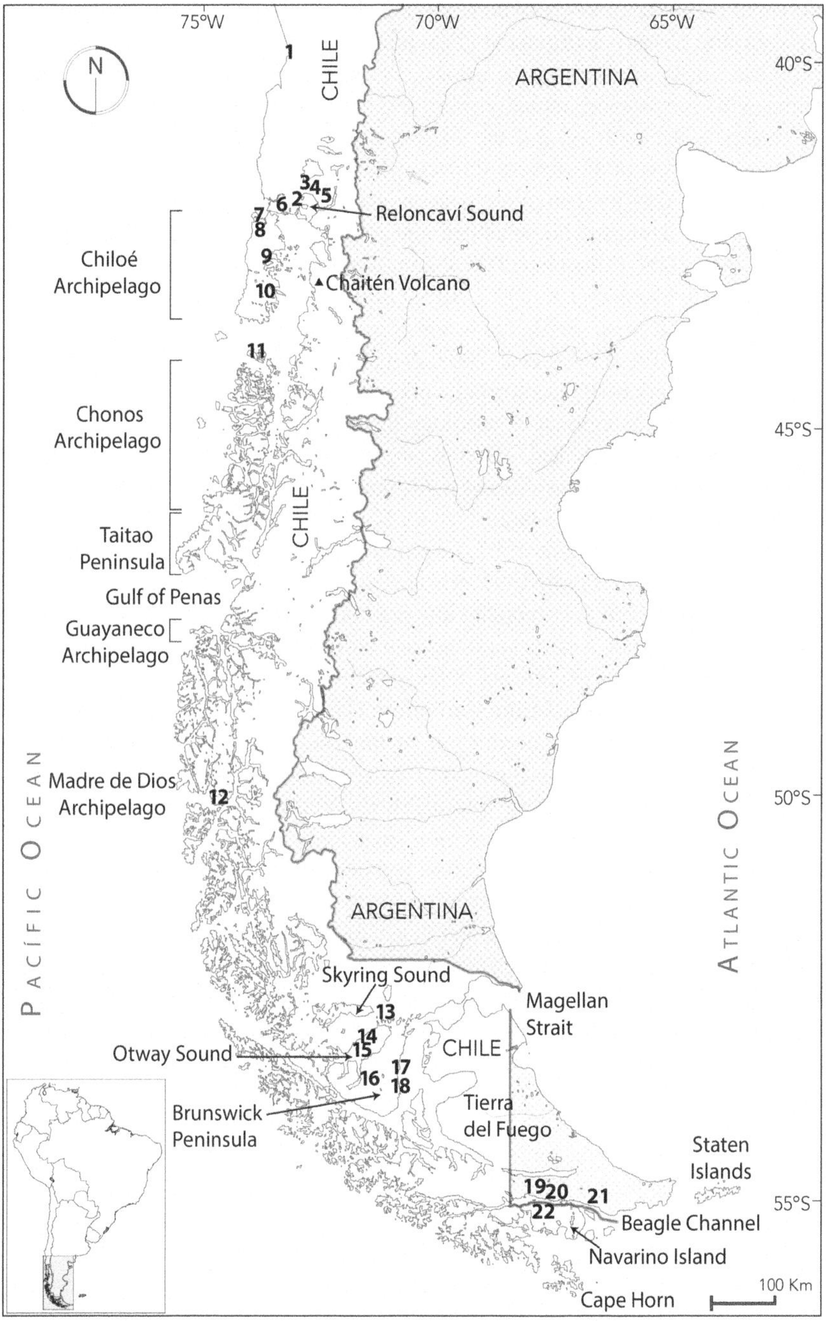

Fig. 1 Western Patagonia. Principal geographical features and archaeological sites mentioned in the text. 1—Chan-Chan 18; 2—Puntilla Tenglo; 3—Piedra Azul; 4—Bahía Ilque 1; 5—Centro Acuicultura Metri; 6—Monte Verde; 7—Puente Quilo; 8—Chepu; 9—Conchal Gamboa; 10—Yaldad 2; 11—GUA-010 Terraza; 12—Cueva Ayayema; 13—Ponsonby; 14—Englefield and Pizzulic Sites; 15—Bahía Colorada; 16—Punta Baja; 17—Bahía Buena and KM 44; 18—Punta Santa Ana; 19—Lancha Packewaia; 20—Tunel 1; 21—Imiwaia; 22—Áridos Guérrico

and faunal components of these contexts and their capacity to gather resources and survive in the coastal environment (Zangrando 2018).

Beyond these preliminary questions, we explore other, more specific areas dealing with subsistence, mobility, the local paleoenvironment, the circulation of raw materials, technology and the biological characteristics of the populations. These are just a few of the diverse subjects which enable us to reconstruct fragments of information that explain both general and particular aspects of the marine hunter-gatherer groups who inhabited the Austral Archipelagos over time (see Álvarez 2004, 2006, Barberena 2008, García-Bour et al. 2004, Legoupil et al. 2011, Martinoli 2015, Morello et al. 2002, 2015, Orquera 2005, Orquera et al. 2006, Piana et al. 2006, San Román 2010, 2014, San Román et al. 2009, Tessone et al. 2003, Torres and Ruz 2011, Yesner et al. 2003, Zangrando 2009a, b, Zangrando and Tivoli 2015). The intensity of archaeological work in the southern tip of western Patagonia has developed into a regional archaeology that supports different research lines.

At the other geographical extreme, in the northern part of the Patagonian Archipelago, regional archaeological fields have been investigated for decades, but with discontinuous work in a large territory (Dillehay et al. 2008, 2015, Gaete et al. 2004, Legoupil and Fontugne 1997, Legoupil 2005, Massone et al. 2016, Ocampo and Rivas 2004, 2008, Porter 1993, Rivas et al. 1999, Sierralta et al. 2019). In this approach, site data were incorporated into previously existing questions, such as the date of the peopling of America and the date and location of the earliest adaptation to littoral and marine environments. Thus inductive reconstruction of the process to be explained was based on fragmented information obtained from a few sites with initial sequences of human occupation in the Mid-Holocene in Chiloé and the Reloncaví Sound (e.g. Puente Quilo and Piedra Azul), and on the equally important presence of the Late Pleistocene site of Monte Verde, less than 50 km from the coast of Reloncaví. The results were not "anchored" in any regional archaeology, and tended to provide chronological or techno-typological data in line with the discussion of the peopling and dispersal of canoe groups in western Patagonia (Ocampo and Rivas 2004, Orquera and Piana 2006, 2009, Piana and Orquera 2007, Rivas and Ocampo 2006). Thus general explanations were proposed for the means of subsistence of the hunter-gatherer groups, who inhabited this portion of the archipelago (e.g. ecotone nuclei and diachronic existence of different cultural traditions), to account for the use and occupation of island spaces (Aspillaga et al. 2006, Massone et al. 2016, Ocampo and Rivas 2008).

The search for material evidence of human occupations in the Patagonian Archipelago should not become separated from the central issues with which it is associated. Although it is a well-founded starting point, from the beginning of the investigation we must consider a prior material characterisation that will allow us to lay the foundations of minimum knowledge on which to construct and contribute to regional issues. An exploratory study is, therefore, needed of very large areas where there is little archaeological documentation, in which we can first assess what was happening in this space by measuring the intensity of certain variables (occupation, chronology, contexts of finds, types of sites, technology, etc.), according to the different analytical scales used previously, and how these perform in a specific territory. Perhaps the most basic work is to create a database of sites and finds. This is essential for answering how and when a territory was occupied, by whom, under what conditions, and whether there were changes in the occupation of this territory over time.

Our study area was the Chonos Archipelago (43°–47°S), an extensive island system (sensu Fitzpatrick et al. 2016, Keegan et al. 2008) with features that make it suitable for an archaeological investigation. The archipelago is geographically isolated. Geographical barriers exist (Borrero 1999, 2001a, 2018, Borrero and Franco 1997, Morello et al. 2012), creating a continuous barrier (Barberena 2008:87) which would have affected the "hierarchisation" of human use of the spaces occupied (Belardi 2003, Borrero 1989–90, 1994–95, 2004). The distances of between 20 and 40 km by sea to reach the mainland coast, the Andes Mountains, dense montane forests, volcanoes and ice fields which bar access to the rest of the continent, all suggest the existence of practically impermeable barriers to overland contact with other cultural groups living beyond the Andes or on the steppes (such as did occur in the northern and southern extremes of the Patagonian Archipelago (Borrero 1997, Gaete et al. 2004, Morello et al. 2012). Different archaeological projects carried out in the Aysén Region over more than a decade, explicitly investigating the possibility of contacts between the valleys of the Andean steppes and the archipelagos off the coast, have been unable to produce archaeological documentation of any such interaction (Méndez and Reyes 2008, 2015 Méndez et al. 2012, Reyes et al. 2006a, b, 2007, 2009, 2015). We believe, therefore, that the Chonos Archipelago is a singularly suitable space for this type of investigation. One advantage of the isolated and marginal nature of human occupation in this area is that it allows us to measure its pulses, which can be understood as moments of presence alternating with periods of abandonment. This, combined with the high sensitivity of the archaeological signal, offers an ideal scenario for evaluating the possible conditioners of the occupation and abandonment of the archipelago. This marginal condition (in terms of the frequency/density of the archaeological record) is apparently not uniformly expressed, but—as we propose—must have been discontinuous and at no time intense. In this process, understanding the landscape as a context and frame of reference (sensu Binford 2001) not only refers to the environmental and geographical reconstruction associated with different moments, but also incorporates elements relating to decision-making and the modes of material production of the human population.

In the investigations summarised in this book, our archaeological work was guided by the following hypotheses: (i) The human occupation of the Chonos Archipelago occurred from the Mid-Holocene onward, which was when the current environmental conditions of this space were established. The frequency of the archaeological evidence complies with two essential characteristics: it is limited, and would extend from ca. 6200 cal. BP onwards (Porter 1993, Reyes et al. 2009, 2015). (ii) In view of the bio-geographical characteristics of the area, which restrict access, the human occupation of this archipelago occurred independently of events in the adjacent extra-Andean and montane zones. (iii) The settlers were specialist hunter-gatherer groups; in other words, the whole occupation sequence was carried out exclusively by marine hunter-gatherer-fisher groups with a subsistence economy based on obtaining resources from the littoral and marine environments (Orquera and Piana 1999b, 2009, Perlman 1980, Yesner 1980). This was made possible by a mobility strategy based on sea-travel technologies (Ames 2002), which gave them access to both the islands of the archipelago and the mainland coast. In parenthesis, it may be noted that the technological specialisation of sea travel, inferred geographically from the exclusive capacity of access to, from and between island territories, is far from being a truism. It is based on a similar line of argument to that projected, for example, for the archaeological occupations detected in Navarino Island, south of the Beagle Channel (Orquera and Piana 2009, Piana and Orquera 2009), Staten Island in the South Atlantic (Chapman 1987, Horwitz and Weissel 2011) or the Cape Horn Archipelago (Legoupil 1994, Legoupil and Fontugne 1997). To return to our guiding hypotheses: (iv) During the Late Holocene there was a gradual increase in the population of the Chonos Archipelago. At macro-regional and meso-regional level (sensu Dincauze 2000), the human occupation of this territory from the Mid-Holocene by marine hunter-gatherer groups would be neither continuous nor intense (Borrero 2004). It is only in the Late Holocene that a stronger archaeological signal is found, with a higher intensity of occupation associated with a population increase in the area. Finally, understanding that the distribution of archaeological sites obeys both geomorphological (e.g. Bailey and Flemming 2008, Borrero 2014, Bird 1988, Clark et al. 2014) and cultural factors, related with the type of settlement (Reyes et al. 2018) and the use given to certain areas, we hypothesise that: (v) Abrupt, permanent geomorphological changes in the region had no impact on the pattern of coastal settlement by marine hunter-gatherers.

Consequently, the general archaeological objectives of our investigation focused on describing the human occupation of the Chonos Archipelago within the regional context; we explored human mobility over time and its relation with the occupation of adjacent areas, and evaluated the relation of the human groups with their environment, in other words with the exploitation of marine resources, the technology used and the areas occupied during the Holocene. To understand the dynamics of this process, we considered the paleoenvironmental and geomorphological factors which affected the archaeological record, evaluating the formation of the existing record and the particular conditions under which we can record the various cultural contexts today. This allowed us to devote our efforts to understanding the archaeological record in terms of spatial distribution and chronology. Population flows were assessed by analysis of the material remains, together with characteristics of subsistence and the technological assemblages ascribable to specialist groups adapted to the marine and insular environment.

The Peopling of a Region by Specialist Hunter-Gatherers, the Canoe Peoples

Starting from the situation outlined above, in our characterisation of the occupation of the Chonos Archipelago, we considered the minimum variables and definitions which would enable us to assess the peopling of an island system by marine hunter-gatherer groups. There is a rich bibliography on these subjects, and we will only highlight a few aspects. Specialist adaptation to the coastal environment refers, in general terms, to exclusive or majority dependence on the littoral and marine resources of an ecosystem for the subsistence and development of a human group (e.g. Borrero 2001b, Erlandson 2001, Perlman 1980, Yesner 1980). This specialisation in littoral resources refers not only to the diet ingested—predominantly or exclusively marine or pericoastal (e.g. Llagostera 1979, Osborn 1977, Yesner 1980)—but also to the development of specific technologies for obtaining these efficiently (Arnold 1995, Fagan 2004, Llagostera 1982, Orquera and Piana 2005). For example, this would include the use of canoes and fishing nets (inferred from the presence of weights), the creation of instruments for hunting and fishing marine fauna, the intensive procurement of exclusively littoral raw materials (Orquera and Piana 2005), and the specific spatial organisation and ranges of action of the human groups occupying the area (sensu Ames 2002, Borrero and Barberena 2006, Erlandson 2001, Llagostera 1989, Richardson 1981). A strategy for obtaining resources which meets these conditions differs clearly from those of other groups who may have access to the shoreline but complement their diet with resources obtained inland (e.g. Barberena 2002, Gómez-Otero 2007, Gómez-Otero et al. 2000, Gómez-Otero and Novellino 2011, Orquera and Gómez-Otero 2007, Panarello et al. 2006, Yesner 1980, 2003). As Orquera and colleagues say, this would differentiate *real littoral adaptation from simple opportunist exploitation* (Orquera et al. 2006, 268). We know how complex it is to establish the importance, frequency and complementary selection of dissimilar resources obtained from continuous coastal-inland environments in human groups of differing cultural traditions with variable ranges of action, as may occur in the northern and southern extremes of the Western Channels or on the Atlantic coast of Patagonia (e.g. Barberena 2008, Borrero and Barberena 2006, Dillehay et al. 2008, Gómez-Otero 2007, Moreno et al. 2011, Zilio et al. 2014).

We understand the peopling of a region as a concept combining different moments, including the initial occupation by human groups of an uninhabited area. These moments may be understood as phases characterised by different interactions between operative parameters such as settlement, mobility, subsistence and technology. These enable us to discern the human occupation of a particular space, based on the logic of keeping social networks active, and maintaining a "hierarchisation" of spaces in the landscape (sensu Belardi 2003) in a continuous peopling process (Borrero 1989–90, 1999, 2004, Borrero and Franco 1997).

To understand the phases of human settlement which may guide us on the occupation of the archipelago, we can use the model developed by Borrero (1989–

90, 1994–95, 2001a) to explain the use and occupation of space. The model considers theoretical moments that characterise the influx of human groups into a territory according to their knowledge of that territory, and we believe that it is applicable to the Chonos Archipelago. Very briefly, the *exploration* phase of an area involves initial displacement into unoccupied spaces following natural routes and not necessarily making use of the best localities. In this phase, we can expect to find discontinuous occupation at the regional and local scale; there is a greater possibility that the use of the localities will alternate with non-human agents, with the resulting impacts on site integrity. Thus archaeological sites from this phase are more likely to be destroyed or hidden, and the chances of finding them are more limited. The first occupations of some sites, more restricted and of lower intensity than shallower or later occupations, are included in this phase. The occupants probably had very wide ranges of action, consistent with ignorance of their surroundings. Consequently, the technology expected in these populations incorporates the systematic use of local resources to facilitate the group's high mobility (Borrero and Franco 1997, Franco 2002). The next phase, *colonisation*, implies the initial consolidation of human groups in certain sectors of the space, with more specific and restricted ranges of action. It implies more repetitive use of sites in preferred locations, and a more discrete record now that ranges of action no longer overlap. In these conditions, archaeological visibility should increase. Effective occupation of the space represents the moment when the whole of the desirable space is being used. This may occur under two different strategies: *stable occupation of the space* and *saturation of the space*. The first implies a population distribution that does not fluctuate greatly over the generations, while the second projects populations that are on the border of sustainability (e.g. competition for resources, cultural drift).

To characterise the occupation of the space of the archipelago, we must also understand that the displacement of human populations can be understood from flows of materials, especially raw materials used for producing artefacts by knapping (e.g. Gamble 1999), and dietary preferences identified by isotope analysis (e.g. Barberena 2002, 2008, Borrero and Barberena 2006, Zangrando et al. 2004). Movements do not occur as linear displacements in a flat space, but along routes constructed under social and technological variables (Gamble 1996, 1999, Pallo 2011, Pallo and Borrero 2015), as well as bio-geographical variables which permit or limit access to other regions (Borrero 2018). This is the approach that we have used in central Patagonia, in the Aysén region of Chile, in the extra-Andean valleys and in the continental area adjacent to the Chonos Archipelago to analyse flows of materials (distribution of obsidian) and relate them to the bio-geographical conditioners observed (Méndez et al. 2008/9, 2012, Méndez and Reyes 2008, 2015, Reyes et al. 2009). Information on the provenance of lithic resources found in the coastal sites of the archipelago (Méndez et al. 2018, Reyes 2017, Reyes et al. 2006a, b, 2007, Stern 2018, Stern and Porter 1991, Stern and Curry 1995, Stern et al. 2002, 2012) and the inland valleys (Méndez et al. 2012, 2018, Reyes et al. 2009) enabled us to develop this line of investigation, evaluating terrestrial hunter-gatherers' knowledge of the structure of resources in the region and the exchange and circulation networks at specific moments of the Holocene sequence.

The mobility of human groups is considered to be one of the distinctive characteristics that define hunter-gatherers, and one of the determinants of their way of life (e.g. Lee and De Vore 1968). In archaeology, the study of the mobility of hunter-gatherers has been addressed through investigation of their settlement systems and the types of archaeological sites formed; the mobility of hunter-gatherer groups is also related with the concrete distribution of resources in the habitat of a human group (e.g. Binford 1980, 1982, Kelly 1983, 1995, Mandryck 1993), and the spaces used will depend on this pattern of distribution and also on the mobility of the resources in the area and the area's bio-geographical characteristics (Borrero 2018, Butzer 1971).

Based on the study of terrestrial hunter-gatherer groups, and using an ethnoarchaeological approach, Binford (1980, 1982) defined general models of settlement and mobility associated with unequal distribution of resources. Thus *collector* and *forager* societies were defined in terms of their strategies for obtaining resources, their mobility and the expected material record. Some societies use one adaptation strategy or the other, but more complex systems also exist in which both options are used, so a broad variety can be expected in the patterns of the archaeological record (Binford 1980). The difference between the two strategies lies in the organisational relationship between individual displacement and group movements (Kelly 1983, 1992), which also involves social and economic variables like subsistence, mobility, technology, territoriality, etc. (Kelly 1995). This approach is enriched by arguing the importance of transport technology, in this case the use of canoes, in the evolution of mobility strategies around residential bases (Binford 1990). The spatial particularities of the organisation of any mobile group are adaptively related to the spatial and temporal distribution and the density of the resources available in the environment (e.g. Binford 1980, 1981, Dincauze 2000, Foley 1981a, Kelly 1995).

Although the environmental structure of the study area presents moderate temperatures and high productivity, propitious for *forager* strategies, the distribution of resources is not the same in all cases, leading to the development of logistical strategies. Thus archaeological sites formed by canoe groups with *forager*-type mobility strategies would present very low archaeological visibility. The existence of boats suggests an efficient technological solution, with high mobility, which reduces the problems of transporting catches, equipment and artefacts; it also allows various activities (e.g. hunting, fishing and even cooking) on the water, leaving no material trace. The archaeological signal is therefore very different from that of hunter groups who come down to the coast from inland (Ames 2002, Bjerck 2017, Emperaire 1963, Fagan 2004, Orquera and Piana 1999b). In other words, archaeological remains can be expected to be organised around evidence of residential camps where general activities were carried out, and which also contain evidence indicating logistical voyages to carry out specific tasks (Binford 1980, 1981). At the same time, the dynamic imposed by the high mobility of these groups suggests that, in certain cases, the sites where base camps were set up and the locations of specific activities could overlap in different occupation events, which would mean that there is not necessarily exact superposition of later occupations on the original use (Binford 1982). Furthermore, the variability of the archaeological evidence also responds to the formative history of the material record (Gifford 1981, Roksandic et al. 2014, Schiffer 1996, Stein 1992, Stein et al. 2003, Wandsnider 1992). We believe, therefore, that the characteristics of the mobility of hunter-gatherer groups can be inferred from the archaeological record, considering the general ecological characteristics and the different phases of the peopling of a region, as well as the general characteristics of the artefactual and ecofactual assemblages and the features present.

We further believe that the spatial and temporal distribution of archaeological sites in the Chonos Archipelago were influenced by both environmental and geomorphological factors (e.g. Bailey and Flemming 2008, Belardi 2003, Bintliff et al. 1988, Butzer 2008, Erlandson and Moss 1999, Sandweiss et al. 2004, Scheinsohn 2001, Waters and Kuehn 1996) in a "taphonomically active" zone (sensu Borrero 2014) and by cultural factors related with the type of site location and the use given to certain areas (Bailey 2004, Erlandson 2001). In terms of intensity of use, these areas have been classified as "marginal" (Borrero 2004), "empty" (Laming-Emperaire 1972), "circulation" (Borrero 1982). A final aspect would be the availability of resources and/or the feasibility of access to certain locations and of permanent or seasonal occupation (Ames 2002, Bailey and Parkington 1988, Pallo 2011). We maintain, therefore, that the archaeological record does not take the form of habitat nuclei or concentrations limited to benign "ecotone" zones, or with better resource productivity (Massone et al. 2016, Ocampo and Rivas 2008), but is dictated by different, archaeologically explainable variables ranging from the quantity and types of archaeological contexts identified in a distributional analysis of sites and the intensity of areas prospected (e.g. Belardi et al. 1998, Borrero and Lanata 1992, Ebert 1992, Dincauze 2000) to the differential preservation of areas and the treatment of absence of evidence as a datum (sensu Blintliff 2000, Foley 1981b). These are elements of judgement which contribute to the generation of a more precise archaeological landscape (Borrero 2013).

Finally, to explain the problems associated with the peopling of the Chonos Archipelago, i.e. timings, identification of data (sites) and contexts of finds, we must incorporate basic information including the reconstruction of past environments and the modelling of contemporary landscapes. We, therefore, believe that paleogeography and the paleoenvironment offer a way of calculating the possibilities of finding evidence through directed systematic search (Bailey and Flemming 2008, Gusick and Faught 2011, Punke and Davis 2006, Reeder-Myers 2015, Rick et al. 2006, 2013, Zangrando et al. 2016). This concept situates human settlements in their context, providing better arguments for understanding man in island environments (Fitzpatrick et al. 2016, Keegan et al. 2008). This strategy solves two central problems. Firstly, learning about past and present scenarios will help us to look for and find the best sites to control and examine our expectations; the current diversity of the landscape constitutes a frame of reference (sensu Binford 2001) for establishing general and qualitative environmental differences for contrasting against the archaeological evidence. And secondly, it allows us to reconstruct the conditions under which the human occupation of a region took place, providing references through which we can understand groups' decisions on space selection, site functions, settlement patterns and the relations between human groups and their environment. To construct the scheme, in the first place we must use information from areas where records of human action have been recovered, understanding the conditions of the process, the circumstances of the find and the preservation of the remains. In a second step we must generate expectations of our searches and investigate in areas where we can reasonably expect to identify data. This procedure leads not only to finds, but to finds contextualised in such a way that the interaction of human beings with the medium can be discussed.

Note

[1] The dates cited in the text will be presented in "years BP" (before the present), unless it is specified that they are calibrated, in which case they will be cited in "years cal. BP".

Bibliography

Álvarez M (2004) ¿A qué responde la diversidad instrumental? Algunas reflexiones a partir del análisis funcional de materiales líticos de la costa norte del Canal Beagle. En Contra viento y marea. Arqueología de Patagonia, In: M. Civalero, P. Fernández y A. Guráieb (eds):29–43. Sociedad Argentina de Antropología. Buenos Aires

Álvarez (2006) Organización tecnológica en la costa norte del Canal Beagle: Estrategias de uso de materiales líticos de los cazadores litorales del extremo sur de Patagonia. Relaciones (2005):33–57

Ames K (2002) Going by boat. In: Beyond Foraging and Collecting. B. Fitzhugh and J. Habu (eds) Springer, pp 19–52

Arnold J (1995) Transportation Innovation and Social Complexity among Maritime Hunter-Gatherer Societies. Am Anthropol 97(4):733–747

Aspillaga E, Castro M, Rodríguez M, Ocampo C (2006) Paleopatología y estilo de vida: El ejemplo de los Chonos. Magallania (Punta Arenas), 34(1):77–85<ParaRef">Bailey G (2004) World prehistory from the margins: The role of coastlines in human evolution. J Interdisc Stud Hist Archaeol 1:39–50

Bailey G, Flemming N (2008) Archaeology of the continental shelf: marine resources, submerged landscapes and underwater archaeology. Quat Sci Rev 27(23):2153–2165

Bailey G, Parkington J (1988) The archaeology of prehistoric coastlines. Cambridge University Press

Barberena R (2002) Los límites del mar: Isótopos estables en Patagonia meridional. Sociedad Argentina de Antropología. Buenos Aires

Barberena R (2008). Arqueología y biogeografía humana en Patagonia Meridional. Sociedad Argentina de Antropología. Buenos Aires

Belardi J (2003) Paisajes Arqueológicos: Un estudio comparativo de diferentes ambientes patagónicos. Tesis doctoral inédita Facultad de Filosofía y Letras, Universidad Nacional de Buenos Aires. Buenos Aires

Belardi J, García M, Campán P (1998) Las distribuciones artefactuales y las poblaciones humanas. En Arqueología de la Patagonia Meridional, In: Borrero L (eds) Búsqueda de Ayllu, Buenos Aires, pp 53–71

Binford L (1980) Willow smoke and dogs' tails: hunter-gatherer settlement systems and archaeological site formation. Am Antiq 45(1):4–20

Binford L (1981) Bones: Ancient Men and Modern Myths. Academic Press, New York

Binford L (1982) The Archaeology of Place. J Anthropol Archaeol 1: 5–31

Binford L (1990) Mobility, Housing, and Environment: a Comparative Study. J Anthropol Res 46 (2):119–152

Binford L (2001) Constructing Frames of Reference. An Analytical Method for Archaeological Theory Building Using Ethnographic and Environmental Data Sets. University of California Press, California.

Bintliff J (2000) The concepts of'site'and'offsite'archaeology in surface artefact survey. Non-destructive techniques applied to landscape archaeology 4

Bintliff J, Davidson D, Grant E (1988) Conceptual issues in Environmental Archaeology. University Press, Edinburgh

Bird, J. 1938. The antiquity and migration on the early inhabitants of Patagonia. The Geographical Review 28:250-275.

Bird J (1988) Travels and Archaeology in South Chile (J. Hyslop, ed.).University of Iowa Press, Iowa City

Bjerck H (2017) Settlements and seafaring: Reflections on the integration of boats and settlements among marine foragers in Early Mesolithic Norway and the Yámana of Tierra del Fuego. J Island Coast Archaeol 12(2):276–299

Borrero L (1982) Un modelo de ocupación humana de la región del Seno de la Ultima Esperanza (Magallanes, Chile). Publicaciones del Instituto de Antropología (Universidad Nacional de Córdoba) 38–39:155–171.

Borrero L (1989–90) Evolución cultural divergente en la Patagonia austral. Anales del Instituto de la Patagonia Ser Cs Hs 19:133–139

Borrero L (1994–95) Arqueología de la Patagonia. Palimpsesto. Revista de Arqueología 4: 6–69, Buenos Aires

Borrero L (1997) The origins of ethnographic subsistence patterns in Fuego-Patagonia. In Patagonia. Natural History, Prehistory and Ethnography at the Uttermost End of the Earth, C McEwan, LA Borrero y A Prieto (eds) pp. 60–81. British Musem Press, London

Borrero L (1999) The prehistoric exploration and colonization of Fuego-Patagonia. J World Prehist 13(3):321–355

Borrero L (2001a) El poblamiento de la Patagonia. Toldos, milodones y volcanes. Emecé editires, Buenos Aires

Borrero L (2001b) Regional taphonomy: the scales of application to the archaeological record. Animals and Man in the Past, 17–20. H. Buitenhuis, W. Prummel (Eds) ARC-Publicatie 41. Groningen, Netherlands

Borrero L (2004) The Archaeozoology of Andean "Dead Ends" in Patagonia: Living near the Continental Ice Cap. In: En Colonisation, Migration, and Marginal Areas. A Zooarchaeological Approach, M Mondini, S Muñoz, S. Wickler (eds) Oxbow Books, Oxford, pp 55–61

Borrero L (2013) Paisajes desconocidos, geografía cultural y tafonomía total. Anuario de Arqueología, Rosario, 5:17–30

Borrero L (2014) Multi-service taphonomy. Shells, garbage, and floating palimpsests. In: Taphonomic Approaches to the Archaeological Record. Intersecciones en Antropología. Special Issue 1:13–20. Argentina

Borrero L (2018) Barreras en la construcción de la arqueología patagónica y más allá… Boletín de la Sociedad Chilena de Arqueología. 48

Borrero L, Barberena R (2006) Hunter-gatherer home range and marine resources. Curr Anthropol 47:855–867

Borrero L, Franco N (1997) Early patagonian hunter-gatherer: subsistence and technology. J Anthropol Res 57:219–239

Borrero L, Lanata J (1992) Análisis Espacial en la Arqueología Patagónica. (compiladores) Ediciones Ayllu, Buenos Aires

Butzer K (1971) Environment and archaeology. 2a edición, Aldine-Atherton, Chicago

Butzer K (2008) Challenges for a cross-disciplinary geoarchaeology: the intersection between environmental history and geomorphology. Geomorphol 101(1):402–411

Camus P (2001) Biogeografía marina de Chile Continental. Revista Chilena de Historia Natural. 74:587–617

Chapman A (1987) La Isla de los Estados en la prehistoria. Primeros datos arqueológicos. EUDEBA, Buenos Aires

Clark J, Mitrovica J, Alder J (2014) Coastal paleogeography of the California-Oregon-Washington and Bering Sea continental shelves during the latest Pleistocene and Holocene: implications for the archaeological record. J Archaeol Sci 52:12–23

Dillehay T, Ramírez C, Pino M, Collins M, Rossen J, Pino-Navarro J (2008) Monte Verde: Seaweed, Food, Medicine, and the Peopling of South America. Sci 320:784–786

Dillehay T, Ocampo C, Saavedra J, Sawakuchi A, Vega R, Pino M (2015) New Archaeological Evidence for an Early Human Presence at Monte Verde, Chile. PLoS ONE 10(11): e0141923. doi:10.1371/journal.pone.0141923

Dincauze D (2000) Environmental archaeology. Principles and practice. Cambridge University Press

Ebert J (1992) Distributional archaeology. University of Utah Press

Emperaire J (1963) Los nómades del mar. Ediciones de la Universidad de Chile. Santiago

Erlandson J (2001) The archaeology of aquatic adaptations: Paradigms for a new millennium. J Archaeol Res 9(4):287–350

Erlandson J, Moss M (1999) The systematic use of radiocarbon dating archaeological survey in coastal and other erosional environments. Am Antiq 64(3):431–443

Fagan B (2004) The house of the sea: An essay on the antiquity of planked canoes in southern California. Am Antiq 7–16

Fitzpatrick S, Thompson V, Poteate A, Napolitano M, Erlandson J (2016) Marginalization of the Margins: The Importance of Smaller Islands in Human Prehistory. J Island Coast Archaeol 11 (2):155–170

Foley R (1981a) A model of regional archaeological structure. In: Proceedings of the Prehistoric Society. vol 47, Cambridge University Press, pp 1–17

Foley R (1981b) Off-site archaeology: an alternative approach for the short-sited. Pattern of the past: studies in honour of David Clarke, 157–183

Franco N (2002) ¿Es posible diferenciar los conjuntos líticos atribuidos a la exploración de un espacio de los correspondientes a otras etapas del poblamiento? Werken 3:119–132

Gaete N, Navarro X, Constantinescu F, Mera R, Selles D, Solari M, Vargas L, Oliva D, Duran L (2004) Una mirada al modo de vida canoero del mar interior desde Piedra Azul. Chungara. Tomo I:333–346

Gamble C (1996) Making tracks. Hominid networks and the evolution of the social landscape. En: The archaeology of human ancestry. Power, sex and tradition. Steele J, Shennan S (eds) Routledge, London y New York, pp 253–277

Gamble C (1999) The Palaeolithic societies of Europe. Cambridge University Press, Cambridge

García-Bour J, Pérez-Pérez A, Alvarez S, Fernández S, López A, Arroyo E, Turbón D (2004) Early population differentiation in extinct aborigines from Tierra del Fuego-Patagonia: ancient mtDNA sequences and Y-Chromosome STR characterization. Am J Phys Anthropol 123:361–370

Gifford D (1981) Taphonomy and Paleocology: a critical review of Archaelogy's Sister Disciplines. En Advances in Archaeological Method and Theory (ed. por M. Schiffer) 4, Academic Press, New York, pp 365–438

Gómez-Otero J (2007) Isótopos estables, dieta y uso del espacio en la costa atlántica centro-septentrional y el valle inferior del río Chubut (Patagonia, Argentina). In Arqueología de Fuego-Patagonia. Levantando piedras, desenterrando huesos... y develando arcanos (pp 151–161). CEQUA. Punta Arenas, Chile

Gómez-Otero J, Belardi J, Tykot R, Grammer S (2000) Dieta y poblaciones humanas en la costa norte del Chubut (Patagonia Argentina). Desde el País de los Gigantes. Perspectivas arqueológicas en Patagonia, Vol.1 Universidad Nacional de la Patagonia Austral, Río Gallegos, Argentina

Gómez-Otero J, Novellino P (2011) Diet, nutritional status and oral health in hunter-gatherers from the central-northern coast of Patagonia and the Chubut river lower valley, Argentina. Int J Osteoarchaeol 21(6):643–659

Gusick A, Faught M (2011) Prehistoric archaeology underwater: Anascent subdiscipline critical to understanding early coastal occupations and migration routes. In: Bicho N, Haws J, Davis L (eds) Trekking the Shore: Changing Coastlines and the Antiquity of Coastal Settlement, Springer, New York, 27–50

Horwitz V, Weissel M (2011) Arqueología de Isla de los Estados: la frontera de la abundancia. Los cazadores-recolectores del extremo oriental fueguino. Arqueología de Península Mitre e Isla de los Estados. Sociedad Argentina de Antropología, Buenos Aires

Keegan W, Fitzpatrick S, Sealey K, Le Febvre M, Sinelli P (2008) The role of small islands in marine subsistence strategies: Case studies from the Caribbean. Hum Ecol 36(5):635–654

Kelly R (1983) Hunter gatherer mobility strategies. J Anthropol Res 39:227–306

Kelly R (1992) Mobility/sedentism: concepts, archaeological measures and effects. Ann Rev Anthropol 21:43–66

Kelly R (1995) The foraging Spectrum: Diversity in Hunter - Gatherer Lifeways., Smithsonian Institution Press, Washington/Londres

Laming-Emperaire A, Lavallée D, D`humbert R (1972) Le site de Marazzi en Terre de Feu. Objects et Mondes XII (2):225–244

Lee R, De Vore I (1968) Problems in the study of hunters and gatherers. In: Lee e R, DeVore I (eds) Man the hunter, Aldine, Chicago, pp 3–12

Legoupil D (1994) El archipiélago del Cabo de Hornos y la costa sur de la isla Navarino: poblamiento y modelos económicos. Anales del Instituto de la Patagonia, Serie Ciencias Humanas 22:101–121

Legoupil D (1997) Bahía Colorada (île d'Englefield): Les Premiers Chasseurs de Mammifères Marines de Patagonie Australe. Éditions Recherche sur les Civilisations, Paris

Legoupil D (2003) Cazadores-recolectores de Ponsonby (Patagonia austral) y su paleoambiente desde VI al III milenio AC Magallania 31, tirada especial

Legoupil D (2005) Recolectores de mariscos tempranos en el sureste de la isla de Chiloé. Magallania 33(1):51–61

Legoupil D, Fontugne M (1997) El poblamiento marítimo en los archipiélagos de Patagonia: núcleos antiguos y dispersión reciente. Anales del Instituto de la Patagonia, Serie Ciencias Humanas 25:75–87

Legoupil D, Béarez P, Lefèvre C, San Román M, Torres J (2011) Estrategias de subsistencia de cazadores recolectores de Isla Dawson (Estrecho de Magallames) durante la segunda mitad del Holoceno: primeras aproximaciones. Magallania (Punta Arenas), 39(2):153–164

Llagostera A (1979) 9,700 years of maritime subsistence on the Pacific: an analysis by means of bioindicators in the north of Chile. Am Antiq 44(2):309–324

Llagostera A (1982) Tres dimensiones en la conquista prehistórica del mar: Un aporte para el estudio de las formaciones pescadoras de la costa surandina. Actas del VIII Congreso Nacional de Arqueología Chilena, Kultrún, Santiago, 217–245

Llagostera A (1989) Caza y Pesca marítima (9.000 a 1000 a.C) En Prehistoria, 57–80. Editorial Andrés Bello, Chile

Mandryk C (1993) Hunter gatherer social costs and the nonviability of submarignal environments. J Anthropol Res 49:39–71

Martinoli M (2015) Procesamiento y consumo de pinnípedos: el caso de las ocupaciones canoeras tempranas del sitio Imiwaia I (Tierra del Fuego, República Argentina). Intersecciones en Antropología 16:367–381

Massone M, Morello F, Borrero L, Legoupil D, Mena F, Prieto A, Ocampo C, Rivas P, San Román M, Martin F, Méndez C, Reyes O, Munita D (2016) Cazadores-Recolectores en la Patagonia chilena desde 11.000 años a.p. a la colonización occidental. In: Falabella F, Uribe M, Sanhueza L, Aldunate C, Hidalgo J (eds) Prehistoria en Chile. Desde sus primeros habitantes hasta los Incas. Editorial Universitaria, Santiago, Chile

Méndez C, Reyes O (2008) Late Holocene human occupation of Patagonian forests: a case study at Cisnes River basin (44° S, Chile), Antiq 82:560–570

Méndez M, Reyes O (2015) Archaeology near the southern ice-end. Current advances in human interdisciplinary research in central western Patagonia. SAA Archaeol Rec. 15(3):21–26

Méndez C, Stern C, Reyes O (2008/9) Transporte de obsidianas a lo largo de los Andes de Patagonia Central (Aisén, Chile). Cazadores-Recolectores del Cono Sur. Revista de Arqueología 3:51–68. Mar del Plata

Méndez C, Stern C, Reyes O, Mena F (2012) Early Holocene long-distance obsidian transport in Central-South Patagonia. Chungara Revista de Antropología Chilena 44(3):363–375

Méndez C, Stern C, Nuevo A, Reyes O, Gutierrez F, Mena F (2018) Spatial and temporal distributions of exotic and local obsidians in Central Western Patagonia, southernmost South America. Quatern Int 468:155–168

Morello F, San Román M, Prieto A (2002) Puntas de proyectil lanceoladas en Patagonia meridional y Tierra del Fuego. Anales del Instituto de la Patagonia, Serie Ciencias Humanas 30:155–166

Morello F, Borrero L, Massone M, Stern C, Garcia-Herbst A, McCulloch R, Martinez I (2012) Hunter-gatherers, biogeographic barriers and the development of human settlement in Tierra del Fuego. Antiq 86(331):71–87

Morello F, Stern C, San Román M (2015) Obsidiana verde en Tierra del Fuego y Patagonia: caracterización, distribución y problemáticas culturales a lo largo del Holoceno Intersecciones en Antropología. Volumen especial 2:139–153

Moreno E, Zangrando F, Tessone A, Castro C, Panarello H (2011) Isótopos estables, fauna y tecnología en el estudio de los cazadores recolectores de la costa norte de Santa Cruz. Magallania 39:265–276

Ocampo C, Rivas P (2004) Poblamiento temprano de los extremos geográficos de los canales patagónicos: Chiloé e isla Navarino 1. Chungara. Volumen Especial, tomo 1:317–331

Ocampo C, Rivas P (2008) Manuscrito. El poblamiento marítimo de los canales occidentales de la Patagonia: desde Chiloé hasta Cabo de Hornos. pp 57–69

Orquera L (2005) Mid-Holocene littoral adaptation at the southern end of South America. Quatern Int 132:107–115

Orquera L, Gómez-Otero J (2007) Los cazadores-recolectores de las costas de Pampa, Patagonia y Tierra del Fuego. Relaciones. Sociedad Argentina de Antropología, (32):75–100

Orquera L, Piana E (1999b) La vida material y social de los Yámana. Eudeba-IFIC, Buenos Aires

Orquera L, Piana E (2005) La adaptación al litoral sudamericano sudoccidental: qué es y quiénes, cuándo y dónde se adaptaron. Relaciones-Sociedad Argentina de Antropología, (30):11–32

Orquera L, Piana E (2006) El poblamiento inicial del área litoral sudamericana sudoccidental. Magallania 34(2):21–36

Orquera L, Piana E (2009) Sea nomads of the Beagle Channel in southernmost South America: over six thousand years of coastal adaptation and stability. J Island Coast Archaeol 4:61–81

Orquera L, Piana E, Alvarez M, Fiore D, Vásquez M, Zangrando F, Tessone A, Tivoli A (2006) El Proyecto arqueológico Canal Beagle. In: Cruz I, Caracotche S (eds) Arqueología de la costa patagónica: perspectivas para su conservación. UNPA. Río gallegos, Argentina. 267–290

Orquera L, Legoupil D, Piana E (2011) Littoral adaptation at the southern end of South America. Quatern Int 239:61–69

Osborn A (1977) Aboriginal exploitation of marine food resources. Ph.D. diss., University of New Mexico

Pallo C (2011) Condicionamientos de la dinámica ambiental en las decisiones humanas sobre asentamiento y circulación a lo largo del estrecho de Magallanes durante el Holoceno tardío. Magallania 39(2):177–192

Pallo C, Borrero L (2015) ¿Intercambio o Movilidad?: Una Evaluación sobre el Uso de Escalas de Análisis Espaciales y Curvas de Declinación en Patagonia Centro-Meridional (Argentina). Latin Am Antiq 26(3):287–303

Panarello H, Zangrando F, Tessone A, Kozameh L, Testa N (2006) Análisis comparativo de paleodietas humanas entre la región del canal Beagle y península Mitre: Perspectivas desde los isótopos estables. Magallania 34(2):37–46

Perlman S (1980) An optimum diet model, coastal variability, and hunter-gatherer behavior. Adv Archaeol Method Theory, 3:257–310

Piana E, Orquera L (2007) Diferencias regionales y temporales en el litoral sudoccidental de Sudamérica. Arqueología de Fuego-Patagonia: 311 - 326. Ediciones CEQUA. Punta Arenas, Chile

Piana E, Orquera L (2009) The southern top of the world: The first peopling of Patagonia and Tierra del Fuego and the cultural endurance of the Fuegian Sea-Nomads. Arctic Anthropol 46 (1–2):103–117

Piana E, Tessone A, Zangrando F (2006) Contextos mortuorios en la región del canal Beagle: del hallazgo fortuito a la búsqueda sistemática. Magallania 34 (1):103–117

Piana E, Zangrando F, Orquera L (2012) Early occupations in Tierra del Fuego and the evidence from layer S at the Imiwaia I site (Beagle cannel, Argentina). In: Miotti L, Salemme M, Flegenheimer N, Goebel T, (eds) Southbound Late Pleistocene peopling of Latin America. College Station: Center for the Study of the First Americans, 171–175

Porter C (1993) GUA-010, un sitio costero erosionado en una zona sísmica activa. Actas del XII Congreso Nacional de Arqueología Chilena. Boletín del Museo Regional de la Araucanía 4:81–88

Prieto A, Stern C, Estévez J (2013) The peopling of the Fuego-Patagonian fjords by littoral hunter–gatherers after the mid-Holocene H1 eruption of Hudson Volcano. Quatern Int 317:3–13

Punke M, Davis L (2006) Problems and prospects in the preservation of late Pleistocene cultural sites in southern Oregon coastal river valleys: Implications for evaluating coastal migration routes. Geoarchaeol 21(4):333–350

Reeder-Myers L, Erlandson J, Muhs D, Rick T (2015) Sea level, paleogeography, and archeology on California's Northern Channel Islands. Quatern Res 83(2):263–272

Reyes O, Méndez C, Velásquez H, Trejo V (2006a) Distribuciones espaciales y contextos arqueológicos de cazadores recolectores esteparios en alto río Cisnes: (XI Región de Aisén). Magallania. (Punta Arenas), 34(2):75–90

Reyes O, Méndez C, Trejo V, Velásquez H (2006b) De las estepas a los canales septentrionales: uso humano de distintos ambientes. Bosques del valle del río Cisnes (Holoceno tardío, XI Región de Aisén)". Actas del XVII Congreso Nacional de Arqueología Chilena [2010]. Tomo 2:1487–1495. Valdivia

Reyes O, Méndez C, San Román M, Cárdenas P, Velásquez H, Trejo V, Morello F, Stern C (2007) Seno Gala 1; Nuevos Resultados En La Arqueología De Los Canales Septentrionales (44º S, XI Región de Aisén, Chile). Magallania. 35(2):105–119

Reyes O, Méndez C, Maldonado A, Velásquez H, Trejo V, Cárdenas M, Abarzúa A (2009) Uso del espacio de cazadores recolectores y paleoambiente Holoceno en el valle del río Cisnes, Región de Aisén, Chile. Magallania 37(1):91–107

Reyes O, Moraga M, Méndez M, Cherkinsky A (2015) Maritime Hunter-Gatherers in the Chonos Archipelago (43°50'–46°50' S), Western Patagonian Channels. J Island Coast Archaeol vol 10. Issue 2:207–231

Reyes O, Kuzminsky S, Méndez C, San Román M (2016a) Examining biological variation among the maritime hunter-gatherers of the Chonos Archipelago, Western Patagonian Channels, Chile. Libro Resúmenes 81 The Annual Meeting of the Society for American Archaeology, Orlando

Reyes O, San Román M, Morello F (2016b) Search for maritime hunter-gatherer archaeological record in the shifting shorelines of the South Pacific Coast (Chonos and Guaitecas Archipelago, Chile). In: Marine Ventures: Archaeological Perspectives on Human-Sea Relations. Equinox Publishing, United Kingdom, pp 141–155 doi: 10.1558/equinox.24563

Reyes O, Méndez C, San Román M, Francoise J (2018) Earthquakes and coastal archaeology: Assessing shoreline shifts on the southernmost Pacific coast (Chonos Archipelago 43°50′–46° 50′ S, Chile, South America). Quatern Int 463:161–175

Richardson JB III (1981) Modeling the development of sedentary maritime economies on the coast of Peru: A preliminary statement. Annals of the Carnegie Museum 50:139–150

Rick T, Erlandson J, Vellanoweth R (2006) Taphonomy and site formation on California's Channel Islands. Geoarchaeol 21(6):567–589

Rivas P, Ocampo C, Aspillaga E (1999) "Poblamiento Temprano de los Canales Patagónicos: El Núcleo Ecotonal Septentrional". Anales del Instituto de la Patagonia, vol 27:221–230

Rivas P, Ocampo C (2006) La adaptación humana al bosque en la isla de Chiloé. Estrategias adaptativas en el litoral septentrional de los canales patagónicos. Actas del XVII Congreso Nacional de Arqueología Chilena (2010). T2: 1449–1460. Valdivia

Roksandic M, de Souza S, Eggers S, Burchell M, Klokler D (2014) The cultural dynamics of shell-matrix sites. UNM Press

Sandweiss D, Maasch K, Chai F, Andrus C, Reitz E (2004) Geoarchaeological evidence for multidecadal natural climatic variability and ancient Peruvian fisheries. Quatern Res 61 (3):330–334

San Román M (2010) La explotación de recursos faunísticos en el sitio Punta Santa Ana 1: estrategias de subsistencia de grupos de cazadores marinos tempranos de Patagonia meridional. Magallania 38(1):183–198

San Román (2014) Sea-level changes and coastal peopling in southernmost Pacific South America: Marine hunters from Patagonia. In: Encyclopedia of Global Archaeology (pp 6515–6525). Springer, New York

San Román M, Fontugne M, Salas K (2009) Primeros avances en la reconstrucción de secuencias de ocupación de cazadores recolectores marinos en el estrecho de Magallanes, Patagonia meridional. In: Salemme M, Álvarez M, Piana E, Vázquez M, Mansur M (eds) Arqueología de Patagonia: una mirada desde el último confín, Ushuaia: Editorial Utopías

San Román M, Reyes O, Elgueta J, Morello F (2016a) Archaeology of Maritime hunter-gatherers from Southernmost Patagonia, South America: discussing timing, changes and cultural traditions during the Holocene). In: Marine Ventures: Archaeological Perspectives on Human-Sea Relations. Equinox Publishing, United Kingdom

San Román M, Reyes O, Méndez C, Cárcamo J, Torres J (2016b) Western Patagonia subsistence strategies: Zooarchaeological studies of marine hunter-fisher-gatherers of the Chonos archipelago, Chile. Libro Resúmenes 81 The Annual Meeting of the Society for American Archaeology, Orlando

Scheinsohn V (2001) 2001: Odisea del espacio: paisajes y distribuciones artefactuales en arqueología. Relaciones de la Sociedad Argentina de Antropología, 26

Schiffer M (1996) (1987) Formation processes of the archaeological record. University of Utah Press

Sierralta S, Delgado A, Kelly P, Rebolledo S (2019) Cronología absoluta en los canales septentrionales, el mar interior y la costa pacífica austral. In: Arqueología de la Patagonia: El pasado en las arenas. Actas de las X Jornadas de Arqueología de la Patagonia. Puerto Madryn, Argentina. 167–179

Steffen H (1944) Patagonia occidental: las cordilleras patagónicas y sus regiones circundantes. Universidad de Chile, Santiago

Stein J (1992) Sediment analysis of the British Camp shell midden. In: Stein J (ed) Deciphering a Shell Midden, Academic Press, San Diego, pp 135–162

Stein J, Deo J, Phillips L (2003) Big sites-short time: accumulation rates in archaeological sites. J Archaeol Sci 30(3):297–316

Stern C (2018) Obsidian sources and distribution in Patagonia, southernmost South America. Quatern Int 468:190–205

Stern C, Porter C (1991) Obsidiana en yacimientos arqueológicos de Chiloé y las islas Guaitecas. Anales del Instituto de la Patagonia. Serie Ciencias Humanas 20:205–209

Stern C, Curry P (1995) Obsidiana del sitio Posa Las Conchillas, Isla Traiguén (45º30′S), Archipiélago de los Chonos. Anales del Instituto de la Patagonia. Serie Ciencias Humanas. 23:119–124

Stern C, Navarro X, Muñoz J (2002) Obsidiana gris translúcida del volcán Chaitén en los sitios arqueológicos de Quilo (Isla Grande de Chiloé) y Chanchán (X Región) Chile, y obsidiana del Mioceno en Chiloé. Anales del Instituto de la Patagonia, Serie Ciencias Humanas 30:167–174

Stern C, Caracotche S, Cruz I, Charlin J (2012) Obsidiana gris porfírica calco-alcalina del volcán Chaitén en sitios arqueológicos al sur del río Santa Cruz, Patagonia meridional. Magallania 40 (1):137–144

Tessone A, Zangrando F, Valencio S, Panarello H (2003) Isótopos estables del carbono en restos óseos humanos de la región del canal Beagle, Isla Grande de Tierra del Fuego. Revista Argentina de Antropología Biológica. vol (2):33–43

Torres J, Ruz J (2011) Pescadores de la tradición cultural Englefield: Datos preliminares en la zona del Estrecho de Magallanes y mar de Otway XII Region de Magallanes, Chile. Magallania, 39 (2):165–176

Wandsnider L (1992) The spatial dimension of time. In: Space, Time, and Archaeological Landscapes. Rossignol J, Wandsnider L (eds) Plenum Press, New York, pp 257–282

Waters M, Kuehn D (1996) The geoarchaeology of place: The effect of geological processes on the preservation and interpretation of the archaeological record. Am Antiq pp 483–497

Yesner D (1980) Maritime hunter-gatherers: ecology and prehistory [and comments and reply]. Curr Anthropol, 727–750

Yesner D, Torres M, Guichon R, Borrero L (2003) Stable isotope analysis of human bone and ethnohistoric subsistence patterns in Tierra del Fuego. J Anthropol Archaeol 22(3):279–291

Zangrando F (2009a) Historia evolutiva y subsistencia de cazadores-recolectores marítimos de Tierra del Fuego. Sociedad Argentina de Antropología. Buenos Aires

Zangrando F (2009b) Is fishing intensification a direct route to hunter-gatherer complexity? A case study from the Beagle Channel region (Tierra del Fuego, southern South America). World Archaeol 41(4):589–608

Zangrando F (2018) Poblamiento temprano y arqueología de costas en Patagonia y Tierra del Fuego: vacío de información, preconceptos y perspectivas. Intersecciones en Antropología, 19 (2):63–97

Zangrando F, Ponce J, Martinoli M, Montes A, Piana E, Vanella F (2016) Palaeogeographic changes drove prehistoric fishing practices in the Cambaceres Bay (Tierra del Fuego, Argentina) during the middle and late Holocene. Environ Archaeol 21(2):182–192

Zangrando F, Tessone A, Valencio S, Panarello H, Mansur M, Salemme M (2004) Isótopos estables y dietas humanas en ambientes costeros. En Avances en Arqueometría 2003, Universidad de Cádiz, España, pp 91–97

Zangrando F, Tivoli A (2015) Human use of birds and fish in marine settings of southern Patagonia and Tierra del Fuego in the Holocene: A first macro-regional approach. Quatern Int 373:82–95

Zilio L, Gordón F, Béguelin M, Castro A (2014) Paleodietas humanas en el sur del Golfo San Jorge (provincia de Santa Cruz) a partir del análisis de isótopos estables. Revista argentina de antropología biológica, 16(1):51–64

Prologue

I am grateful for the opportunity to prologue this book. It is a demonstration of Omar Reyes Baez's perseverance and growth, which I have been fortunate enough to witness throughout our long friendship and more recently tutoring his doctoral research. Now, a wealth of new readers will have the chance of learning about his ideas and accessing new data on the maritime culture of the originary people of Western Patagonia.

In this new book, we can learn as to when, how and who peopled the Chonos Archipelago of the northern Patagonian channels of Chile, an area that until recently remained relatively unexplored from an archaeological point of view. These goals were accomplished based on the results obtained from a long-term field and lab project. This has allowed Omar Reyes to generate a large data set by integrating environmental and archaeological information. The archaeological patterns and trends identified, through years of sustained research, has triggered a series of renovated interpretations about the originary canoe people and their adaptation to maritime coastal settings.

A starting point in this book, is an overview of previous archaeological research through the thoughtful reading of the literature on past and present maritime hunter-gatherers. This is followed by the analyses on how changes in the environment and sea-level conditioned human occupations through time. Some issues addressed here are the effect of catastrophic events, isostatic and tectonic processes, as well as the action of tsunamis and great storms as recorded in coastal geomorphology and sediments.

Fieldwork did not prove easy in this region and this is addressed in the book. Collecting field data involved surmounting very challenging practical problems. One was dealing with the constraints imposed by accessibility to the rocky coast, and the other, was the low archaeological visibility due to the dense cover of evergreen forest that characterises the littoral and interior of the research area. Another issue was evaluating site representation by taking into consideration postdepositional agents and processes in their preservation. A multistage fieldwork strategy was designed for exploring and sampling different geomorphological, topographical and ecological zones.

This combined approach was crucial for successfully obtaining representative information of the archaeological temporal and spatial variability of the region. The research design unveiled an unsuspectedly wide diversity of sites. This allowed the identification of features associated with distinct procurement activities, collecting diverse categories of artefacts and animal bone samples during the excavation of different kinds of sites, and, in some cases, the recovery of human remains uncovered at special locations.

The chronological framework results from more than 80 radiocarbon dates that extend from circa 6200 years cal BP to the nineteenth century. The dating strategy focused on establishing both the temporality of cultural deposits as well as their association with the individual landforms considered relevant for archaeological landscape reconstruction. Dates were also useful for defining the chronology of stratigraphic sequences at certain sites and for dating special features or findings (e.g. hearths or burials).

Results are presented in a very detailed way for the whole region, area by area, and site by site, all supported by tables and figures that summarise primary and secondary data. Estimates of the frequency of sites per area were calculated to assess the variability in site distribution in the region. The way of life of the originary people that occupied this sea and landscape for millennia was undertaken by integrating several lines of analytical research. These included lithic technology and faunal analyses, complemented with the study of isotopes from human remains. The quality and quantity of the evidence produced is outstanding and is critically assessed in view of the models and frames of reference presented at the beginning of the book.

Lithic analysis revealed different categories of instruments and the provenance of raw materials used. The great diversity of marine invertebrates and sea and terrestrial vertebrates in the faunal record showed how they were used with different selectivity and intensity depending on the geographical site location and the duration of the occupations. Isotope analysis of food resources and human bones was the key to understanding changes in the types of food consumed through time, and also detected a contrast in diet between pre and post-contact individuals.

The narrative provided by Omar Reyes allows us to travel through time and space. Central to the discussion was to know if there was continuity in the occupation in the long run and if there were any significative changes in the overall cultural interaction with this environment. The questions made at the beginning of the book are all answered. I firmly believe this book will be a major contribution to our present knowledge of the way of life of early and recent inhabitants of these archipelagos. And, at the same time, it also reveals the impact that the contact period had on the originary people that lived in that region.

The ethnographic and historical literature analysed by Omar shows that when exploring the southern coastal region in the sixteenth century, the Europeans encountered the Chono, the name given to the indigenous people they met in the area. In the following centuries, this whole region was incorporated into the economic system of the national state of Chile, and the Chono people were made to move to other areas, were evangelized, and also used as forced labour.

As we all know, archaeologists reconstruct the past grounded in the present and that archaeology is a way of connecting the past with the present. But there is no present without history. This book will certainly be invaluable to other researchers worldwide interested in maritime adaptations and their cultural trajectories. But it also has a great anthropological potential by offering a history of the ancestors of indigenous people that seem to have officially vanished but, in real terms, have probably become invisible.

Guillermo L. Mengoni Goñalons
Facultad de Filosofía y Letras
Instituto de Arqueología
Universidad de Buenos Aires
Buenos Aires, Argentina

Contents

Chapter 1
Study Area

Abstract One of the fundamental aspects is to understand the evolution of the coastline in the Chonos Archipelago. Only by understanding these changes, it is possible to understand the location and placing of archaeological sites, their preservation and the possibility of making finds by designing an archaeological prospection of the coast. In this Chapter, the general characteristics of the study area, the climatic and paleoclimatic history, the coastal geomorphological changes and the factors (world, regional and local) that affect the said geographic configuration are developed, as well as the characterization of the available resources.

Keywords Western patagonia · Chonos archipelago · Holocene · Glacial geomorphology · Sea-level change · Earthquakes · Paleoclimatic history

1.1 Geomorphological Aspects of the Chonos Archipelago

The Chonos Archipelago is located in the northern part of the Patagonian Channels. It consists of a group of more than 150 islands, forming a dense network of narrow, steep-sided channels and fjords. It extends from the Gulf of Corcovado, at the southern end of the Chiloé Archipelago (43°50′S), to the Taitao Peninsula (46°50′S); it measures 360 km from north to south and 150 km from east to west, covering an area of ~54,000 km^2. The northern part includes the Guaitecas Archipelago, which forms part of the same island system (Fig. 1.1).

The geomorphology of this archipelago, and of the western channels of Patagonia in general, presents the typical characteristics of a complex system moulded by glaciation and the retreat of the ice, postglacial isostasy and marine transgressions during the Holocene, with an important contribution of plate tectonics and volcanic activity. These agents of change have contributed not only to the existing geography of the archipelago, but also to frequent changes in the position of the coastline, and therefore to the distribution and preservation of archaeological sites.

The geology of the region is dominated by North Patagonian Batholith, which is composed of granodiorites and tonalites of the Cenozoic and Mesozoic (SERNAGEOMIN 2003). Metamorphic rocks of the Triassic and Miocene define the geology

O. Reyes, *The Settlement of the Chonos Archipelago, Western Patagonia, Chile*,
The Latin American Studies Book Series,
https://doi.org/10.1007/978-3-030-54326-6_1

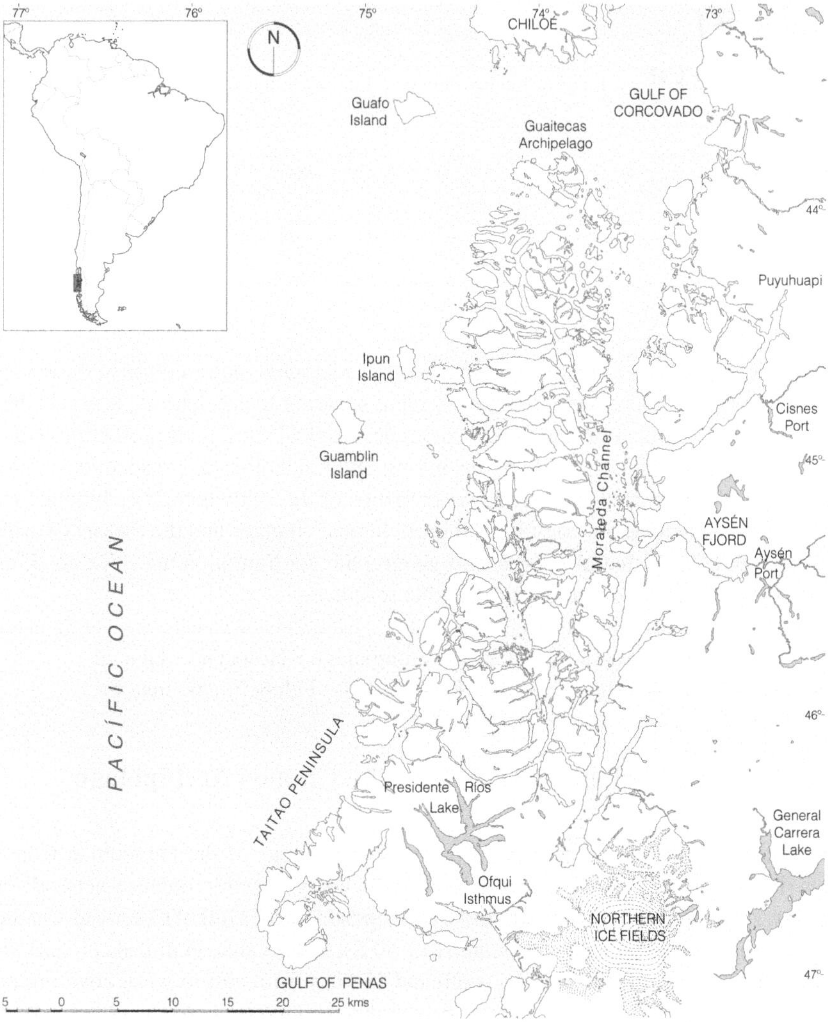

Fig. 1.1 Study area. Location of the Chonos Archipelago in Western Patagonia

of the Taitao Peninsula, the Chonos Archipelago and Chiloé Island (Herve et al. 2010). The crystalline basement and the presence of flooded land and abundant vegetation help to produce highly acid soils or *podzol* throughout the archipelago (Holdgate 1961).

The major morphological feature characterising the archipelago is the disappearance by sinking of the Central Valley, which runs right down Chile from the north of the country. The sea flooded the central plain and lower valleys, giving rise to islands, channels, estuaries, fjords and deltas (Heusser 2003). The evidence suggests that this

Fig. 1.2 Coastal modelling by the ice sheet of the Last Glacial Maximum. Steep coasts and fjords can be observed throughout the Chonos Archipelago. The ground rises hundreds of metres and the sea bed descends to similar depths. Refugio channel

complex system of fjords and channels was formed in the Pleistocene (~125,000–11,000 years BP) due to the advance and retreat of the glaciers and the tectonic sinking of the Central Valley (Borgel 1970; Glasser et al. 2008; Heusser et al. 1992; Vargas et al. 2013). Like the Coastal Range, it has been intensely eroded by glacial action and subduction tectonics (Borgel 1970; Díaz-Naveaz and Frutos 2010; García 2012). The abrupt topography of the channels and islands has steep hillsides rising hundreds of metres, while the fjord and channel bottoms lie at depths of 200 to 700 m (Delgado 2004), resulting in coastlines with long stretches of near vertical cliffs (Fig. 1.2). Although the ice has disappeared, it is thought to have been the principal transforming agent, modelling valleys, fjords and mountains, and depositing huge volumes of moraine material (Glasser et al. 2008). Nevertheless, the general forms of the existing fjords and valleys were defined before the first great invasion of ice, and during the ice age they were deepened by the glacial masses (Glasser et al. 2004, 2008).

1.2 Changes in the Coastline

One of the fundamental aspects of this work is to understand the evolution of the coastline in the Chonos Archipelago. Only by understanding these changes, it is possible to understand the location and placing of archaeological sites, their preservation and the possibility of making finds by designing an archaeological prospection of the coast.

Changes in the coastline during the postglacial result from a series of processes and factors which operate and interact on various spatial scales (i.e. global, regional and local) and have repercussions for the changes in sea level. They include the total volume of ice in the world, regional glacial hydro-isostatic adjustments and regional/local tectonics (Milne and Shennan 2007; Murray-Wallace 2007; Nelson 2007; Shennan 2007). These changes may be abrupt or gradual depending on regional/local tectonic behaviour (Reyes et al. 2018). Furthermore, catastrophic events like earthquakes associated with subduction zones can cause major changes in coastlines through subsidence or uplift due to the deformation of the earth's crust during the rupture associated with the seismic event (Nelson 2007). Finally, phenomena such as tsunamis and great storms may also be responsible for modifying coastlines through erosion or by bringing in new material.

1.3 Tectonic Activity

In the Chonos Archipelago, coastline changes are associated not only with changes in sea level but also with vertical deformation caused by an isostatic rebound and the subduction of the Nazca Plate under the South American Plate (Atwater et al. 1992; Hervé and Ota 1993; Lara et al. 2008; Thomson 2002). This process is complemented by local subsidence and uplift caused by major earthquakes, generally also associated with tsunamis (Cisternas et al. 2005).

The zone between 42°S and 47°S is near the meeting point of three tectonic plates (Nazca, South American and Antartic), known as the Chilean Triple Junction, which lies just off the Taitao Peninsula (48°S). The most important tectonic feature is the Liquiñe-Ofqui fault system. The principal fault, more than 900 km long from north to south, runs through the whole length of the archipelago (Delgado 2004). It starts in the south near the Triple Junction and reaches the area of Liquiñe in the Andes Mountains (38°S) in the north (Thomson 2002). This fault system is largely responsible for the morphology of the fjords and channels (Cembrano et al. 1996; Glasser and Ghiglione 2009). It also determines the location of a regional centre of volcanic activity (Stern 2004) known as the Southern Volcanic Zone (SSVZ, 42°–46°S), with 13 volcanoes which have erupted repeatedly during the Holocene (Naranjo and Stern 2004). On the mainland facing the northern part of the archipelago are Hornopirén, Huequi, Michimahuida, Chaitén, Corcovado, Llanteles, Melimoyu, Mentolat, Maca, Cay and Hudson volcanoes. One of the most recent events recorded led to landslides and tsunamis on the shore of Aysén Fjord (Lara 2008; Naranjo et al. 2009).

This active margin runs parallel and close to the Peru–Chile subduction trench, characterised by the frequency and high magnitude of the earthquakes caused by coseismic tectonic movements (Lomnitz 1970). These also contribute to modelling of the coastlines in the archipelago, which may rise and/or fall (Vidal Gormaz 1901). Major earthquakes and tsunamis have been recorded along the coast. The best-known example is the greatest contemporary subduction event recorded in the world, the Chilean earthquake of 1960, between 37° and 48°S, magnitude 9.5 MW (SHOA 2000). In the Chonos Archipelago, it caused the islands of Guafo, Guamblin and Ipún (offshore islands and Pacific border of the archipelago) to rise by more than 5.7 m, while the inshore area sank by between 1 and 2 m (Cisternas et al. 2005; Plafker and Savage 1970; Reed et al. 1988; Vita-Finzi 1996). Associated with this major event occurred one of the most destructive tsunamis recorded, which affected much of the archipelago and swept across the whole Pacific Basin (Abe 1979; Cisternas et al. 2005; SHOA 2000). There are historical references to land sinking due to earthquakes, like the Valdivia earthquake of 1737 which affected the whole northern archipelago (Darwin 1845; Fitz-Roy 1839; Lomnitz 1970). A century later, in the Valdivia earthquake of 1837, the coast of the island of Lemu in the Chonos Archipelago rose by 2.4 m.

These tectonic processes are generally associated with big tsunamis which generate important sedimentary movements and transformations along the shoreline. For example, sediment studies in Huelde Lake, on the west coast of Chiloé Island, recorded 17 tsunami events over the past 5500 years cal. BP (Kempf et al. 2017). Earthquakes can trigger other phenomena, such as the drastic redistribution of fauna (Castilla 1988) due to the rapid and profound remodelling of the coast. Reed and collaborators (1988) propose that sinking of the land can be observed in the line of dead trees below sea level in the area of San Rafael Lake and the Ofqui Isthmus in the Chonos Archipelago, which would indicate a difference of between 2 and 2.5 m in elevation between the healthy trees and the well-preserved dead trunks (Fig. 1.3).

Some authors suggest that large magnitude earthquakes can occur in the same area at a frequency of approximately 100 to 130 years (Abe 1979; Lomnitz 1970), and that earthquakes of magnitudes like that of Valdivia in 1960 (M_W 9.5) are recorded approximately every 500 years (SHOA 2000).

1.4 Melting of the Ice, Rising Sea Levels and Rising Land Levels Due to Glacial Isostasy

The relief and the general landscape of the northern part of the Patagonian Channels have been moulded by the morphogenetic factors and agents typical of a subduction zone, suggesting a general tendency for the land to sink; and by the effects of glaciation and deglaciation, when glacio-eustatic response caused the continent to rise as it was freed from the weight of the Patagonian ice sheet (Milne and Shennan 2007; Murray-Wallace 2007; Nakada and Lambeck 1988).

Fig. 1.3 Estero Puelma in the Taitao Peninsula, in the south of the Chonos Archipelago, where the sunken remains of forest can be seen below the present sea level, probably due to phenomena associated with vertical tectonics

The level of the world's oceans varies in direct relation to the volume of ice in the world (Chappell et al. 1996). Together with the melting of enormous sheets of ice, and their incorporation into the system as large volumes of water, the change in the ice mass also causes an isostatic rebound of the underlying earth (Dietrich et al. 2010).

At the time of the Last Glacial Maximum (LGM), the Patagonian ice sheet covered an area of nearly 500,000 km^3 to a thickness of 2,000 m (Doake 2001; Glasser et al. 2008; Hulton et al. 2002; Mercer 1976). During the early Holocene, changes in atmospheric conditions, with higher temperatures and lower humidity, resulted in the rapid melting of the polar and continental ice caps. Sea levels around the world rose quickly (marine transgressions), with varying effects on coastlines depending on their latitude and morphology (Isla 1989; Long 2001). Different records of the maximum postglacial marine transgression are found with different radiocarbon datings during the mid-Holocene (*ca.* 7000 to 5000 years BP approx.) and different heights above present sea level (average between 2 and 8 m). During the late Holocene, conditions became more stable, with much smaller rates of change (Milne et al. 2009). Thus much of the geomorphological evidence related to deposits along coastlines from the early Holocene are underwater or have been eliminated due to these global changes (Nelson 2007). Many geomorphological studies indicate glacial advances and retreats during the Postglacial (Francoise 2014; Glasser et al. 2005, 2006, 2008; Heusser 1960). There is also evidence that there was a resurgence of glacial activity during the Mid to Late Holocene (Araneda et al. 2007; Harrison et al. 2008; Winchester and Harrison 2000; Winchester et al. 2001). This phenomenon can be seen in the chronology of glacial advances in Patagonia (Aniya 1999, 2013; Aniya et al. 1997;

Aniya and Sato 1995; Mercer 1982); it would appear to be directly related with the increase in precipitations and the fall in temperatures recorded during this period.

In the southernmost extremes of the Patagonian Channels, in the area of the Magellan Strait and the Otway and Skyring Sounds, geo-archaeological studies have established changes in the coastline throughout the Postglacial, and the location of archaeological sites on an elevation gradient (McCulloch and Morello 2009; Porter et al. 1984; San Román 2014; Santiago et al. 2007), or related with a change in the presence of faunal resources at the site (e.g. Zangrando et al. 2016). In particular, these studies indicate a direct relationship between the elevation of archaeological sites and the chronology associated with changes in regional sea level.

For the Chonos Archipelago in contrast, it is impossible to model the search for archaeological sites based on the geoform of the coast where they may have existed (i.e. a direct relation between the elevation of the archaeological site and its antiquity). Although isostatic rebound and marine transgressions can be modelled, the important vertical tectonics of the area and its effects (rises and falls in level ranging from millimetres to metres) cannot be deduced, eliminating geomorphological projections and models (Reyes et al. 2018). These changes may be abrupt or gradual depending on regional/local tectonic behaviour.

1.5 History of Postglacial Vegetation

Pollen records in the Los Lagos Region and on Chiloé Island indicate changes experienced by terrestrial ecosystems and the climate of the archipelago region since the LGM. They show the development of the North Patagonian forests, related to a colder, wetter climate than at present, between 18,000 and 11,500 years BP (Abarzúa et al. 2004; Francoise 2014; Moreno 2004; Moreno and León 2003; Pesce and Moreno 2014). Pollen records in the Taitao Peninsula and the Chonos Archipelago also indicate a proliferation of hygrophilous taxa like *Pilgerodendron* during the transition from the LGM (Bennett et al. 2000; Haberle and Bennett 2004).

During the transition from the Pleistocene to the Holocene and the early part of the Holocene, thermophilous temperate forest taxa predominate (Valdivian forest), followed by an increase in the percentages of conifers and *Nothofagus spp.* during the Mid-Holocene (North Patagonian forest). These changes in the pollen spectrum, and therefore in the vegetation, suggest a change from warm/dry conditions to colder, wetter conditions around 8000 years cal. BP (Haberle and Bennett 2004). Precipitation during the early Holocene appears to have been lower than in the Mid to Late Holocene.

The information obtained from the record at Stibnite Lake, on the Taitao Peninsula (Lumley and Switsur 1993; Mássaferro and Brooks 2002), shows warm, dry conditions between 11,000 and 8000 years cal. BP, and wet conditions between 8000 and 4000 years cal. BP. The most remarkable change in this record is the decline of *Pilgerodendron uviferum* and *Weinmannia trichosperma* and the expansion of the Podocarpaceae and *Nothofagus spp.* around 2700 years cal. BP. This change indicates an increase in precipitations and/or a fall in temperatures in the area, and may be linked to the neoglacial activity recorded (Aniya 2013; Mercer 1982). Other pollen records from the same archipelago, e.g. Miranda Lake, Fácil Lake and Oprasa Lake (Haberle and Bennett 2004; Haberle et al. 2000), show the expansion of *Tepualia stipularis* and *Weinmannia trichosperma* between approximately 12,000 and 7000 years cal. BP, which has been interpreted as a warm, dry moment. The present floral assembly of *Nothofagus*, *Pilgerodendron* and *Tepualia* is thought to have formed around 7000 years cal. BP, suggesting an increase in precipitation in the area. During the past 3000 years BP, there has been an increase in the concentration of carbon particles (indicating fire). This has been interpreted by Haberle and Bennett (2004) as a period of more variable climate associated with the El Niño Southern Oscillation (ENSO) and possibly anthropic intervention. Another characteristic registered in these pollen records are brief expansions of thermophilous taxa during the Mid and Late Holocene, suggesting warmer, drier moments in this basically colder, wetter period.

It may be added that, consistent with the pluviometric gradient, evergreen forest formations dominate to the west, turning to deciduous forest on the other side of the Andes Mountains and open relief with steppe vegetation further east. The paleoenvironmental information available suggests that this phytogeographic distribution has not varied substantially during the Holocene, except for minor fluctuations in the location of the forest/steppe margin (Bennett et al. 2000; de Porras et al. 2012, 2014; Markgraf et al. 2007; Reyes et al. 2009).

1.6 Present Climate and Vegetation

The climate of the archipelago region is oceanic; it is strongly dictated by the influence of the sea and the variations along the northern edge of the Southern Westerly Wind Belt, which brings heavy precipitation regimes to the west of the Andes (Garreaud et al. 2009). Low ground and coastal areas show higher mean annual temperatures, with smaller variation over the course of the year. The mean annual temperatures in the archipelago are around ~10°C, with little seasonal variation. In the Andes Range on the other hand, where the highest elevations in the region are found (>1,000 masl), there is a resulting reduction in mean annual temperatures allowing the persistence of masses of ice, remnants of the last glacial period. The largest of these are the Northern (4,200 km^2) and Southern (13,000 km^2) Ice Fields. The Northern Ice Field, with some 70 glaciers larger than 0.5 km^2 (Rivera et al. 2007), lies opposite the southern part of the Chonos Archipelago. Precipitations are heavier and persistent south of 45°S, around 4,000 mm per year. The maximum precipitations occur in winter, tending to diminish in the summer but with no dry months (12 months of wet climate); these climatic conditions give the area of the channels a temperate climate with an oceanic tendency (Di Castri and Hajek 1976). Further east, in contrast, in the Andean zone, the level of precipitations (carried in from the west) declines as a result of the rain shadow effect of the Andes Mountains (Garreaud et al. 2009).

The current vegetation, typical of cold, rainy regimes, closely follows these climatic parameters (Gajardo 1994; Luebert and Pliscoff 2006) and is controlled latitudinally and altitudinally by strong temperature and precipitation gradients (Abarzúa et al. 2004; Sepúlveda 2009). Viviani (2004:6) says: "*The plant formations do not respond only to the rainfall, but to a combination of the rainfall distribution, temperature, wind, relative humidity, soil, solar radiation, and other factors such as the volcanic ash which fertilises the soil, the geographical exposure of the slopes, the age of the soil since its release from the ice of the last glaciation (distance to glaciers still moving actively towards the sea) and the maturation of the ecological succession towards the climax state of the biome. The very wet oceanic climate, with remarkably stable temperatures year-round, accompanied by abundant precipitation in the form of rain, allows a multi-stratified evergreen forest to flourish, with abundant epiphytes, climbing plants and ferns among trees of great girth. These are not cold forests of conifers or deciduous trees with snow underfoot in winter and spring flowers, as are common in other fjord systems of the world*".

The archipelago has been characterised in simpler and more general descriptive terms as the evergreen forest and peat bogs (Fig. 1.4) consisting of two forest subregions: evergreen forest with conifers like *Fitzroya cupressoides* and *Pilgerodendron uvifera*; and evergreen forest of species like *Nothofagus nitida* and *Drimys winteri* with Chonos peat bogs (Álvarez et al. 2010; Gajardo 1994; SERPLAC 2005).

Fig. 1.4 North-west coast of Traiguén Island. Multi-stratified evergreen forest projected to the high tide mark

1.7 The Fauna in the Archipelago

The faunal richness present in the northern archipelagos shows a marked contrast between terrestrial and marine resources. This is explained by the combination of oceanographic, climatic and geomorphological factors described above, which make up an environmental mosaic. Among marine fauna, there is a direct relationship between the type of fjord and the energy of the waves which beat on the shore. While high energy prevails in more exposed sections (open fjords), low energy is found in the protected inner marine spaces of the archipelagos (Paskoff 2010). Extreme gradients are recorded of salinity, temperature, oxygenation, nutrients, tidal currents, oxygen, pH and suspended sediments, among other factors that combine in different proportions in different places (Castillo et al. 2006). Every fjord and channel is a different system, where atmospheric, edaphic, vegetational, hydrological, geomorphological, tidal and oceanographic processes interact in different ways and with different intensities (Sievers and Silva 2006). The proximity of glaciers, the shoreline and narrow channels, and the orientation of those channels determine the distribution and quantity of biotic resources. "*The tides then are the fundamental driver for renewal of the marine waters of the complicated littoral system of Western Patagonia*" (Viviani 2004:14).

The sea and the littoral environment are home to a vast biodiversity of marine mammals. They include 18 species of cetaceans (among them *Eubalaena australis, Caperea marginata, Balaenoptera physalus, Balaenoptera borealis, Balaenoptera edeni, Balaenoptera bonaerensis, Balaenoptera acutorostrata* and *Megaptera novaeangliae,* and the biggest mammal of all, the blue whale *Balaenoptera musculus*, as well as three species of cachalots: *Physeter macrocephalus, Logia breviceps* and *Logia sima)*; more than 10 species of dolphins, the orca (*Orcinus orca*), porpoises and beaked whales (Bastida and Rodríguez 2003). There are also two species of pinnipeds (*Otaria flavescens,* with 37 rookeries currently recorded, and *Arctocephalus australis*, with one known rookery); other Antarctic otariids can be observed occasionally such as the elephant seal (*Mirounga leonina*) and the leopard seal (*Hydrurga leptonyx)* (Aguayo et al. 2006; Lozano 2006; Zamorano et al. 2010).

The extensive avifauna includes both terrestrial and marine birds, with around 46 species from 40 genera and 21 families (Vuilleumier 1985). It includes Humboldt (*Spheniscus humboldti*) and Magellanic penguins (*Spheniscus magellanicus*), 5 species of grebe, 11 albatrosses (among them the royal albatross *Diomedea epomophora*), 21 petrels and shearwaters, 3 storm petrels, a pelican (*Pelecanus thagus*), one booby (*Sula variegata*), 6 cormorants, 9 plovers, 16 taxa of waders, marine Anatidae (like the black-necked swan *Cygnus melanocorypha*), kelp goose (*Chloephaga hybrida*), the Fuegian steamer duck (*Tachyeres pteneres*), the snowy sheathbill (*Chionis alba*), 6 species of skuas, 15 gulls, 6 herons and one species of ibis. There is a variety of species of goose (*Chloephaga spp.*); the Anatidae is the best represented family with 24 species. The ringed kingfisher (*Ceryle torquata*) is one of the most handsome of the huge variety of birds in the archipelago (Lozano 2006).

Among benthic species of biological, environmental and economic interest are crustaceans, molluscs, annelids, echinoderms and fish. Twenty-nine species of fish have been recorded, grouped in 25 genera from 23 families (Navarro and Pequeño 1979; Pequeño 2000; Pequeño and Riedemann 2002). There is also a great diversity of malacological fauna with a very broad bio-geographical distribution and formations in different marine substrates (e.g. sea beds, rocky and sandy substrates). Forty-eight species of molluscs (Polyplacophora, gastropods, bivalves and scaphopods), crustaceans and echinoderms have been recorded (Cárcamo 2015; Osorio and Reid 2004).

In contrast, terrestrial resources are almost non-existent in the islands; only a few rodents like the coypu (*Myocastor coypus*) and mustelids like otters (*Lontra provocax* and *Lontra felina* (Aguayo et al. 2006)). On the mainland coast facing the archipelago, there are pudú (*Pudu pudu*, the world's smallest deer) and huemul (*Hippocamelus bisulcus*, a large deer); puma (*Felis concolor*) are also seen occasionally (González et al. 2009; Jiménez 2010; Mella et al. 2002). The range of action of this fauna is not completely clear, due to the major disruptions to the habitat caused by depredation (Flueck and Smith-Flueck 2012).

References

Abarzúa A, Villagrán C, Moreno P (2004) Deglacial y postglacial climate history in east-central Isla Grande de Chiloé, southern Chile (43°S). Quatern Res 62:49–59

Abe K (1979) Size of great earthquakes of 1837–1974 inferred from tsunami data. J Geophys Res 84(B4):1561–1568

Aguayo A, Acevedo J, Vargas R (2006) Diversidad de mamíferos marinos en las aguas del Archipiélago de los Chonos. Ciencia y Tecnología del Mar 29(2): 129–145

Álvarez M, San Martin C, Novoa C, Toledo G, Ramírez C (2010) Diversidad florística, Vegetacional y de hábitats en el Archipiélago de los Chonos (Región de Aisén, Chile). Anales del Instituto de la Patagonia 38:35–56

Aniya M (1999) Recent glacier variations of the Hielos Patagonicos, South America and their contribution to sea-level change. Arct Antarct Alp Res 31(2):165–173

Aniya M (2013) Holocene glaciations of Hielo Patagónico (Patagonia Icefield), South America: a brief review. Geochem J 47(2):97–105

Aniya M, Sato H (1995) Holocene glacier variations at Tyndall Glacier area, southern Patagonia. Bolletin Glacier Res 13:97–109

Aniya M, Sato H, Naruse R, Skvarca P, Casassa G (1997) Recent glacier variations in the Southern Patagonian Icefield, South America. Arctic Alp Res, 1–12

Araneda A, Torrejón F, Aguayo M, Torres L, Cruces F, Cisternas M, Urrutia R (2007) Historical records of San Rafael glacier advances (North Patagonian Icefield): another clue to 'Little Ice Age' timing in southern Chile? The Holocene 17(7):987–998

Atwater B, Núñez H, Vita-Finzi C (1992) Net Late Holocene emergence despite earthquake-induced submergence, south-central Chile. Quatern Int 15–16:77–85

Bastida R, Rodriguez D (2003) Mamíferos marinos de la Patagonia y Antártica. Vazquez Manzini editores, Buenos Aires, Argentina

Bennett K, Haberle S, Lumley S (2000) The last glacial-holocene transition in Southern Chile. Science 290:325–328

Borgel R (1970) Geomorfología de las regiones australes de Chile. Revista Geográfica de Chile 'Terra Australis' 20:135–140

Cárcamo J (2015) El consumo de moluscos en el conchal GUA-10, Archipiélago de las Guaitecas, Patagonia septentrional: *Un Estudio Arqueomalacológico*. Universidad de Tarapacá, Memoria de Título

Castilla J (1988) Earthquake-caused coastal uplift and its effects on rocky intertidal kelp communities. Science 242:440–443

Castillo M, Bello M, Reyes H, Guerrero Y (2006) Patrones de corrientes y distribución vertical de temperatura y salinidad en la entrada oceánica del canal Darwin en invierno y primavera de 2002. Ciencia y Tecnología del Mar. 29(2):5–21

Cembrano J, Hervé F, Lavenu A (1996) The Liquiñe Ofqui fault zone: a longlived intra-arc fault system in southern Chile. Tectonophysics 259(1–3):55–66

Chappell J, Omura A, Esat T, McCulloch M, Pandolfi J, Ota Y, Pillans B (1996) Reconciliation of late Quaternary sea levels derived from coral terraces at Huon Peninsula with deep sea oxygen isotope records. Earth Planet Sci Lett 141(1–4):227–236

Cisternas M, Atwater B, Torrejon F, Sawai Y, Machuca G, Lagos M, Eipert A, Youlton C, Salgado I, Kamataki T, Shishikura M, Rajendran C, Malik J, Rizal Y, Husni M (2005) Predecessors of the giant 1960 Chile earthquake. Nature 437(7057):404–407

Darwin Ch (1845) The Voyage of the Beagle (Journal of researches) into the natural history and geology of the countries visited during the voyage of H.M.S. Beagle round the world, under the Command of Capt. Fitz Roy, R.N. 2ª edición, Londres: John Murray, Cap. 13:273–290

Delgado S (2004) Relación entre el perfil de basamento en fiordos y canales y la morfoestructura regional en norpatagonia. Universidad de Chile, Santiago

de Porras ME, Maldonado A, Abarzúa A, Cárdenas M, Francois JP, Martel A, Stern C, Méndez C, Reyes O (2012) Postglacial vegetation, fire and climate dynamics at Central Chilean Patagonia (Lake Shaman, 44°S). Quarter Sci Rev 50:71–85

de Porras ME, Maldonado A, Quintana F, Martel-Cea A, Reyes O, Méndez C (2014) Environmental and climatic changes in central Chilean Patagonia since the Late Glacial (Mallín El Embudo, 44° S). Clim Past 10(3):1063–1078

Díaz-Naveas J, Frutos J (2010) Geología Marina de Chile. Comité Oceanográfico Nacional de Chile—Pontificia Universidad Católica de Valparaíso—Servicio Nacional de Geología y Minería de Chile.

Dietrich R, Ivins E, Casassa G, Lange H, Wendt J, Fritsche M (2010) Rapid crustal uplift in Patagonia due to enhanced ice loss. Earth Planet Sci Lett, 22–29

Doake C (2001) Ice-shelf stability. British Antarctic Survey, 1282–1290

Fitz-Roy R (1839) Voyages of the adventure and beagle. Volume II. Proceedings of The Second Expedition, pp 1831–1836. Henry Colburn (ed. London.)

Flueck W, Smith-Flueck J (2012) Huemul heresies: beliefs in search of supporting data. Historical and zooarcheological considerations. Animal Product Sci 52(8):685–693

Francoise J (2014) Antecedentes (paleo) ambientales de la Patagonia Noroeste (41°-48°S). Manuscrito Fondecyt, 1130151

Gajardo R (1994) La vegetación natural de Chile. Editorial Universitaria, Santiago

García J (2012) Late Pleistocene ice fluctations and gacial geomorphology of the Archipielago de Chiloe, Southern Chile. Geografiska Annaler: series a, physical geography 94(4):459–479

Garreaud R, Vuille M, Compagnucci R, Marengo J (2009) Present-day South American climate. Palaeogeogr Palaeoclimatol Palaeoecol 281(3–4):180–195

Glasser N, Jansson K, Harrison S, Kleman J (2008) The glacial geomorphology and Pleistocene history of South America between 38°S and 56°S. Quatern Sci Rev 27:365–390

Glasser N, Harrison S, Winchester V, Aniya M (2004) Late pleistocene and holocene paleoclimate and glacier fluctuations in Patagonia. Global and Planet Chan, 79–101

Glasser N, Jansson K, Mitchell W, Harrison S (2006) The geomorphology and sedimentology of the 'Tempanos' moraine at Laguna San Rafael Chile. J Quat Sci 21(6):629–643
Glasser N, Jansson K, Harrison S, Rivera A (2005) Geomorphological evidence for variations of the North Patagonian Icefield during the Holocene. Geomorphology 71(3–4):263–277
Glasser N, Ghiglione M (2009) Structural, tectonic and glaciological controls on the evolution of fjord landscapes. Geomorphology 105(3–4):291–302
Gonzalez G, Torres-Mura J, Munoz-Pedreros A (2009) Orden Artiodactyla. In Mamíferos de Chile (A. Munoz-Pedreros y J. Yanez, eds.): 231–250. Valdivia: CEA Ediciones
Haberle S, Bennett K (2004) Postglacial formation and dynamics of North Patagonian Rainforest in the Chonos Archipelago Southern Chile. . Quatern Sci Rev 23(23):2433–2452
Haberle S, Szeicz J, Bennett K (2000) Late Holocene vegetation dynamics and lake geochemistry at Laguna Miranda, Regi6n XI Chile. Rev Chil Nat Hist 73:655–669
Harrison S, Glasser N, Winchester V, Haresign E, Warren C, Duller G, Bailey R, Ivy-Ochs S, Jansson K, Kubik P (2008) Glaciar Leon, Chilean Patagonia: late-Holocene chronology and geomorphology. Holocene 18(4):643–652
Hervé F, Ota Y (1993) Fast Holocene uplift rates at the Andes of Chiloé, southern Chile. Andean Geol 20(1):15–23
Herve F, Mpodozis C, Davidson J, Godoy E (2010) Observaciones estructurales y petrográficas en el basamento metamórfico del Archipiélago de los Chonos, entre el canal King y el canal Ninualac, Aisén. Andean Geol, 13–14
Heusser C (1960) Late-pleistocene environments of the Laguna de San Rafael Area Chile. . Geogr Rev 50(4):555–577
Heusser C (2003) Ice Age Southern Andes. A chronicle of paleocological events. Amsterdam, The Netherlands, Elsevier B.V.
Heusser C, Heusser L, Hauser A (1992) Paleoecology of late quaternary deposits in Chiloé Continental, Chile. Revista Chilena de Hist Natur 65:235–245
Holdgate M (1961) Vegetation and soils in the South Chilean Islands. J Ecol 49(3):559–580. https://doi.org/10.2307/2257223
Hulton N, Purves R, McCulloch R, Sugden D, Bentley M (2002) The last glacial maximum and deglaciation in southern South America. Quat Sci Rev, 233–241
Isla F (1989) Holocene sea-level fluctuation in the southern hemisphere. Quatern Sci Rev 8:359–368
Jiménez J (2010) Southern pudu, Pudu puda (Molina 1782). Neotropical cervidology: biology and medicine of latin American Deer. Funep y IUCN, Jaboticabal y Gland, 140–150
Kempf P, Moernaut J, Van Daele M, Vandoorne W, Pino M, Urrutia R, De Batist M (2017) Coastal lake sediments reveal 5500 years of tsunami history in south central Chile. Quatern Sci Rev 161:99–116
Lara L (2008) Holocene submarine volcanoes in the Aysén fjord, Patagonian Andes (44°S): relations with the Liquiñe-Ofqui Fault Zone. In: 7th International Symposium on Andean Geodynamics (ISAG 2008, Nice), Extended Abstracts, pp 285–288
Lara L, Cembrano J, Lavenu A (2008) Quaternary vertical displacement along the Liquine-Ofqui fault zone: differential uplift and coeval volcanism in the Southern Andes. Int Geol Rev 50(11):975–993
Lomnitz C (1970) Major Earthquakes and Tsunamis in Chile during the period 1535 to 1955. Mexico: Geologische Rundschau 59(3):938–960
Long A (2001) Mid-Holocene sea-level change and coastal evolution. Prog Phys Geogr 25(3):399–408
Lozano P (2006) Manuscrito. Descripción y caracterización biogeográfica de las grandes unidades ambientales de la Patagonia chilena
Luebert F, Pliscoff P (2006) Sinopsis bioclimática y vegetacional de Chile. Editorial Universitaria, Santiago
Lumley S, Switsur R (1993) Late quaternary chronology of the Taitao Peninsula, southern Chile. J Quat Sci 8(2):161–165

Markgraf V, Whitlock C, Haberle S (2007) Vegetation and fire history during the last 18,000 cal yr B.P. in Southern Patagonia: Mallín Pollux, Coyhaique, Province Aisén (45°41"30"" S, 71°50"30"" W, 640 m elevation). Palaeogeogr Palaeoclimatol Palaeoecol 254:492–507
Massaferro J, Brooks S (2002) Response of chironomids to Late Quaternary environmental change in the Taitao Peninsula, southern Chile. J Quat Sci 17(2):101–111
McCulloch R, Morello F (2009) Evidencia glacial y paleoecologica de ambientes Tardiglaciales y del Holoceno temprano. Implicaciones para el poblamiento temprano de Tierra del Fuego. In: Salemme M, Santiago F, Alvarez M, Piana E, Vásquez M, Mansur E, Utopias U (eds.), Arqueología de la Patagonia. Una mirada desde el último confín, pp 119–136
Mella J, Simonetti J, Spotorno A, Contreras L (2002) Mamíferos de Chile. In: Ceballos, G, Simonetti J (eds.), Diversidad y conservación de los mamíferos neotropicales. CONABIO-UNAM. México, D.F.
Mercer J (1976) Glacial history of southernmost South America. Quatern Res 6:125–166
Mercer J (1982) Holocene glacier variations in southern South America. Striae 18:35–40
Milne G, Gehrel W, Hughes C, Tamisiea M (2009) Identifying the causes of sea-level change. Nat Geosci 2(7):471–478
Milne G, Shennan I (2007) Sea level studies. Isostasy. In: Elias S (ed), Encyclopedia of Quaternary Science, pp 3043–3051. Oxford, Elsevier
Moreno P (2004) Millennial-scale climate variability in northwest Patagonia over the last 15000 yr. J Quat Sci 19(1):35–47
Moreno P, León A (2003) Abrupt vegetation changes during the last glacial to Holocene transition in mid-latitude South America. J Quat Sci 18:787–800
Murray-Wallace C (2007) Sea level studies. Eustatic Sea-Level Changes Since the Last Glaciation. In: Elias SA (ed), Encyclopedia of quaternary science, pp 3034–3043. Oxford, Elsevier
Nakada M, Lambeck K (1988) The melting history of the late Pleistocene Antarctic ice sheet. Nature 333:36–40
Naranjo J, Stern C (2004) Holocene tephrochronology of the southernmost part (42°30'-45°S) of the Andean Southern Volcanic Zone. Rev Geol Chile 31(2):225–240
Naranjo J, Arenas M, Clavero J, Muñoz O (2009) Mass movement-induced tsunamis: main effects during the Patagonian Fordland seismic crisis in Aisén (45°25'S), Chile. Andean Geology 36:137–145
Navarro J, Pequeno G (1979) Peces litorales de los archipielagos de Chiloe y los Chonos, Chile. Revista Biología Marina 16:205–309
Nelson A (2007) Sea-levels, late quaternary. Tectonic Locations. In: Elias SA (ed), Encyclopedia of quaternary science, pp 3072–3087. Oxford, Elsevier
Osorio C, Reid D (2004) Moluscos marinos intermareales y submareales entre la Boca del Guafo y el estero Elefantes, sur de Chile. Investigaciones Marinas 32(2):71–89
Paskoff R (2010) Geomorfología costera. Geología marina de Chile. Comité Oceanográfico Nacional, Valparaíso
Pequeño G (2000) Delimitaciones y relaciones biogeograficas de los peces del pacifico suroriental. Estud Oceanol 19:53–76
Pequeño G, Riedemann A (2002) Los peces bentónicos de los canales orientales de Aysén (XI Región, Chile), con nuevos avances sobre su distribución geográfica. Resultados del Crucero CIMAR 7:149–154
Pesce O, Moreno P (2014) Vegetation, fire and climate change in centraleast Isla Grande de Chiloé (43°S) since the Last Glacial Maximum, northwestern Patagonia. Quatern Sci Rev 90:143–157
Plafker G, Savage JC (1970) Mechanism of the Chilean Earthquakes of May 21 and 22, 1960. Geol Soc Am Bull 81(4):1001–1030
Porter S, Stuiver M, Heusser C (1984) Holocene sea-level changes along the Strait of Magellan and Beagle channel, southernmost South America. Quatern Res 22(1):59–67
Reed D, Wood, R, Best J (1988) Earhquakes rivers and ice: scientific research at the Laguna San Rafael, southern Chile. The Geograph J, 392–405

Reyes O, Méndez C, Maldonado A, Velásquez H, Trejo V, Cárdenas M, Abarzúa A (2009) Uso del espacio de cazadores recolectores y paleoambiente Holoceno en el valle del río Cisnes, Región de Aisén Chile. Magallania 37(1):91–107

Reyes O, Méndez C, San Román M, Francoise J (2018) Earthquakes and coastal archaeology: assessing shoreline shifts on the southernmost Pacific coast (Chonos Archipelago 43°50′–46°50′ S, Chile, South America). Quatern Int 463:161–175

San Román M (2014) Sea-level changes and coastal peopling in southernmost Pacific South America: Marine hunters from Patagonia. In: Encyclopedia of global archaeology, pp 6515–6525. Springer, New York

Santiago F, Bujalesky G, Salemme M (2007) Prospección arqueológica en la cuenca del río Chico. Tierra del Fuego, Argentina. Arqueología de Fuego-Patagonia. Levantando piedras, desenterrando huesos... y develando arcanos, 357–363

Sepúlveda J, Pantoja S, Hughen K, Bertrand S, Figueroa D, León T, Drenzek N, Lange C (2009) Late Holocene sea-surface temperature and precipitation variability in northern Patagonia, Chile (Jacaf Fjord, 44°S). Quatern Res 72:400–409

SERNAGEOMIN (2003) Mapa Geológico de Chile: versión digital. Servicio Nacional de Geología y Minería, Publicación Geológica Digital N°4, Santiago

SERPLAC (2005) Atlas de Aysén. Ediciones LOM, Chile

Shennan I (2007) Sea level studies. Overview. In: Elias S (ed), Encyclopedia of quaternary science, pp 2967–2974. Oxford, Elsevier

SHOA (2000) El maremoto del 22 de mayo de 1960 en las costas de chile. Ediciones Servicio Hidrográfico y Oceanográfico de la Armada de Chile, Valparaíso

Sievers H, Silva N (2006) Masas de agua y circulación en los canales y fiordos australes. In: Silva N, Palma S (eds), Avances en el conocimiento oceanográfico de las aguas interiores chilenas, Puerto Montt a cabo de Hornos, pp 53–58. Comité Oceanográfico Nacional—Pontificia Universidad Católica de Valparaíso, Valparaíso

Stern C (2004) Active Andean volcanism: its geologic and tectonic setting. Revista geológica de Chile 31:161–206

Thomson S (2002) Late Cenozoic geomorphic and tectonic evolution of the Patagonian Andes between latitudes 42°S and 46°S: an appraisal based on fission-track results from the transpressional intra-arc Liquiñe-Ofqui fault zone. Geol Soc Am Bull 114(9):1159–1173

Vargas G, Rebolledo S, Sepúlveda S, Lahsen A, Thiele R, Townley B, Padilla C, Rauld R, Herrera MJ, Lara M (2013) Submarine earthquake rupture, active faulting and volcanism along the major Liquiñe-Ofqui Fault Zone and implications for seismic hazard assessment in the Patagonian Andes. Andean Geol 40:141–171

Vidal Gormaz F (1901) Hundimientos y solevantamientos verificados en las costas chilenas. Revista Chilena de Historia Natural 10:213–224

Vita-Finzi C (1996) Paleoseismology in coastal Chile. J Geophys Res Solid Earth (1978–2012) 101(B3):6109–6114

Viviani C (2004) Manuscrito. Caracterización del litoral de la Región de Aisén. Universidad de Magallanes, Punta Arenas

Vuilleumier F (1985) Forest birds of Patagonia: ecological geography, speciation, endemism, and faunal history. Ornithol Mono 36:255–304

Winchester V, Harrison S (2000) Dendrochronology and lichenometry: colonization, growth rates and dating of geomorphological events on the east side of the North Patagonian Icefield Chile. Geomorphology 34(3–4):181–194

Winchester V, Harrison S, Warren CR (2001) Recent Retreat Glaciar Nef, Chilean Patagonia, Dated by Lichenometry and Dendrochronology. Arct Antarct Alp Res 33(3):266–273

Zamorano J, Gibbons J, Capella J (2010) Diversity and summer distribution of cetaceans in inlet waters of northern Aisen, Chile. Anales del Instituto de la Patagonia 38(1):151–157

Zangrando F, Ponce J, Martinoli M, Montes A, Piana E, Vanella F (2016) Palaeogeographic changes drove prehistoric fishing practices in the Cambaceres Bay (Tierra del Fuego, Argentina) during the middle and late Holocene. Envir Archaeol 21(2):182–192

Chapter 2
Background

Abstract We present a summarised view of the main historical antecedents on the disarticulation of the way of life of one of the ethnic hunter-gatherer groups of Patagonia, known generically as Chono and who occupied the northern area of the Patagonian archipelago. This cultural dismemberment occurred in barely 200 years after European contact (18th century). The records of the first sightings, although imprecise, coincide in describing a subsistence pattern based on their ability to navigate and on obtaining marine resources. From the archaeological point of view, the main known antecedents of the areas adjacent to the archipelago are reviewed, presenting the main topics discussed around the antiquity of the coastal adaptation manifestations and the technology used. The lines of work carried out around other sites discovered are also mentioned, which relate a sequence of human occupation from 6,000 cal. BP to the late ceramic period. Topics are explored that transcend regional archeology and refer to the possible origins of coastal adaptation in Patagonia, the dispersion process and the possible population influences in relation to the lithic technological components (mainly lanceolate points of medium and large size) characteristic of the northern archipelago.

Keywords Western patagonia · Chonos archipelago · Chono natives · Regional archaeology · Coastal archaeology · Marine hunter-gatherers · Archipelagic settlements · Biogeographic barrier · European contact · Cultural dismantling

2.1 Historical Information on the Last Inhabitants of the Chonos Archipelago

The indigenous group known in ethnohistory as Chono (e.g. Byron 1901 [1746]; Cooper 1917, 1946; Ibar 1959; Latcham 1928) inhabited the northern part of the Patagonian archipelagos, between the south of Chiloé and the Taitao Peninsula (43°S–47°S); however ethnohistorical sources for the northern archipelagos mention a great diversity of indigenous factions and ethnic denominations (Álvarez 2002; Urbina 2016). Of the various indigenous groups known as canoe peoples for whom

O. Reyes, *The Settlement of the Chonos Archipelago, Western Patagonia, Chile*, The Latin American Studies Book Series,
https://doi.org/10.1007/978-3-030-54326-6_2

we have the slightest written and material records, the Chono was the first to disappear (Emperaire 1963; Martinic 2004, 2005; Quiroz and Olivares 1988; Urbina 1988, 2014). The principal direct references to the Chono are to be found in the first hand sources of Bibar (1966 [1558]); Byron (1901 [1746]); De Ovalle (1969 [1646]); De Goicueta 1879 [1557–1558]; De Vea (1886 [1675]) and García (1889 [1766]), while other secondary sources of interest can be found in Álvarez (2002), Cárdenas et al. (1991), Cooper (1917, 1946), Emperaire (1963), Latcham (1928), Martinic (2005), R. Urbina (1983, 1988), X. Urbina (2010, 2013, 2016) and other authors.

The first sightings and contacts started in 1553 in the early days of the Colony, during the voyage of Captain Francisco de Ulloa, accompanied by Hernán Gallego and Francisco Cortez de Ojea, to reconnoitre the archipelagos (De Goicueta 1879[1557–1558]). This first encounter off the Taitao Peninsula (47°S) turned into a fight, creating an impression which would never change. From then on, most of the expeditions to reconnoitre, conquer and evangelise the region sailed from Chiloé Island (Hanisch 1982; Urbina 1983, 1988). They give eloquent descriptions of human groups who lived off the sea, hunting sea-lions and birds, eating the abundant shellfish of the archipelago and travelling in canoes known as *dalcas* (Fig. 2.1) which received high praise for their seaworthiness (Bibar 1966 [1558]; de Ovalle 1969 [1646]; García 1889 [1766]; Urbina 2010, 2013); they differed from bark canoes seen south of the Taitao Peninsula.

Fig. 2.1 Artist's impression of the activities of a group of marine hunter-gatherer-fishers on the shoreline

In parallel with these expeditions of exploration and conquest, the *encomienda* system was introduced in mainland Chile. This was the principal economic driver of the colonial period, based on obtaining permanent slave labour for productive work on the new and growing Spanish *haciendas*. In the south of the colony, the War of Arauco meant that local indigenous groups could not be used to supply labour for this system in central and northern Chile, so landowners turned their eyes towards the newly explored archipelagos. The result was forced transportation of the canoe peoples captured by Spanish *conquistadors* in raids to obtain the slave labour required by their economy, seizing entire families and shipping them first to Chiloé Island and then as far north as Conception, Santiago, La Serena and even Lima (Mansilla 2002; Urbina 2016).

At the same time, the early Jesuit missions of the sixteenth and seventeenth centuries, spreading across from Paraguay [1], tried to evangelise these *"gentiles"* and to stop the scandalous Spanish slaving raids from Chiloé Island (Moreno 2011; García 1889 [1766]; Urbina 2016). They also instigated mass transport and settlement, although peacefully, to people missions like those on the islands of Huar, Cailín and Chaulinec in the Chiloé Archipelago, resulting not only in profound changes in the culture of the Chono, but also in miscegenation with Huilliche and Mapuche individuals, as well as Europeans (Harb et al. 1995; Moreno et al. 1999; Munizaga 1978a, b). These more numerous indigenous groups had already converted to a sedentary, agricultural way of life. Other individuals, fleeing these forced and/or pacific movements, migrated to the area south of the Taitao Peninsula—away from the routes frequented by the Spanish—where they mixed with Kawésqar groups (De Vea 1886 [1675]).

The movement and transportation of large numbers of Chono, the massacres, the new diseases introduced by contact with other populations and acculturation, for example, through the introduction of agricultural practices and new commercial strategies (Urbina et al. 2020) resulted in the dismantling of their culture (Reyes et al. 2019a); with their low population density, they were unable to absorb so many changes in such a short time and recover. By the end of the eighteenth century the Chono, described with different emphasis as hunters, gatherers, fishermen and sailors of the archipelago, had disappeared as a culture (subsistence techniques, language, rituals) (Moraleda 1888 [1786]; Simpson [1871, 1875]). In his voyage of 1675–1676, the sailor Antonio de Vea (1886) recorded how the Chono avoided the routes commonly used by the Spanish, while Antonio Machado, in an exploration in 1769, found only a few Indians who lived in the mission of Cailín, in the Chiloé Archipelago, but no evidence of other Indians in the area (Machado 1889 [1769]). FitzRoy says: *"Indians do not frequent that quarter; indeed, no traces of them whatever were found by Mr. Stokes, or any of our party, among the Chonos islands"* (FitzRoy 1839: 375). There were not even any drawings or scientific expeditions which could give a better idea of the indigenous groups of the archipelago, as occurred further south (e.g. Chapman 2012; Darwin 1845; FitzRoy 1839; Martial et al. 2007) with the increasing hydrographical surveying and scientific expeditions of the nascent republic. The collapse of this world took barely 200 years from the first unfriendly encounter in 1553.

One of the most eloquent testimonies of the depopulation of the original inhabitants of the area was written a century later by the seaman and explorer Enrique Simpson: "*All the remains and vestiges of this race have already almost disappeared at the hands of the brutal executioners; who consider it merit to destroy all that they find of these, to them, abominable gentiles. At the same time, the great tidal waves caused by earthquakes, which have visited the archipelago in the last century, passing over the low beaches frequented by the Chonos, have greatly helped to cause the memorials left by them to disappear; and also the dense vegetation has covered once more the sites that they cleared... Their dwellings were caves and sometimes circular huts, the stakes of which I have seen. They often buried their dead close to these dwellings; but generally they preferred to place them in caves, covering them with branches. In some of these, seamen, in times gone by, found mummies preserved in coffins of cypress bark in the shape of eggs; but they have all been removed or destroyed*" (Simpson 1871: 192).

As may be appreciated, when we examine the logs, voyage reports, diaries and publications of primary sources, or historical studies derived from these, a series of doubts arise about accounts containing no more than sketches or superficial mentions. They are irregular, mediatised by the historical and religious context and by their authors (see Orquera and Piana 1995). By way of example, the limits of their marine travels are estimated to be between the south of Chiloé and the Taitao Peninsula, depending on the different factions or ethnic groups into which they were divided at the time, either by dialect, geographical location or references of others (Álvarez 2002). Some accounts mention vegetable plots and cultivation of potatoes (e.g. Bibar 1966 [1558]), while others only say that these groups lived exclusively by hunting, without undertaking any kind of agricultural work such as they were taught in the missions (e.g. García 1889 [1766]). There are very few descriptions of their technology in terms of the use of tools (stone, bone, leather and baskets); however, the absence of pottery is cited to highlight their cultural backwardness. Mentions of their everyday and social life are limited to reports of certain anecdotes, superstitions, clothing, food and sometimes funerary practices, as well as descriptions of hunting expeditions. The wreck of the English frigate *Wager* in the Guayaneco archipelago south of the Taitao Peninsula, and the journey to the safety of the survivors across the Ofqui Isthmus to Chiloé Island thanks to the guidance of indigenous Chono Indians, is perhaps one of the best-known references today (see Byron 1901 [1746]; Carabias 2009; García 1889 [1766]; Urbina 2013, 2014).

The technological item most widely mentioned and described, admired, used and reformulated by the Spanish and other Europeans, due to its versatility in stormy waters, is the *dalca* used by the Chono, consisting of three planks and able to carry a variable number of passengers (see Bibar 1966 [1558]; Byron 1901 [1746]; García 1889 [1766]). In the report of the first expeditionary voyage to the Patagonian archipelagos of Francisco de Ulloa in 1553, we read: "*On land we found a canoe made of three planks very well sewn of 24 to 25 feet, and on the stitching they had daubed a paste that they prepare. Its shape was like that of a shuttle, with very high ends...*" Bibar 1966: 179 [1558]. Later studies propose that this canoe was probably a technology derived from Huilliche territory further north, in the northern part of the

Chiloé Archipelago, since these forest-dwelling groups were specialists in the use of wood; and that it only later spread to the Chonos Archipelago (Medina 1984). After European contact and the forced migrations of many Chono, or their flight to safer areas south of the Taitao Peninsula (47°S), the *dalca* would have gradually replaced the bark canoe used by Kawésqar groups (Carabias et al. 2007; Cooper 1917; Finterbusch 1934; Latcham 1930; Lira et al. 2015; Lothrop 1932; Medina 1984; Trivero 2018).

This information allows us to draw up a general ethnographical panorama (see Álvarez 2002; Cooper 1917, 1946; Emperaire 1963; Latcham 1928; Martinic 2005) in which the first sailors and explorers, and later scientific expeditions of the nineteenth century, defined a limited range for the Chono people based on their physical appearance, the island environment in which they moved and their way of obtaining resources, describing them as canoe Indians like the Yámana and Kawésqar. They were clearly distinguished from the land-based tribes of the eastern Andes and the steppe, like the Aonikenk and the Selknam (see Casamiquela 1973; Latcham 1928; Martinic 1989; Skottsberg 1913), and differed also from the more northerly groups like the Huilliche and Mapuche (Cárdenas et al. 1991; Cooper 1946).

This group of indigenous canoe peoples, known to ethnohistory as Chono, has been invisibilised in the general ambit of Patagonian cultures. In the compilation *La memoria olvidada. Historia de los pueblos indígenas de Chile*, they are relegated to a footnote (Bengoa 2004: 578). The same situation is repeated in the *Informe de la Comisión Verdad Histórica y Nuevo Trato con los Pueblos Indígenas* (2008: 516); in this case, it is incomprehensible that a vanished people should merit no more than a footnote. It must be recognised that even the accounts of the slaving expeditions gave much more information and recognised their existence as a people or nation, in contrast to some historians and the State of Chile today.

2.2 The Archaeological Areas Adjacent to the Chonos

The northern part of the Patagonian Archipelago (approx. 41°30′ to 47°S), including the Reloncaví Sound, the Chiloé Archipelago and the mainland coast opposite, and the Guaitecas and Chonos Archipelagos, forms a large geographical zone 600 km long. It interrupts the continuity of the Pacific coast, which breaks up into an island system with strong bio-geographical barriers: the channels, fjords, gulfs, mountain ranges and ice fields which characterise the west coast of the southern tip of South America (Camus 2001). The first signs of human occupation in this oceanic environment (Garreaud 2009), modelled mainly by tectonic activity (see Plafker and Savage 1970) and glacial action (see McCulloch et al. 2000) go back to ~6500 years cal. BP. These would be groups of marine hunter-gatherers, as is shown by datings obtained in different deposits discovered in the Reloncaví Sound, Chiloé Island and the Guaitecas Archipelago (Legoupil 2005; Ocampo and Rivas 2004; Porter 1993; Reyes et al. 2016b; Rivas et al. 1999).

The archaeology of hunter-gatherer groups adapted to the littoral and marine environment begins on the mainland, in the Reloncaví Sound. This is recognisably the start of this way of life in the western channels of Patagonia, differing from the consumption of coastal resources evidenced in Monte Verde (Dillehay et al. 2008). The first controlled excavations were carried out in the shell-midden of Capera Huapi Island, opposite Calbuco, by J. Bird in the 1930s and 40s (1988); these were the first stratigraphic descriptions in which large lanceolate stone projectile points and stone axes were found. Bird mentioned, in his tour through the western channels of Patagonia, the existence of large shell-middens in eastern Chiloé (the inland sea). He also appreciated the geomorphological changes of the coastline through observation of the base deposits of the shell-middens on Chiloé, their erosion profiles and the sunken forests of the Taitao Peninsula. However, decades would pass before specific archaeological excavations would be carried out maintaining a record of the materials recovered and interpretations of the culture which used them (e.g. Díaz and Garretón 1972/73; Vásquez de Acuña 1963).

It was only in the 1990s and early years of the twenty-first century that new investigations were carried out under a series of funded projects in the Chiloé area (e.g. Aspillaga et al. 1995; Ocampo and Rivas 2004, 2008; Rivas and Ocampo 2005, 2006; Rivas et al. 1999). Stress was laid on the older finds from the end of the Mid-Holocene, like the Puente Quilo site located on the north coast of Chiloé Island, in the Gulf of Quetalmahue; the area of Chiloé/Reloncaví Sound was proposed as a possible point of origin of the groups specialising in marine resources (Ocampo and Rivas 2004). In general terms, this site was a settlement and workshop occupied by hunter-gatherer groups with a maritime subsistence strategy. In its formation in the Mid-Holocene (level 110–120 from 5030 years BP, ~6000 years cal. BP) (Ocampo and Rivas 2004), a sector of the site can be identified devoted to the production of stone tools (leaf-shaped, lanceolate, double point and subtriangular bifacial points); this situation is replicated in higher levels, where occupations associated with ceramic technology are also recorded. The remains of seven individuals of both sexes were recovered at different levels. Funerary deposition in right lateral flexed position was identified, with the presence of red body paint (Aspillaga et al. 1995; Ocampo and Rivas 2004); these are characteristic of the funerary pattern found in the area (Munita et al. 2011).

Subsequently, Legoupil (2005) added more evidence of occupation of Chiloé during the Mid-Holocene by systematic prospections in Chadmo, Huildad and Yaldad Fjords on the coast of the southern part of the island. A series of archaeological shell-middens were recorded by the Yaldad, Huildad and Chadmo streams. In general, these shell-middens are located on the shoreline (less than 20 m from the sea) at an elevation of between 1 and 9 masl. The most interesting of these big shell-middens is the Yaldad 2 site, with evidence of occupation from the Mid-Holocene. Tests in this shell-midden enabled its depth to be estimated as at least 3.80 m; crushed shells at this level were dated to 5950 ± 80 BP. The size of these shell-middens and the low presence of associated lithic and osseous material, documented by Bird (1988), led the author to wonder whether the formation of these monumental middens can be attributed to an emphasis on gathering practices over other subsistence activities:

"So the inhabitants of south-east Chiloé appear to have been more shell gatherers than true marine hunters" (Legoupil 2005: 60).

Important information on specialised marine adaptation has also been obtained from the study of a series of archaeological sites distributed in the area of the Reloncaví Sound as part of environmental impact studies. These have led to the recording of a series of archaeological shell-middens with occupation from around the Mid-Holocene (~6400 to 5200 years cal. BP), such as Piedra Azul, Puntilla Tenglo, Centro Acuicultura Metri and Ilque 1 (Campbell 2012; Flores and Lira 2006; Flores 2007; Flores and Correa 2011; Gaete et al. 2004; Gaete and Navarro 2004; Ladrón de Guevara et al. 2003; Munita 2017). So far the lowest stratigraphic components of the Piedra Azul and Ilque 1 shell-middens present the oldest littoral occupations on the coast of the mainland, at the start of the northern channels, with dates of ~6400 years cal. BP. The stratigraphic succession terminates in coastal occupations with pottery, and shows evidence of an industry producing lanceolate and bifacial points related to the consumption of marine resources obtained by hunting, fishing and gathering (Gaete et al. 2004; Gaete and Navarro 2004; Munita 2017), as well as human burials. Centro Acuicultura Metri (Flores and Correa 2011; Flores and Lira 2006) and Puntilla Tenglo present initial occupation from ~5500 to 5200 years cal. BP (Álvarez et al. 2008; Ladrón de Guevara et al. 2003), and the same bifacial technology and manufacture of lanceolate stone projectile points are observed (Galarce 2005a, b, c) in a context of hunting, gathering and fishing for resources from the littoral and marine environments.

Thus, the material record of this specialist maritime adaptation has been detected both on islands of the Chiloé Archipelago and on the mainland north of Reloncaví Sound (Flores and Lira 2006; Flores 2007; Massone et al. 2016; Munita 2007; Munita et al. 2011; Navarro 1995, 2001; Gaete et al. 2004; Gaete and Navarro 2004; Quiroz and Sánchez 2004). In terms of the Mid-Holocene component, we note a recurrence in the remains of lithic technology associated with large lanceolate or leaf-shaped points, some manufactured on rhyolithic obsidian from Chaitén Volcano. This raw material is widely distributed along the Pacific coast, from the Chan-Chan 18 site, on the coast of the south Chilean mainland, to Goñi Island in the Chonos Archipelago (Munita 2007; Stern and Curry 1995; Stern and Porter 1991; Stern et al. 2002). Similar records from south of the Chiloé Archipelago include the GUA-010 archaeological deposit from the Mid-Holocene (~6200 years cal. BP) associated with the presence of lanceolate points similar to those described for the Chiloé area, where we also recorded rhyolithic obsidian from Chaitén Volcano (Porter 1993; Reyes et al. 2016b; Stern and Porter 1991). The Seno Gala 1 site, on the mainland coast of the Chonos Archipelago, also presented lanceolate points found in eroded sectors of the coast; the geochemical source of these tools was identified as Chaitén in mainland Chiloé (Reyes et al. 2007).

Thus, the archaeological record shows the extensive spatial distribution of this rhyolithic obsidian. The recorded distribution has been modified significantly in recent years to include data from the north-east coast of the Magellan Strait and the Atlantic coast. This obsidian, collected from the surface, can be geochemically identified on a stretch of coastline over 600 km long from Chan-Chan 18, in the Los

Lagos Region, to Goñi Island (Stern and Curry 1995) and the south coast of Traiguén Island (Reyes et al. 2011) in the Chonos Archipelago. It has also been identified at M11 Zona Norte and A1 Cabeza de León, two terrestrial hunter sites on the Atlantic coast of the province of Santa Cruz, Argentina; and at the Alero Frailes 2 site in the volcanic zone of Pali Aike on the border between Argentina and Chile, 1,000 km from the source. It was, therefore, transported and dispersed over a range of some 1,400 km (Stern et al. 2012; Stern 2018).

As a result of these finds, the archaeological record of the transition to the Late Holocene and the historical period in Reloncaví Sound and the Chiloé area began to be recognised for its diversity of deposits and types of sites (Flores et al. 2010), for example, large, multicomponent, mound-shaped shell-middens, ephemeral deposits with ceramic materials, rock paintings and carvings in caves, fishing pens and funerary sites, among other types of evidence (Álvarez et al. 2008, 2019; Labarca et al. 2016; Ladrón de Guevara et al. 2003; Mena et al. 2011; Munita et al. 2005, 2011, 2016; Munita 2007; Rodríguez et al. 2010; Sáez 2008). The appearance of pottery attributed to Huilliche populations of Chiloé and southern Chile has not been analysed in detail in some of the sites where it is recorded, beyond TL (thermoluminescence) dating, for example, in the upper levels of Puente Quilo and Piedra Azul (Gaete et al. 2004; Rivas et al. 1999); the Centro Acuicultura Metri, Yaco Alto 1 and Ilque 1 sites are exceptional in this respect (Flores and Correa 2011; Munita 2017; Munita et al. 2011). It is estimated that the presence of pottery is an important marker in the archaeological record as it indicates changes in the systems of settlement and space organisation, particularly associated with semi-sedentary agricultural settlements and the arrival of groups with different cultural roots to the base occupations, which would have been hunter-gatherer-fisher groups (Flores et al. 2010). Information on the presence of agriculture is scarce, and we have no solid evidence; however, it appears to have developed in historical times (Rivas and Ocampo 2006), although some authors propose that it may have been present for approximately the past 1,000 years BP (Aldunate 1996; Cooper 1917, 1946; Menghin 1962).

It was only with the unpublished reports of Galarce (2005a, b and c) of the analysis of lithic material recovered from sites in Chiloé (Puente Quilo and Chepu) and Reloncaví Sound (Punta Metri; PM005, PM009 and PM012), and with a time sequence covering both the Mid and Late Holocene, that a model of occupation and use of space was proposed which identified differences in subsistence and settlement strategies based on the particularities of the lithic technology recovered in the series of levels excavated. To summarise, during the Mid-Holocene the sites identified would be logistical camps with specialised (bifacial) lithic tools for hunting marine fauna, as well as fishing and gathering. The sites formed in the Late Holocene would be residential camps with fewer specialised lithic tools and greater diversity for obtaining different faunal resources. This explanation developed by Galarce (2005a, b and c) was taken up by Rivas and Ocampo (2006), who proposed it as an explanation of the human occupation of the area of the Chiloé Archipelago and the Reloncaví Sound.

The Chonos Archipelago extends more than 600 km to the south, as far as the Taitao Peninsula and the Gulf of Penas (47°S). This forms the central portion of the extended Patagonian Archipelago, and is largely isolated from settled territories on the mainland by the Southern Ice Field. Archaeological work in this part of the archipelago is even more scarce and scattered. It includes the identification and excavation of late shell-middens (approx. 2200 years BP) in the area of the Guayaneco Archipelago (e.g. Stuven 1 site, Legoupil et al. 2007), the identification of a series of sites around the Messier Channel (Bird 1988; Emperaire 1963), with chronologies later than 1800 years BP (Curry 1991) and the record of funerary caves and rock art in the Madre de Dios Archipelago (see Jaillet et al. 2009). The highpoint of the latter context is the Ayayema cave burial, with human bones dated to 4500 years BP (Legoupil and Sellier 2004). This is the oldest testimony of indigenous presence in this area. It was in this portion of the Patagonian channels that Joseph Emperaire (1963) carried out his anthropological and linguistic work with the last groups of Kawésqar settled in the town of Puerto Edén in the Messier channel, between 1946 and 1948.

2.3 Discussion of Some Regional Issues

After this brief historical summary of the archaeological investigations carried out in the northern channels, we can go on to contextualise the guiding themes of this work, largely influenced by older, systematic work carried out in the southern extremes of western Patagonia.

Since the start of archaeological investigations in the tip of South America (Bird 1938, 1946; Emperaire and Laming 1961; Lothrop 1928), a consensus has existed that western Patagonia was occupied by terrestrial hunter groups who might have originated in one of two regions: eastern Patagonia or south-central Chile, both with initial peopling documented from around 13,000 years BP (e.g. Borrero 2001; Dillehay 2009; Dillehay et al. 2008, 2015; Jackson et al. 2011; Massone 2004; Méndez 2012; Miotti and Salemme 2003; Prates et al. 2013). Nevertheless, discussion centres on when marine adaptation originated, and the area(s) from which this specialised process initiated and spread.

The archaeological investigations carried out to date suggest that the earliest evidence of marine adaptation occurred in two possible areas of the southern end of the archipelagos. The first is in Chile (54°50′S), where there is a series of sites distributed in the western third of the Magellan Strait (Punta Santa Ana sites, dated *ca.* 6500 years BP, and Bahía Buena, with 5895–5210 years BP [Ortiz-Troncoso 1975, 1979; San Román 2010]), and in the Otway Sound (Ponsonby level C, 5380–4430 BP [Legoupil 2003; Legoupil and Fontugne 1997]; the Englefield sites, 6100 years BP and Bahía Colorada, 5900 years BP [Legoupil 1988, 1997]; and Pizzulic 1 and 2, 5595 and 6055 years BP [San Román 2005]). The second is located 150 km further south in Tierra del Fuego (54°50′S), on the north shore of the Beagle Channel (in Argentina), which is also the source of the oldest datings (of samples not conditioned for the Reservoir Effect [Alberó et al. 1986, 1988]) of hunter-gatherers with a maritime tradition, starting around 6400 years BP (or 7200 years cal. BP) [Orquera and Piana

2009; Orquera et al. 2011; Piana and Orquera 2007; Piana et al. 2012]. This evidence comes from the second component of the Túnel I site (6400–4300 years BP) and from Imiwaia I, 6400–5800 years BP [Orquera and Piana 1999, 2009]. On the south coast of the Beagle Channel in Navarino Island (Chile), rather more recent datings are found (Seno Grandi, 6160 years BP [Legoupil 1994] and Áridos Guerrico, *ca.* 6400 years BP in a little reported site [Ocampo and Rivas 2000]).

Nevertheless, some authors agree that there were coastal occupations a thousand years older from *ca.* 7000 years BP, such as Ponsonby level D, 6700 and 7400 years BP, in Otway Sound (Legoupil 2003; Legoupil and Pigeot 2009), the first component of Túnel I, 6680 years BP (Orquera and Piana 1999) and layer S of Imiwaia, 7840 years BP (Orquera and Piana 2009), on the north coast of the Beagle Channel. The latter however are interpreted as discrete and/or non-diagnostic terrestrial hunter occupations (although with early dates for the region [see Massone 2004; Laming-Emperaire et al. 1972]) and not as the type of marine adaptation described for later moments (see Zangrando 2009a and b). Moreover, both areas, the Magellan Strait/Otway Sound and the south coast of the Beagle Channel, are located in ecotonal environments of forest/steppe transition, allowing hunting of both marine and terrestrial species and communication with completely terrestrial environments (Legoupil and Fontugne 1997; Orquera 2005; Orquera and Piana 1999). Thus, one hypothesis suggests that the process of maritime transformation and adaptation, consolidated more than 6000 years BP (Piana and Orquera 2009), originated in the area of Otway Sound/western third of the Magellan Strait, and then spread to the rest of the archipelagos (Legoupil and Fontugne 1997; Orquera and Piana 2006).

The second hypothesis on the origin and dispersal of this specialisation in littoral environments suggests that groups of specialised marine hunter-gatherers originated and spread from the north of the Patagonian Archipelago around the area of Chiloé/Reloncaví Sound (Legoupil and Fontugne 1997; Ocampo and Rivas 2004; Orquera and Piana 2006; Orquera et al. 2011; Rivas et al. 1999; Rivas and Ocampo; 2006). However, the dates obtained so far in the Puente Quilo and Yaldad 2 sites on Chiloé Island [Legoupil 2005; Ocampo and Rivas 2004; Rivas et al. 1999], the Piedra Azul and Ilque 1 sites on the north shore of Reloncaví Sound [Gaete and Navarro 2004; Gaete et al. 2004; Mera et al. 2016]) and the GUA-10 Terraza site in the Guaitecas Archipelago (included in the Chonos Archipelago) [Porter 1993; Reyes et al. 2016b] are rather younger than those recorded in the southern extreme of the archipelagos.

Despite this marked absence of early sites, one of the arguments of Ocampo and Rivas (2004) in favour of marine adaptation originating in the northern area is that it would have been available for occupation during the early Holocene, due to earlier deglaciation here than in the far south (Ashworth et al. 1991; Heusser 1998, 2002; Mc Culloch et al. 2000). The availability of forest would have been an essential condition for obtaining wood to make canoes. Moreover, the evidence of peopling of the Monte Verde site at the end of the Pleistocene (*ca.* 12,600 years BP, Dillehay 2009; Dillehay et al. 2008, 2015), less than 50 km from the coast where the northern archipelagos begin, indicates early human occupation of the area and the first use of the coastal space. All this indicates the possibility that older archaeological sites may exist containing material and chronological evidence of this process of marine specialisation. There is a direct relation with the biases of visibility and preservation according to the geomorphological behaviour found in the archipelago, not to mention the lack of systematic investigation (see Zangrando's Discussion Forum in the Intersecciones en Antropología journal, 2018).

The final aspect of the date and area(s) of origin of the marine adaptation process that is still debated is the dispersion of techno-typological elements. Stone projectile points of lanceolate design are found throughout the archipelago, of medium to large size and thick cross section (San Román et al. 2016). This technology is associated with specialised hunting of marine fauna and is widely distributed through the northern part of the Patagonian Archipelago, in the area of Reloncaví Sound/Chiloé (Díaz and Garretón 1972/3; Flores and Lira 2006; Gaete and Navarro 2004; Gaete et al. 2004; Galarce 2005a, b, c; Ladrón de Guevara et al. 2003; Ocampo and Rivas 2004), the Guaitecas (Porter 1993; Reyes et al. 2015, 2016b) and the Chonos Archipelago (Reyes et al. 2007, 2011, 2015) in contexts dated from ~6400 years cal. BP. In the area of the Magellan Strait/Otway Sound (Legoupil 2003; Morello et al. 2002; San Román 2005) and the Beagle Channel (Orquera et al. 1977), they are found mainly in contexts associated with dates around ~5000 to 4000 BP. The investigations propose that there are technological and subsistence aspects of the stone projectile points which support the presence of a second cultural group ("the Ponsonby tradition", Legoupil and Pigeot 2009) which arrived from the northern area and occupied the inland seas at least between 5000 and 4000 years BP, and may have continued into the following millennium (Morello et al. 2002, 2004). Technological and temporal similarities to finds of stone projectile points in the north, in Chiloé/Reloncaví Sound (Flores and Lira 2006), have led some authors to presume a northern influx differentiated from the earlier contexts of maritime adaptation recorded in the Mid-Holocene in Tierra del Fuego/Patagonia, defined as the "Englefield Culture" (between 6500 and 5000 years BP; Emperaire and Laming 1961; Legoupil 1997; Ortiz – Troncoso 1975, 1979; San Román 2005, 2010). On the contrary, and distinct from the position on the north shore of the Beagle Channel, these lithic designs should be seen as merely the dissemination of a technological innovation, with no influence of population replacement or migration (Piana and Orquera 2007), in archaeological assemblages which "*create an impression of homogeneity, continuity and stability*" (Orquera et al. 2011: 68).

2.4 Meanwhile, in the Chonos Archipelago …

A few archaeological notes from the Chonos Archipelago (Ocampo and Aspillaga 1984, 1991) reported two datings interpreted in the light of the few references available for Gran Guaiteca Island (Site Ch11 Rep 02, shell-midden mound and burials dated to 2430 ± 80 years BP, a sample from the base of the shell-midden) and Benjamín Island (Site Ch11 BEN 01, shell-midden in cave dated 410 ± 70 years BP, cypress bark sample). In both sites, there are unreported excavations. It is understood that the presence of shell-middens indicates the late population of this area of the archipelago, and that the funerary record in caves reflects an even later practice, associated with the historical accounts which described them as part of the last records of the prehistoric sequence (Ocampo and Aspillaga 1991). These first works registered various ossuaries from different archaeological funerary sites (caves and rock shelters) and excavations of shell-middens (with no known archaeological record beyond the name of the island or sector where they were obtained); this was the basis of the Chonos Osteological Collection of the University of Chile. The studies carried out on this collection led to the first publications on paleopathologies, aspects of subsistence and ways of life of marine hunter-gatherer groups (Aspillaga et al. 1990; Constantinescu and Aspillaga 1991; Linossier et al. 1994; Rodriguez 2011), and to the proposal that the canoe groups of this archipelago suffered an adaptive crisis, reflected in an increase in morbidity and the incidence of infectious diseases due to contact with other populations prior to contact with Europeans (Aspillaga et al. 2006; Sáez 2008).

Porter (1993) recorded a site in Gran Guaiteca Island, GUA-010, in an area of the island recognised archaeologically; it was dated from a speck of carbon in a forested terrace 5 m above the present sea level to ~5020 years old (5583–5918 years cal. BP). This is the first formal archaeological record which describes a number of features: soil acidity *versus* differential preservation of remains; lithic materials (obtained principally from the intertidal zone) characterised by the use of bifacial lanceolate stone projectile points similar to those recovered in Chiloé, stone axes and net weights; malacological fauna (obtained by testing a shell-midden on the lower terrace adjacent to the intertidal); and for which the occupations are interpreted in the light of the geomorphological changes produced by the tectonics of the region. The same investigator started cultural projections on long-distance mobility and exchange, with geochemical identification of the sources of lithic raw materials used by the canoe groups in this archipelago (Stern 2004, already glimpsed by Díaz and Garretón 1972/73), such as the rhyolithic obsidian from Chaitén Volcano in mainland Chiloé, more than 150 km north-east of the Guaitecas Archipelago (Porter 1993; Stern and Porter 1991). The age of this site, the typology of the lanceolate projectile points and the use of rhyolithic obsidian supported arguments on the origin from which the northern archipelagos were peopled and the dispersal of marine hunter-gatherer groups around the fifth millennium (Rivas et al. 1999; Ocampo and Rivas 2004). Despite this information, however, other investigators maintained; *"In GUA-10, Guaitecas Islands,* Porter (1993)*obtained a dating of 5020 $\pm$ 90 BP. However, as Porter remarks, the carbon analysed was taken from a test pit in the forested area, away from the shell-midden. The very acid soil and the size of the test pit (only 1m^2) may explain the lack of genuinely associated bone instruments and faunal remains, but the fact remains that the materials demonstrably related with the dating include nothing which can be assigned with certainty to littoral specialisation. It is not impossible that this existed (and to reach the Guaitecas some kind of boat was certainly necessary) but the data available are insufficient to demonstrate it"* (Orquera and

Piana 2005: 20). A question, therefore, hung over this site with respect to the specialisation of hunter-gatherer groups in the island and marine environments; evidently further investigation of this context was required (Reyes et al. 2016b, 2018, 2019b).

A few years later, Patricia Curry (1997), in the framework of Operation Raleigh, carried out the first systematic coastal prospection in the central area of the Chonos Archipelago. She identified around 40 sites and isolated finds, principally shell-middens, and collected some 50 pieces of lithic material from the shoreline and intertidal (the nucleus of the Curry Collection deposited in the CEHA, Patagonian Institute, Punta Arenas), mainly of thick cross section, lanceolate stone projectile points, axes and net weights. Although her report contained no datings, it recorded a variety of archaeological sites in varying eroded situations, with the presence of cultural and bio-anthropological material. This valuable data on the distribution of archaeological sites was incorporated into our investigations, re-specifying the location of some sites and analysing the lithic collection (Reyes et al. 2015). Curry's work also provided new information on the range of distribution of the rhyolithic obsidian from Chaitén Volcano, as she recorded pieces in shell-midden sites in the south of Traiguén Island and on Goñi Island—nearly 400 km south of the source—where it was used to make lanceolate stone projectile points (Stern and Curry 1995). Later, when the geochemical characterisation of this obsidian was developed, the range of distribution was extended to the area of Chiloé (Puente Quilo site) and 400 km north of the source, at Chan-Chan 18 on the coast near Valdivia (Stern et al. 2002), dated to the fifth millennium (Navarro 1995, 2001; Navarro and Pino 1999; Pino and Navarro 2005), marking a dispersal range of 800 km along the mainland coast and in the archipelagos. As stated above, this distribution has been modified significantly in recent years to include finds from the north-east coast of the Magellan Strait and the Atlantic coast (Stern et al. 2012; Stern 2018). Existing information suggests that this raw material was first used in the northern channels from the Mid-Holocene and continued thereafter, but with no precise datings. So far we have only a series of estimates from contextual datings (Reyes et al. 2016b, 2019b).

Our own investigations in the Chonos Archipelago started with "steppe problematics" (FONDECYT Project 1050139), in general terms dedicated to characterising the occupations of terrestrial hunter-gatherers in the Cisnes River valley throughout the Holocene (~44°S, Méndez and Reyes 2008, 2015; Reyes et al. 2006a, 2009). This valley, extending over ~160 km, crosses a diversity of environments and is one of the few places offering a connection between the high-level extra-Andean steppe and the Pacific coast; in the rest of the Aysén Region, there are major bio-geographical barriers to the connection between the steppe and the archipelago (mountains, ice fields, volcanoes, etc.). This geographical characteristic made a good argument for assessing the mobility and communication routes of the past. The cultural units defined occupied the space discontinuously, with differential hierarchisation based on the paleoenvironmental variability of the axis between coast and interior (Méndez and Reyes 2008; Reyes et al. 2006b, 2009).

Sites were identified in the first prospection along 60 km of coastline in the Melimoyu Peninsula, on the mainland coast of the Chonos Archipelago. A site of particular interest was Seno Gala 1 (Reyes et al. 2007), a shell-midden

(~1300 years cal. BP) associated with a dispersion of lithic material in the intertidal, as suggested for GUA-010 (Guaitecas Islands, Porter 1993). Tectonic sinking, together with tidal action, had exposed artefacts with a marked patina in the intertidal. The presence of large, thick cross sectioned, lanceolate stone projectile heads, in various stages of knapping, suggested local manufacture. A diversity of raw materials was observed, including rhyolithic obsidian from Chaitén Volcano (Reyes et al. 2007), which agreed with wide ranges of movement, to be expected in groups with a maritime tradition. Excavation of the shell-midden adjacent to the intertidal showed evidence of gathering of malacological fauna from the local environment (*Concholepas concholepas, Choromytilus chorus, Fissurella* sp.), consumption of fish, pudú (*Pudu pudu*) and otariids; other finds included knapping waste and bone technology (a punch made from a bird's tibiotarsus), in what was interpreted as occupation by marine hunter-gatherers with the exploitation of resources from the littoral and marine media.

According to several investigations carried out along the whole Cisnes River valley, and later in the valleys of the Ñirehuao and Simpson Rivers in the eastern Andes, we concluded that there were no arguments, either technological or based on the distribution of raw materials, for relating the two cultural spheres, maritime and terrestrial. This was proposed at one time by Bate and Mena (2003) based on the geographical location of the El Toro rock shelter site, ~37 km in a straight line from Puyuhuapi Channel (~60 km along the river) in the same valley. We drew a different interpretation from the absence of points of contact (techno-typological criteria, types of lithic raw materials, archaeofaunal record), arguing that—beyond the problems associated with the record (e.g. visibility)—it resulted from the effective action of the various bio-geographical barriers (sensu Barberena 2008; Borrero 2004, 2018) which made contact (e.g. circulation of raw materials) and technology transfer impossible between two worlds that had previously been completely separated (Méndez et al. 2006, 2008/9, 2012, 2018; Méndez and Reyes 2008, 2015; Reyes 2017; Reyes et al. 2006b, 2009, 2015, 2016a, 2019a and b).

Archaeological studies carried out in the neighbouring areas further north in the Aysén Region, on the extra-Andean slope, in Comarca Andina and in the Manso River valley (41–43°S), likewise support the barrier effect of the Andes, which restricts the dispersion and mobility of populations to a north–south axis (Bellelli et al. 2008). Further south, in the mid-course of the Ibáñez River valley (area of General Carrera/Buenos Aires Lake), occupations of terrestrial hunter-gatherers detected in the montane deciduous forest decline conspicuously to the west; further into the forest and the Andes foothills no records are found; the valley is effectively a cul-de-sac (García and Mena 2016). Another line of evidence suggesting that these two cultural spheres were completely separate is bio-anthropological information. Mitochondrial DNA analyses carried out to date on various human samples obtained in the Aysén Region also tend to present the population flows traced by the determination of the Pacific coast and the extra-Andean steppe peoples as independent haplogroups (Moraga et al. 2009; Reyes et al. 2011, 2012). Information on the first craniometric studies of human remains recovered in the Chonos Archipelago, with varying chronological sequences, show that the set of crania recovered present greater biological affinity with Pacific coast groups than with samples from the steppe area (Reyes et al. 2016a; Kuzminsky et al. 2018). Thus we argue, based on an extensive archaeological record and a diverse set of analyses, that both in the northern channels and in the east Andean valleys of Aysén there is a clear difference between the hunter-gatherer cultural traditions operating on the two sides of the barrier formed by the Andes, in this latitude at least. If there were contacts, a possibility that we do not deny, the material record does not represent them, and they might have been limited to sporadic situations without temporal continuity or cultural projection.

Continuing with the development of investigation in the archipelago, a FONDECYT Project (1100643) has been executed more recently (2010), the main theme of which was a mitochondrial DNA study of the canoe groups in the Chonos Archipelago. Its object was to provide a genetic characterisation of the Chono group and determine differences or similarities with the known genetics for the Huilliche groups of Chiloé and the Kawésqar groups south of the Taitao Peninsula (Lalueza et al. 1997; Moraga et al. 2000, 2009, 2010). This work is still in progress, and therefore not yet published. It may be remarked, however, that from a genetic-population perspective the mitochondrial markers of the late marine hunter-gatherers of the Chonos Archipelago present a certain homogeneity, appearing not to have changed significantly over nearly 2000 years. This situation is characterised by the presence of only two haplogroups (C, 45% and D, 55%) in very similar frequencies, and a fairly restricted repertoire of sub-haplogroups. In this context, archaeological prospections were carried out to record contextualised bone remains in the central-south area of the Chonos Archipelago between Aysén Fjord, Traiguén Island and the Darwin Channel. A number of funerary archaeological sites (ossuaries and shell-middens with burials) were discovered in contexts of erosion associated with changes in sea level, as described for the sites identified on the north-west coast of Traiguén Island (Reyes et al. 2011). A series of samples were also obtained from the Chonos Collection of the University of Chile (Ocampo and Aspillaga 1984, 1991), and of human remains deposited in the CIEP at Coyhaique after judicial investigations carried out in the archipelago. These samples were used to create a preliminary chronology of the Chonos Archipelago, with isotope analysis of the human remains (Reyes et al. 2013, 2015). The results showed that the funerary contexts in rocky sites were formed synchronically and/or diachronically (revisited), with dates within the Late Holocene (~2000 years cal. BP to ~ 300 years cal. BP), in addition to what were initially interpreted as contexts formed at the end of the occupation sequence around the moment of European contact (Ocampo and Aspillaga 1984, 1991). These studies produced specific information for the first time on the diet of these hunter-gatherers. They are thought to have formed a fairly homogenous group, with a diet consisting principally of marine resources (e.g. marine mammals, molluscs, fish, sea-birds). There were no obvious differences when the $\delta^{13}C$ and $\delta^{15}N$ data were crossed with the geographical, temporal and inhumatory variables, except in samples recovered from the post-contact historical period (Reyes et al. 2013, 2015, 2019a).

Finally, with FONDECYT Project 1130151 we intensified our archaeological prospections throughout the length and breadth of the Guaitecas and Chonos Archipelagos and the offshore islands (Guafo). We recorded and described a large number of sites discovered in new prospections and test pits, establishing their geomorphological situation (site location, distribution and position with respect to the shoreline) and evaluating them in the context of global and regional changes in sea level. We confirmed the dates associated with the base deposits of human formation since around ~6200 years cal. BP (Reyes et al. 2016b, 2018, 2019b) and determined more precisely the technological and subsistence components (archaeofauna, isotopes) and the spatial and temporal distribution of the canoe peoples who travelled around this archipelago throughout their history.

Note

[1] In 1608, the Jesuit Fathers Melchor Venegas and Juan Bautista Ferrufino embarked on a ship that sailed to Valparaiso every year with a cargo of Indian slaves from the Chonos Archipelago and the area of the Gulf of Penas. After 2 years in Chiloé, they had been summoned to Santiago. Father Venegas returned to Chiloé with a new companion, Father Mateo Esteban, who learnt the language of the Chono Indians, translated prayers and wrote a catechism for the use of the Indians scattered among the islands; these are the only existing texts in the Chono language. The fathers carried out three missions among the Chono between 1611 and 1614, in the Chiloé, Guaitecas and Chonos Archipelagos. They travelled in the canoe of converted Indians who accompanied them, made of three planks of cypress wood stitched together with creepers (Pedro Lozano, *Historia de la Compañía de Jesús en la provincia del Paraguay* (1755) in Emperaire 1963; Moreno 2011; Urbina 2016).

References

Alberó M, Angiolini F, Piana E (1986) Discordant ages related to reservoir effect of associated archaeological remains from Tunel site (Beagle Channel, Argentine). Radiocarbon 28(2A):748–753

Alberó M, Angiolini F, Piana E (1988) Holocene 14C reservoir effect at Beagle Channel (Tierra del Fuego, Argentine Republic). Quarter South Am Antarc Penins 5(1987):59–71

Aldunate C (1996) Mapuche: gente de la tierra. In Nueva Historia de Chile desde los orígenes hasta nuestros días. Editorial Zig-Zag. Santiago

Álvarez M, Vázquez M, Piana E (2008) Prácticas mortuorias entre los cazadores-recolectores del canal Beagle: el caso de Shamakush Entierro. Magallania 38(2):7–124

Álvarez R (2002) Reflexiones en torno a las Identidades de las Poblaciones Canoeras, Situadas entre los 44° y 48° de Latitud Sur, Denominadas "Chonos." Anales del Instituto de la Patagonia 30:79–86

Álvarez R, Munita D, Mera R, Borlando I, Ther F, Núñez D, Hidalgo C, Hayward P (2019) rebounding from extractivism: The history and re-assertion of traditional weir-fishing practices in the Interior Sea of Chiloé. Shima 13(2):155–173

Ashworth A, Markgraf V, Villagran C (1991) Late quaternary climatic history of the chilean channels based on fossil pollen and beetle analyses, with an analysis of the modern vegetation and pollen rain. J Quat Sci 6(4):279–291

Aspillaga E, Castro M, Ocampo C (1990) Paleopathology and life-style: the Chono and Fuegian examples (Abstract). Am J Phy Anthropol, Supplement

Aspillaga E, Castro M, Rodríguez M, Ocampo C (2006) Paleopatología y estilo de vida: El ejemplo de los Chonos. Magallania (Punta Arenas) 34(1):77–85

Aspillaga E, Ocampo C, Olivares J, Arensburg B, Meyer J (1995) Una visita a los canoeros de Quetalmahue. Revista Museos 20:18–20

Barberena R (2008) Arqueología y biogeografía humana en Patagonia Meridional. Sociedad Argentina de Antropología, Buenos Aires

Bate F, Mena F (2003) Alero El Toro: un campamento indígena en el bosque siempreverde cercano al litoral en Aisén. En Actas del XVI Congreso Nacional de Arqueología Chilena, Tomé (pp. 675–680). Museo de Historia Natural de Concepción

Bellelli C, Scheinsohn V, Podestá M (2008) Arqueología de pasos cordilleranos: un caso de estudio en Patagonia Norte durante el Holoceno Tardío. Boletín del museo chileno de arte precolombino 13(2):37–55

Bengoa J (2004) (compilador). La memoria olvidada. Historia de los pueblos indígenas de Chile, p 768. Publicaciones del Bicentenario, Santiago

Bibar G (1966 [1558]) Crónica y relación copiosa y verdadera de los Reynos de Chile. Tomo II. Fondo Historico y Bibliográfico José Toribio Medina. Santiago

Bird J (1938) The antiquity and migration on the early inhabitants of Patagonia. Geograph Rev 28:250–275

Bird J (1946) The archaeology of Patagonia. In: Steward J (ed.), Handbook of South American Indians, vol I, pp 17–24. Smithsonian Institution, Washington

Bird J (1988) Travels and Archaeology in South Chile. Hyslop J(ed.). Iowa City: University of Iowa Press

Borrero L (2001) El poblamiento de la Patagonia. Toldos, milodones y volcanes. Emecé editires, Buenos Aires

Borrero L (2004) The archaeozoology of andean “Dead Ends” in Patagonia: living near the continental ice cap. In: Mondini M, Muñoz S, Wickler S (eds.), Colonisation, migration, and marginal areas. A Zooarchaeological approach, pp 55–61. Oxbow Books, Oxford

Borrero L (2018) Barreras en la construcción de la arqueología patagónica y más allá... Boletín de la Sociedad Chilena de Arqueología. 48

Byron J (1901 [1746]) Relato del Honorable John Byron que contiene una esposicion de las grandes penurias sufridas por él i sus compañeros en la costa de la Patagonia desde el año 1740 hasta su arribo a Inglaterra en 1746.... Imprenta Cervantes. Santiago

Campbell R (2012) Ojo al fechado. Un ejercicio de evaluación cronológica y re-calibración para tres sitios de los canales patagónicos septentrionales. Boletín de la Sociedad Chilena de Arqueología. N° 41–42:5–10

Camus P (2001) Biogeografía marina de Chile Continental. Revista Chilena de Historia Natural. 74:587–617

Carabias D (2009) Encuentro de dos mundos. Naufragio del HMS Wager en la Patagonia. Santiago: Consejo Nacional de la Cultura y las Artes, FONDART. Chile

Carabias D, Lira N, Chapanoff M (2007) Documentación, análisis e interpretación de los restos de una dalca procedentes del lago Chapo, Región de los Lagos. In: Arqueología de Fuego-Patagonia. Ediciones CEQUA. Punta Arenas

Cárdenas R, Montiel D, Hall K (1991) Los Chonos y los Veliche de Chiloé. Editorial Olympo

Casamiquela R (1973) Alacalufes, canoeros occidentales y pueblos marginales o metamórficos. Relaciones de la Sociedad Argentina de Antropología, 7

Chapman A (2012) Yaganes del Cabo de Hornos. Encuentros con los europeos antes y después de Darwin. Ediciones Pehuén, Santiago, Chile

Cooper J (1917) Analytical and critical bibliography of the tribes of Tierra del Fuego and adjacent territory. Smithsonian Institution, Washington

Cooper J (1946) The Chono. Handbook of South American Indians, vol. 1: The Marginal Tribes. Editado por J. Steward. Bureau of American Ethnology Bulletin. 143. Smithsonian Institution. Washington, DC

Constantinescu F, Aspillaga E (1991) Paleopatalogia y anatomia funcional de la columna de una muestra de indigenas Chonos. Santiago, Chile. Actas del XI Congreso Nacional de Arqueologia Chilena (pp. 237–41)

Curry P (1991) Distribución de sitios e implicaciones para la movilidad de los canoeros en el canal Messier. En Anales del Instituto de la Patagonia. 20:145–154

Darwin Ch (1845) The Voyage of the Beagle (Journal of researches) into the natural history and geology of the countries visited during the voyage of H.M.S. Beagle round the world, under the Command of Capt. Fitz Roy, R.N. 2ª edición, Londres: John Murray, Cap. 13: 273–290

De Goicueta M (1879 [1557–1558]) Viaje del capitán Juan Ladrillero al descubrimiento del Estrecho de Magallanes. Anuario Hidrográfico de la Marina de Chile. Año V. Imprenta Nacional. Santiago

De Ovalle A (1969 [1646]) Histórica Relación del Reyno de Chile y de las Misiones y Ministerios que ejercita en él la Compañía de Jesús. Universitaria, Santiago de Chile

De Vea A (1886) Espedicion de Antonio de Vea (1675–1676). Anuario Hidrográfico de la Marina de Chile. Año XI. Imprenta Nacional, Santiago

Díaz C, Garretón M (1972/73) El poblamiento prehispánico del área insular septentrional chilena. In Actas del VI Congreso de Arqueología chilena, pp 559–584

Dillehay T (2009) Probing deeper into first American studies. PNAS 106:971–978

Dillehay T, Ramírez C, Pino M, Collins M, Rossen J, Pino-Navarro J (2008) Monte verde: seaweed, food, medicine, and the peopling of South America. Science 320:784–786

Dillehay T, Ocampo C, Saavedra J, Sawakuchi A, Vega R, Pino M (2015) New archaeological evidence for an early human presence at monte verde Chile. . PLoS ONE 10(11):e0141923. https://doi.org/10.1371/journal.pone.0141923

Emperaire J (1963) Los nómades del mar. Ediciones de la Universidad de Chile, Santiago

Emperaire J, Laming A (1961) Les gisements des îles Englefield et Vivian dans la mer d'Otway (Patagonie australe). Journal de la Société des Américanistes 40:7–77

Finterbusch C (1934) Las Dalcas de Chiloé y los Chilotes. Revista Chilena de Historia y Geografía, Tomo LXXV, 82, 412–433. Santiago, Chile

Fitz-Roy R (1839) Voyages of the adventure and beagle. Volume II. Proceedings of The Second Expedition, 1831–1836. Henry Colburn (ed. London.)

Flores C (2007) Estrategias de aprovechamiento de la diversidad ecológica del intermareal prehistórico conchales del seno de Reloncaví, X Región, Chile. Arqueología de fuego-Patagonia, pp 33–38

Flores C, Broitman B, Rivas P (2010) Changes in the subsistence strategy of prehistoric intertidal gathering: the pre-ceramic and ceramic coastal huntergatherers of Reloncaví Sound, Chile. Comparative Perspectives on the Archaeology of Coastal South America. Co-published: Center for Comparative Archaeology, University of Pittsburgh, Pontificia Universidad Católica del Perú, Ministerio de Cultura del Ecuador

Flores C, Lira N (2006) Aspectos comunes para sitios de cazadores recolectores costeros en los canales patagónicos septentrionales. En Boletín de la Sociedad Chilena de Arqueología 39:103–112

Flores C, Correa I (2011) Explotación de ambientes costeros, intensificación de recursos y transformaciones culturales en la transición holoceno medio a holoceno tardío en la costa de los canales patagónico septentrionales, Chile. Actas XVII Congreso Nacional de Arqueología Argentina. Tomo I:321–326

Gaete N, Navarro X (2004) Estrategias de vida de canoeros cazadores pescadores recolectores del seno de Reloncaví. Entre el bosque siempreverde y el mar interior. Región de Los Lagos, Chile. In: Contra viento y marea. Arqueología de Patagonia, pp 217–235. INAPL-Sociedad Argentina de Antropología, Buenos Aires

Gaete N, Navarro X, Constantinescu F, Mera R, Selles D, Solari M, Vargas L, Oliva D, Duran L (2004) Una mirada al modo de vida canoero del mar interior desde Piedra Azul. Chungara. Tomo I:333–346

Galarce P (2005a) Manuscrito. Tecnología lítica de grupos cazadores-recolectores-pescadores holocénicos en el Seno de Reloncaví (Puerto Montt, X Región). Informe de análisis lítico entregado a Ciprés Consultores

Galarce P (2005b) Manuscrito. Análisis de los instrumentos líticos de los sitios PM005, PM009 y PM012, Seno de Reloncaví (Puerto Montt, X Región). Proyecto Mejoramiento Ruta 7 (Carretera Austral) entre las localidades de Quillaipe—Caleta La Arena. Ciprés Consultores Ltda. Santiago de Chile

Galarce P (2005c) Manuscrito. Análisis de los instrumentos líticos de los sitios Puente Quilo y Chepu 005, costa de Chiloé. X Región. Informe de análisis lítico entregado a Ciprés Consultores

García C, Mena F (2016) ¿La frontera del oeste?: Prospecciones arqueológicas en el bosque montano del extremo occidental del valle medio del río Ibáñez (Andes patagónicos, Chile). Intersecciones en antropología 17(1):49–61

García J (1889 [1766]) Diario del viaje y la navegación hechos por el padre José Joseph García de la Compañía de Jesús, desde su misión de Cailin, en Chiloé, hacia el sur, en los años 1766–1767. Anuario hidrográfico de la Marina de Chile, XIV. Imprenta Nacional Santiago, Chile

Garreaud R, Vuille M, Compagnucci R, Marengo J (2009) Present-day South American climate. Palaeogeogr Palaeoclimatol Palaeoecol 281(3–4):180–195

Hanisch W (1982) La isla de Chiloé, capitana de rutas australes. Academia Superior de Ciencias Pedagógicas de Santiago, Santiago

Harb Z, Llop E, Moreno R, Aspillaga E (1995) Evolución de la mezcla amerindio europea en el Archipiélago de Chiloé. Un análisis etnohistórico y bioantropológico. Actas del Segundo Congreso Chileno de Antropología. Vol. 1.Valdivia

Heusser C (1998) Deglacial paleoclimate of the American sector of the Southern Ocean: Late Glacial- Holocene records from the latitude of Canal Beagle (55°S), Argentine Tierra del Fuego. Palaeogeogr Palaeoclimatol Palaeoecol 141:277–301

Heusser C (2002) On glaciation of the southern Andes with special reference to the Peninsula de Taitao and adjacent Andean cordillera (46°30'S). J S Am Earth Sci 15:577–589

Ibar J (1959) Ensayo sobre los indios Chonos e interpretación de sus toponimias. Anales de la Universidad de Chile, Santiago, Chile

Informe de la Comisión Verdad Histórica y Nuevo Trato con los Pueblos Indígenas (2008 [2003]) Editado por el Comisionado Presidencial para Asuntos Indígenas Rodrigo Egaña Baraona

Jackson D, Maldonado A, Carré M, Seguel R (2011) Huentelauquén cultural complex: the earliest peopling of the pacific coast in the South-American southern cone. In: Vialou D (ed) Peuplements et préhistoire en Amériques. CTHS, Paris, pp 221–232

Jaillet S, Fage L, Maire LR, Tourte B (2009) Découverte d'une grotte ornée dans l'archipel de Patagonie chilienne: la grotte du Pacifique. Spelunca Mémoires34:173–183

Kuzminsky S, Reyes O, Arriaza B, Mendez C, Standen V, San Roman M, Muñoz I, Duran A (2018) Investigating Cranial Morphological Variation of Early Human Skeletal Remains from Chile: A 3D Geometric Morphometric Approach. Am J Phys Anthropol. https://doi.org/10.1002/ajpa.23344

Labarca R, Mena F, Prieto A, Dupradou T, Silva E (2016) Investigaciones arqueológicas en torno a los primeros registros de Arte Rupestre en Morro Vilcún. Boletín Museo Chileno de Arte Precolombino 21(2)

Ladrón de Guevara B, Gaete N, Morales S (2003) El patrimonio como fundamento para el desarrollo del capital social: el caso de un sitio arqueológico y Puntilla Tenglo. Revista Conserva N°7:5–21. Santiago

Lalueza C, Pérez-Pérez A, Prats E, Cornudella L, Turbon D (1997) Lack of founding amerindian mitocondrial DNA lineages in extinct aborigines from Tierra del Fuego-Patagonia. Human Molecular Genetics 6:41–46

Laming-Emperaire A (1972) Los sitios arqueológicos de Patagonia occidental. Anales del Instituto de la Patagonia 3(1–2):87–96
Latcham R (1928) Prehistoria de Chile. Sociedad Impresora y Litográfica Universo, Santiago
Latcham R (1930) La dalca de Chiloé y los canales patagónicos. Boletín del Museo Nacional de Chile, Tomo XIII:63–72. Santiago
Legoupil D (1988) Ultimas consideraciones sobre las dataciones del sitio de isla Englefield (Seno de Otway). Anales del Instituto de la Patagonia, Serie Ciencias Humanas 18:95–98
Legoupil D (2003) Cazadores-recolectores de Ponsonby (Patagonia austral) y su paleoambiente desde VI al III milenio A. C. Magallania 31, tirada especial
Legoupil D (2005) Recolectores de mariscos tempranos en el sureste de la isla de Chiloé. Magallania 33(1):51–61
Legoupil D, Bearez P, Lepetz S, San Román M, Salas K (2007) De la pesca a la caza: evolución económica del sitio Stuven 1, al sur del Golfo de Penas. In: Morello F, Martinic M, Prieto A, Bahamonde G (eds.), Arqueología de Fuego-Patagonia. Levantando piedras, desenterrando huesos… y develando arcanos, pp 279–294. Punta Arenas: Ediciones CEQUA
Legoupil D, Fontugne M (1997) El poblamiento marítimo en los archipiélagos de Patagonia: núcleos antiguos y dispersión reciente. Anales del Instituto de la Patagonia, Serie Ciencias Humanas 25:75–87
Legoupil D, Pigeot N (2009) Les grandes pointes foliacées du type « Ponsonby ». Un traceur culturel en Patagonie australe. Journal de la société des américanistes 95–2
Legoupil D, Sellier P (2004) La sepultura de la cueva Ayayema (isla Madre de Dios, archipiélagos occidentales de Patagonia). Magallania 32:115–124
Linossier A, Aspillaga E, Gajardo M, Olavarria J (1994) Paleomicrobiología oral en poblaciones de Chile: Chonos y Atacameños. Antropología Biológica 2(2):63–73
Lira N, Figueroa V, Braicovich R (2015) Informe sobre los restos de dalca del museo etnográfico de Achao, Chiloé. Magallania 43(1):1–12. Punta Arenas
Lothrop S (1928) The Indians of Tierra del Fuego. Museum of the American Indian, New York
Lothrop S (1932) Aboriginal navigation off the West Coast of South America. J Royal Anthropol Ins Great Britain and Ireland 62:229–256
Machado A (1889 [1769]) Viajes del piloto Don francisco Machado a los archipiélagos occidentales de Patagonia. Anuario hidrográfico de la Marina 14
Mansilla J (2002) La población de Calbuco. Evolución de las cifras Siglos XVII-XX. Revista Austral de Ciencias Sociales. No 6:125–134
Martial L, Hyades P, Deniker J (2007) Etnografía de los indios Yaghan en la misión científica del Cabo de Hornos, pp 1882–1883. Universidad de Magallanes
Martinic M (1989) Los canoeros de la Patagonia meridional. J de la Société des Américanistes 75(1):35–61
Martinic M (2004) Archipiélago Patagónico. La última frontera. Ediciones de la Universidad de Magallanes, Punta Arenas, Chile
Martinic M (2005) De la Trapananda al Aysén. Ediciones Pehuén, Santiago
Massone M (2004) Los cazadores después del hielo. Centro de Investigaciones Barros Arana, DIBAM
Massone M, Morello F, Borrero L, Legoupil D, Mena F, Prieto A, Ocampo C, Rivas P, San Román M, Martin F, Méndez C, Reyes O, Munita D (2016) Cazadores-Recolectores en la Patagonia chilena desde 11.000 años a.p. a la colonización occidental. In: Falabella F, Uribe M, Sanhueza L, Aldunate C, Hidalgo J (eds.), Prehistoria en Chile. Desde sus primeros habitantes hasta los Incas. Universitaria. Santiago, Chile
McCulloch R, Bentley M, Purves R, Hulton N, Sugden D, Clapperton C (2000) Climatic inferences from glacial and palaeoecological evidence at the last glacial termination, southern SouthAmerica. J Quat Sci 15(4):409–417
Medina A (1984) Embarcaciones chilenas precolombinas. La Dalca de Chiloé. Revista Chilena de Antropología. N°4:121–138. Universidad de Chile

Mena F, González E, Labarca R (2011) Primeros registros de arte rupestre en el litoral de Patagonia Septentrional chilena. Magallania 39(2):303–307
Méndez C (2012) Terminal Pleistocene/early Holocene 14 C dates form archaeological sites in Chile: critical chronological issues for the initial peopling of the region. Quatern Int 301:60–73
Méndez C, Velásquez H, Reyes O, Trejo V (2006) Tras los moradores del bosque Conjuntos arqueológicos de Alero El Toro (Valle del río Cisnes, XI Región de Aisén). Revista Werkén 8:101–115. Santiago
Méndez C, Reyes O (2008) Late Holocene human occupation of Patagonian forests: a case study at Cisnes River basin (44 °S, Chile). Antiquity. 82:560–570
Méndez C, Reyes O (2015) Archaeology near the southern ice-end. Current advances in human interdisciplinary research in central western Patagonia. SAA Archaeol Record 15(3):21–26
Méndez C, Stern C, Reyes O (2008/9) Transporte de obsidianas a lo largo de los Andes de Patagonia Central (Aisén, Chile). Cazadores-Recolectores del Cono Sur. Revista de Arqueología 3:51–68. Mar del Plata
Méndez C, Stern C, Reyes O, Mena F (2012) Early Holocene long-distance obsidian transport in Central-South Patagonia. Chungara Revista de Antropología Chilena 44(3):363–375
Méndez C, Stern C, Nuevo A, Reyes O, Mena F (2018) Spatial and temporal distributions of exotic and local obsidians in Central Western Patagonia, southernmost South America. Quatern Int 468:155–168
Menghin O (1962) Estudios de prehistoria araucana. Studia Praehistorica 2. Centro Argentino de Estudios Prehistóricos. Buenos Aires
Mera R, Munita D, Alvarez R, García C, García P (2016) Investigación, conservación y educación patrimonial como compensación de daños en los sitios arqueológicos de bahía Ilque. Boletín de la Sociedad Chilena de Arqueología 46:71–84
Miotti L, Salemme M (2003) When Patagonia was colonized: people mobility at high latitudes during Pleistocene/Holocene transition. Quatern Int 109:95–111
Moraga M, Rocco P, Miquel J, Nervi F, Llop E, Chakraborty R, Rothammer F, Carvallo P (2000) Mitochondrial DNA polymorphisms in chilean aboriginal populations: implications for the peopling of the southern cone of the continent. Am J Phys Anthropol 113:19–29
Moraga M, Aspillaga E, Mena F. (2009) Search for founder mitochondrial lineages in holocene human remains in Patagonia. Am J Phy Anthropol Suppl 48:193–193
Moraga M, de Saint Pierre M, Torres F, Ríos J (2010) Vínculos de parentesco por vía materna entre los últimos descendientes de la etnia Kawésqar y algunos entierros en los canales patagónicos: evidencia desde el estudio de linajes mitocondriales. Magallania. 38:103–114
Moraleda J (1888 [1786]) *Esploraciones jeograficas e hidrográficas…* Imprenta Nacional, Santiago, Chile
Morello F, San Román M, Prieto A (2002) Puntas de proyectil lanceoladas en Patagonia meridional y Tierra del Fuego. Anales del Instituto de la Patagonia, Serie Ciencias Humanas 30:155–166
Morello F, San Román M, Prieto A (2004) Obsidiana verde en Fuego-Patagonia: distribución y estrategias tecnológicas. In: M. Civalero, P. Fernández y A. Guráieb (eds.), Contra Viento y Marea. Arqueología de Patagonia (pp. 149–166). Sociedad Argentina de Antropología
Moreno R (2011) El archipiélago de Chiloé y los jesuitas: El espacio geográfico para una misión en los siglos XVII y XVIII. Magallania, Punta Arenas 39–2:47–55
Moreno R, Llop E, Harb Z (1999) Estudio biomédico de la heterogeneidad en la población chilena. El Archipiélago de Chiloé, un modelo natural. Revista Chilena de Antropología. N°15. Facultad de Ciencias Sociales. Universidad de Chile, Santiago
Munita D (2007) Materias primas líticas en sitios costeros del extremo sur septentrional de chile. Dispersión y aprovisionamiento. Boletín de la Sociedad Chilena de Arqueología 39:103–112
Munita D (2017) Ocupaciones arqueológicas en el borde costero del seno de Reloncaví, el caso de Bahía Ilque. Facultad de Ciencias Sociales, Universidad de Chile, Memoria para optar al título de Arqueóloga
Munita D, Álvarez R, Ocampo C (2005) Corrales de piedra, pesca pasiva en la costa interior de Chiloé. Boletín de la Sociedad Chilena de Arqueología 37:61–74

Munita D, Mera R, Arregui I, Manneschi M (2011) Funebria de grupos canoeros durante el holoceno tardío en la región de los lagos. El conchal de Yaco Alto–1, Calbuco, Chile. Revista cazadores recolectores del cono sur. Argentina

Munita D, Mera R, Alvarez R (2016) Una historia de seis mil años. In *Chiloé*. Edited by Aldunate C. Colección Santander Museo Chileno de Arte Precolombino, pp 58–87

Munizaga J (1978a) Antropología Física de Chiloé. In: Revista Chilena de Antropología. N° 1: 125. Facultad de Ciencias Sociales. Universidad de Chile. Santiago

Munizaga J (1978b) Microevolución en Poblaciones Rurales Contemporáneas de Chiloé, Revista Chilena de Antropología. Vol 1: 143–153. Facultad de Ciencias Sociales. Universidad de Chile. Santiago

Navarro X (1995) Interpretación de ocupaciones precerámicas y cerámicas en los distintos microambientes de la costa de Chan Chan, Valdivia, X Región. Actas del XIII Congreso Nacional de Arqueología Chilena 9(1):127–134

Navarro X (2001) Una mirada desde la arqueología del paisaje para entender las ocupaciones iniciales de la costa norte de Valdivia. XV Congreso Nacional de Arqueología Chilena

Navarro X, Pino M (1999) Estrategias Adaptativas en Ambientes Costeros del Bosque Templado Lluvioso de la Zona Mapuche. Una Reflexión Desde el Precerámico. In: Actas de las Terceras Jornadas de Arqueología de la Patagonia: 65–82

Ocampo C, Aspillaga E (1984) Breves notas sobre una prospección arqueológica en los archipiélagos de las Guaitecas y de los Chonos. Revista Chilena de Antropología 4:155

Ocampo C, Aspillaga E (1991) Problemas del registro arqueológico de los sitios del archipiélago de los Chono y las Guaitecas. Actas del XII Congreso Nacional de Arqueología Chilena 1:17–18

Ocampo C, Rivas P (2000) Nuevos fechados 14C de la costa norte de la isla Navarino, costa sur del canal Beagle, región de Magallanes. Anales del Instituto de la Patagonia 28:197–214

Ocampo C, Rivas P (2004) Poblamiento temprano de los extremos geográficos de los canales patagónicos: Chiloé e isla Navarino 1. *Chungara*. Volumen Especial, tomo 1:317–331

Ocampo C, Rivas P (2008) Manuscrito. El poblamiento marítimo de los canales occidentales de la Patagonia: desde Chiloé hasta Cabo de Hornos, pp 57–69

Orquera L (2005) Mid-Holocene littoral adaptation at the southern end of South America. Quatern Int 132:107–115

Orquera L, Piana E (1995) La imagen de los canoeros magallánico-fueguinos: conceptos y tendencias. Runa: archivo para las ciencias del hombre 22(1):187–245

Orquera L, Piana E (1999) Arqueología de la Región del Beagle (Tierra del Fuego, República Argentina). Sociedad Argentina de Antropología, Buenos Aires

Orquera L, Piana E (2005) La adaptación al litoral sudamericano sudoccidental: qué es y quiénes, cuándo y dónde se adaptaron. Relaciones-Soc Argentina de Antropología 30:11–32

Orquera L, Piana E (2006) El poblamiento inicial del área litoral sudamericana sudoccidental. Magallania 34(2):21–36

Orquera L, Piana E (2009) Sea nomads of the Beagle Channel in southernmost South America: over six thousand years of coastal adaptation and stability. J Island Coastal Archaeol 4:61–81

Orquera L, Sala A, Piana E, Tapia A (1977) Lancha Packewaia. Editorial Huemul, Buenos Aires

Orquera L, Legoupil D, Piana E (2011) Littoral adaptation at the southern end of South America. Quatern Int 239:61–69

Ortiz Troncoso O (1975) Los yacimientos de Punta Santa Ana y Bahía Buena: Excavaciones y fechados radiocarbónicos. *Anales delInstituto de la Patagonia*. Serie Ciencias Humanas 6:93–122

Ortiz Troncoso O (1979) Punta Santa Ana et Bahıa Buena: Deux gisements sur une ancienne ligne de rivage dans le Detroit de Magellan. J de la Societe des Americanistes 66:133–204

Piana E, Orquera L (2007) Diferencias regionales y temporales en el litoral sudoccidental de Sudamérica. *Arqueología de Fuego-Patagonia*, pp 311–326. Ediciones CEQUA. Punta Arenas, Chile

Piana E, Orquera L (2009) The southern top of the world: the first peopling of Patagonia and Tierra del Fuego and the cultural endurance of the Fuegian Sea-Nomads. Arctic Anthropol 46(1–2):103–117

Pino M, Navarro X (2005) Geoarqueología del sitio arcaico Chan-Chan 18, costa de Valdivia: discriminación de ambientes de ocupación humana y su relación con la transgresión marina del Holoceno Medio. Revista Geológica de Chile. 32(1):59–75

Plafker G, Savage JC (1970) Mechanism of the Chilean Earthquakes of May 21 and 22, 1960. Geol Soc Am Bull 81(4):1001–1030

Porter C (1993) GUA-010, un sitio costero erosionado en una zona sísmica activa. En: Actas del XII Congreso Nacional de Arqueología Chilena. Boletín del Museo Regional de la Araucanía 4:81–88

Prates LG, Politis G, Steele J (2013) Radiocarbon chronology of the early human occupation of Argentina. Quat Int 301:104–122

Quiroz D, Olivares J (1988) Nómades canoeros de la Patagonia Occidental Insular Septentrional: El mundo de don Pedro del Agua. Encuentro de Etnohistoriadores. Silva O, Medina E, Téllez E (eds), pp 10–33. Universidad de Chile, Serie Nuevo Mundo: Cinco Siglos, N° 1

Quiroz D, Sanchez M (2004) Poblamientos iniciales en la costa septentrional de la Araucanía (6.500-2.000 a.p.) *Chungara*. Revista de Antropología Chilena Volumen Especial, tomo 1:289–302

Reyes O (2017) El poblamiento del archipiélago de los Chonos (43°–47°S). Patagonia occidental. Chile. Tesis doctoral, Facultad de Filosofía y Letras, Universidad de Buenos Aires, Buenos Aires

Reyes O, Méndez C, Velásquez H, Trejo V (2006a) Distribuciones espaciales y contextos arqueológicos de cazadores recolectores esteparios en alto río Cisnes: (XI Región de Aisén). Magallania. (Punta Arenas) 34(2):75–90

Reyes O, Méndez C, Trejo V, Velásquez H (2006b) De las estepas a los canales septentrionales: uso humano de distintos ambientes. Bosques del valle del río Cisnes (Holoceno tardío, XI Región de Aisén)". Actas del XVII Congreso Nacional de Arqueología Chilena [2010]. Tomo 2:1487–1495. Valdivia.

Reyes O, Méndez C, San Román M, Cárdenas P, Velásquez H, Trejo V, Morello F, Stern C (2007) Seno Gala 1; Nuevos Resultados En La Arqueología De Los Canales Septentrionales (44°S, XI Región de Aisén, Chile). Magallania. 35(2):105–119

Reyes O, Méndez C, Maldonado A, Velásquez H, Trejo V, Cárdenas M, Abarzúa A (2009) Uso del espacio de cazadores recolectores y paleoambiente Holoceno en el valle del río Cisnes, Región de Aisén. Chile. Magallania 37(1):91–107

Reyes O, San Román M, Moraga M (2011) Archipiélago de los Chonos: nuevos registros arqueológicos y bioantropológicos en los canales septentrionales. Isla Traiguén, XI Región de Aisén. Magallania 39(2):295–303

Reyes O, Méndez C, Mena F, Moraga M (2012) The bioanthropological evidence of a ca. 10,000 CALYBP ten-individual group in central Patagonia. Southbound: late pleistocene peopling of latin America, 167–171

Reyes O, Moraga M, Aspillaga E (2013) El registro bioantropológico y las evidencias de ocupación en el Archipiélago de los Chonos (XI Región de Aisén, Chile). Avances en la arqueología de los canales septentrionales del extremo sur. In Tendencias Teórico-Metodológicas y Casos de Estudio en la Arqueología de la Patagonia, pp 227–232. Argentina

Reyes O, Moraga M, Méndez, M, Cherkinsky A (2015) Maritime hunter-gatherers in the chonos archipelago (43°50'–46°50' S), Western Patagonian Channels. J Island Coastal Archaeol 10(2):207–231

Reyes O, Kuzminsky S, Méndez C, San Román M (2016a) Examining biological variation among the maritime hunter-gatherers of the Chonos Archipelago, Western Patagonian Channels, Chile. Libro Resúmenes 81 The Annual Meeting of the Society for American Archaeology. Orlando

Reyes O, San Román M, Morello F (2016b) Search for maritime hunter-gatherer archaeological record in the shifting shorelines of the South Pacific Coast (Chonos and Guaitecas Archipelago, Chile). In Marine Ventures: Archaeological Perspectives on Human-Sea Relations, pp. 141–155. Equinox Publishing, United Kingdom. https://doi.org/10.1558/equinox.24563

Reyes O, Méndez C, San Román M, Francoise J (2018) Earthquakes and coastal archaeology: assessing shoreline shifts on the southernmost Pacific coast (Chonos Archipelago 43°50′–46°50′ S, Chile, South America). Quatern Int 463:161–175

Reyes O, Tessonne A, San Román M, Méndez C (2019a) Dieta e isótopos estables de cazadores recolectores marinos en los canales occidentales de Patagonia Chile . Latin Am Antiquity 30(3):550–568
Reyes O, Méndez C, San Román M (2019b) Cronología de la ocupación humana en los canales septentrionales de la Patagonia occidental Chile. Intersec en Antropol 20(2):195–210
Rivas P, Ocampo C (2005) El antiguo curanto chilote. In: Chile País Oceánico. Badal G, Montanari M, Santa Cruz C, Altamirano C (eds.), pp 70–72
Rivas P, Ocampo, C (2006) La adaptación humana al bosque en la isla de Chiloé. Estrategias adaptativas en el litoral septentrional de los canales patagónicos. Actas del XVII Congreso Nacional de Arqueología Chilena (2010). T2: 1449–1460. Valdivia
Rivas P, Ocampo C, Aspillaga E (1999) Poblamiento Temprano de los Canales Patagónicos: El Núcleo Ecotonal Septentrional. Anales del Instituto de la Patagonia 27:221–230
Rodríguez M (2011) Manuscrito. Análisis intrapoblacional e interpoblacional de paleopatologías en la colección archipiélago de chonos. Memoria de Titulo. Universidad de Chile
Rodríguez M, Aspillaga E, Arensburg B (2010) Colecciones esqueletales del archipiélago de Chiloé: nuevos antecedentes, perspectivas y limitaciones. Actas del XVII Congreso Nacional de Arqueología Chilenas. 1:535–546
Sáez A (2008) Impacto del contacto hispano-indígena en la salud de la población de Chiloé. Un caso de tuberculosis en el cementerio Puqueldón. Magallania 36(2):167–174
San Román M (2005) Nuevos hallazgos de sitios de cazadores recolectores marinos tempranos en isla Englefield. Mar de Otway. Magallania 33(2):173–176
San Román M (2010) La explotación de recursos faunísticos en el sitio Punta Santa Ana 1: estrategias de subsistencia de grupos de cazadores marinos tempranos de Patagonia meridional. Magallania 38(1):183–198
San Román M, Reyes O, Elgueta J, Morello F (2016a) Archaeology of Maritime hunter-gatherers from Southernmost Patagonia, South America: discussing timing, changes and cultural traditions during the Holocene). Archaeological Perspectives on Human-Sea Relations, In Marine Ventures
Simpon E (1871) Memorias Científicas i Literarias. Anales de la Universidad, Agosto
Simpon E (1875) Esploraciones hechas por la Corbeta Chacabuco en 1870 .... Anuario Hidrográfico de la Marina. Tomo I
Skottsberg C (1913) Observations on the natives of the patagonian channel regional. Am Anthropol. New Series. 15(4):578–616
Stern C (2004) Obsidian in Southern Patagonia: review of the current information. Contra viento y marea. Arqueología de Patagonia, 167–176
Stern C (2018) Obsidian sources and distribution in Patagonia, southernmost South America. Quat Int 468:190–205
Stern C, Porter C (1991) Obsidiana en yacimientos arqueológicos de Chiloé y las islas Guaitecas. Anales del Instituto de la Patagonia. Serie Ciencias Humanas 20:205–209
Stern C, Curry P (1995) Obsidiana del sitio Posa Las Conchillas, Isla Traiguén (45°30′S), Archipiélago de los Chonos. *Anales delInstituto de la Patagonia.* Serie Ciencias Humanas. 23:119–124
Stern C, Navarro X, Muñoz J (2002) Obsidiana gris translúcida del volcán Chaitén en los sitios arqueológicos de Quilo (Isla Grande de Chiloé) y Chanchán (X Región) Chile, y obsidiana del Mioceno en Chiloé. Anales del Instituto de la Patagonia, Serie Ciencias Humanas 30:167–174
Stern C, Caracotche S, Cruz I, Charlin J (2012) Obsidiana gris porfírica calco-alcalina del volcán Chaitén en sitios arqueológicos al sur del río Santa Cruz Patagonia meridional. Magallania 40(1):137–144
Trivero A (2018) La Dalca y la navegación en los canales australes. Proyecto Editorial Feyentum
Urbina R (1983) La periferia meridional indiana, Chiloé. Ediciones Universitarias de Valparaíso y Universidad Católica de Valparaíso. Valparaíso
Urbina R (1988) Los chonos en Chiloé: Itinerario y Aculturación. *Revista Chiloé* No 9 – Agosto, pp 29–42

Urbina X (2010) La navegación por los canales australes en la Patagonia occidental insular en los siglos coloniales: la ruta del istmo de Ofqui. Magallania, Punta Arenas 38–2:41–67

Urbina X (2013) Expediciones a las costas de la Patagonia Occidental en el período colonial. Magallania, Punta Arenas 41–2:51–84

Urbina X (2014) Fuentes para la historia de Patagonia occidental en el periodo colonial. Primera parte: Siglos XVI y XVII. Ediciones Universitarias de Valparaíso. Pontificia Universidad Católica de Valparaíso. Chile

Urbina X (2016) Traslados de indígenas de la Patagonia occidental insular a Chiloé en los siglos XVI, XVII y XVIII. In: Valenzuela J (ed), América en diásporas: esclavitudes y migraciones forzadas (siglos XVI-XIX), Santiago, Instituto de Historia, Pontificia Universidad Católica de Chile/RiL editores

Urbina X, Reyes O, Belmar C (2020) Canoeros en Chiloé: de facilitadores de las navegaciones españolas en los archipiélagos de los Chonos y de Guayaneco, a productores y comerciantes. Chungara Revista Chilena de Antropología 52(2):335–346

Vásquez de Acuña I (1963) Arqueología Chiloense: Yacimientos y Material Lítico. *Trabajos de Prehistoria*. Instituto Español de Prehistoria del Consejo Superior de Investigaciones Científicas, Madrid

Zangrando F (2009a) Historia evolutiva y subsistencia de cazadores-recolectores marítimos de Tierra del Fuego. Sociedad Argentina de Antropología. Buenos Aires

Zangrando F (2009) Is fishing intensification a direct route to hunter-gatherer scomplexity? A case study from the Beagle Channel region (Tierra del Fuego, southern South America). World Archaeology 41(4):589–608

Chapter 3
Methodology

Abstract The methodology used included office work, fieldwork and laboratory study focused on discovering, recovering and contextualising archaeological information. The object was to characterise the occupation of the Chonos Archipelago by hunter-gatherers and the impact of natural and cultural processes which have affected spatial and temporal aspects of the archaeological record. We sought to integrate the distribution of the sites, the chronological sequences obtained and their relation with geomorphological information about coastlines with analysis of the different types of material recovered.

Keywords Western patagonia · Chonos archipelago · Maritime hunter-gatherers · Coastal survey · Coastal erosion · Formation processes · Archaeological sites · Sea-level change · Radiocarbon chronology · Stable isotopes

3.1 Office Work

Office work addressed two general areas: documentation, and compilation of existing palaeoenvironmental and archaeological information in order to draw up an archaeological plan subsequently, on the ground, of the record of sites and cultural material and interventions in these. The palaeoenvironmental documentation consisted in compiling data on the types of climate recorded in the Chonos Archipelago throughout the Holocene. We referred to work on the palaeoenvironment in previous investigations on the eastern slope of the Andes and the steppe region at the latitude of the Archipelago, covering the period from the Glacial/Post-glacial transition at the end of the Pleistocene to the late Holocene (de Porras et al. 2012, 2014; García et al. 2019; Méndez and Reyes 2008; Reyes et al. 2009). We obtained data comparable with the various palaeoenvironmental testimonies recorded in the area of the northern archipelagos (see Bennett et al. 2000; Haberle and Bennett 2004; Heusser 1960, 2002). These environmental data were interpreted in the light of their temporal and spatial scope and resolution, evaluating the relation between environmental dynamics and the cultural record of the past (sensu Bailey 2004; Erlandson 2001; Gusick and Faught 2011; Richardson 1981; Sandweiss et al. 2004).

O. Reyes, *The Settlement of the Chonos Archipelago, Western Patagonia, Chile*,
The Latin American Studies Book Series,
https://doi.org/10.1007/978-3-030-54326-6_3

Information was also gathered about the geological and geomorphological records, as an essential strategy for locating sites, predicting finds and analysing the contexts recorded (sensu Butzer 2008; Scheinsohn 2001; Waters and Kuehn 1996). These refer principally to changes in sea level, and therefore to local coastal geography (Porter 1993; Reyes et al. 2016b, 2018). These changes, at different scales, are based on the global, regional and local impact of transforming factors like glacio-eustasy, isostasy, volcanic and tectonic activity, and marine transgressions; they are also directly associated with permanent changes in the coastline, the remodelling of the coastal landscape and the differential preservation of coastal archaeological sites (Bayley and Flemming 2008; Punke and Davis 2006; Reeder-Myers et al. 2015). The use of vertical geomorphological measurements was important for this purpose, due to the subsidence and emergence caused by plate tectonics documented in different sampling points throughout southern Chile (see Díaz-Naveas and Frutos 2010; Kempf et al. 2017; Plafker and Savage 1970; SHOA 2000). These phenomena transform the landscape fairly regularly (Darwin 1845; Lomnitz 1970; Vidal Gormaz 1901).

Regional geological information (SERNAGEOMIN 2003) was a useful tool for discovering sources from which lithic raw materials exploited in the past were obtained (sensu Aragón and Franco 1997; Church 1994; Franco and Borrero 1999; Stern 2004), contributing to the analysis of the circulation of raw materials, as we have seen within the region and on the eastern slope of the Andes (Méndez et al. 2006, 2008/9, 2012; Méndez and Reyes 2008, 2015; Reyes et al. 2007, 2009, 2016b). Contextual information was also documented on the fauna and the forest flora which made up the principal resources present in this island system. This information on the current structure of the environment and supply of available resources served as a frame of reference (sensu Binford 2001) for understanding the environmental dynamics of the past, assessing these as they relate to the archaeological record and materials obtained.

In order to address a wide spectrum of possible activities carried on by marine hunter-gatherer groups, we searched the ethnographical and ethno-archaeological bibliography (e.g. Chapman 2012; Emperaire 1963; Gusinde 1951, 1982; Martial et al. 2007; Orquera and Piana 1999b) for indicators that would help us to establish frames of reference (sensu Binford 2001), and for particular analogous situations to interpret elements of material and social culture, with questions about their origin, establishing the scope of their application to other archaeological examples (Borrero 1991, 1997).

We studied the available archaeological documentation (Curry 1997; Ocampo and Aspillaga 1984, 1991; Porter 1993) as well as reports of our own investigations (Reyes et al. 2006a, b, 2007, 2011, 2013, 2015, 2016a, b, 2018) to plan a series of archaeological prospections and interventions (searches, tests, excavations, surveys of materials, etc.). In the results of this study, we have incorporated not only material archaeological information from before the first contact with Europeans (taking 1553 AD as the first-contact date), but also material and ecofactual information—also archaeological—from the history of the republican period (post-1810) discovered in our archaeological prospection and test work. These are the *cholguero ports* or simply *cholgueros* (created by fishers for the large *cholga* or Magellan mussel, *Aulacomya atra*) present on the coasts of several islands in the archipelago. They are associated with the start of intense industrial extraction of marine resources, beginning in the nineteenth century and continuing until the middle of the twentieth (Marín 2014; Martinic 2005; Mayorga 2016; Nuñez et al. 2016).

These archaeological sites from historical times take us away from the central questions addressing the occupation of this archipelago by marine hunter-gatherers; however, they are indirectly related to the probability of finding archaeological sites and with an evaluation of the state of preservation of the material and ecofactual record on a temporal microscale (sensu Dincauze 2000). Many *cholgueros* overlie and occupy archaeological locations (see Curry 1991; Legoupil et al. 2007), covering them metres deep in shells and disturbing the shallower levels of pre-contact archaeological sites by moving elements in the space that they occupy (cutting down the forest, bringing in cattle and sheep, setting up fishing camps, etc.). For this reason, we searched and took cores at cholguero sites to confirm or discard the presence of archaeological occupations concealed at lower levels. The short time scale of less than 200 years allowed us to make observations and records of the site location of these *cholgueros* and their situation in relation to geomorphological changes in the shoreline, using them as sedimentary witnesses of the coastal dynamic. Using the same temporal microscale, we evaluated the archaeological visibility of the material record after the sites were abandoned.

When designing the prospection plan, various field trips in predefined sectors were planned to ensure a representative geographical sample to cover the 360 km of the north/south dimension and the almost 150 km of the east/west dimension, as well as sampling the mainland coast of the archipelago, the inshore channels and those exposed to the Pacific, and the offshore islands. In each sector, we assessed the type of activity to be carried out, the estimated time and the route, taking into account available information (Aspillaga et al. 2006; Curry 1997; Porter 1993; Ocampo and Aspillaga 1984; Reyes et al. 2007, 2011, 2013) and the general and specific objectives proposed for this investigation. Thus, the Chonos Archipelago was subdivided for operational purposes into five predefined working areas: Northern, Mainland coast, Central, Central-south, and the Offshore Islands and Pacific Border area (Fig. 3.1). This artificial subdivision allowed us to explore different, proportional sectors of the islands, in order to obtain a general archaeological panorama, observing not only the visibility and accessibility of the material record but also its obtrusiveness (Schiffer et al. 1978) and preservation.

A similar intensity of archaeological work (prospection, sampling, excavations) was carried out in order to allow subsequent comparison (see Ammerman 1981; Banning et al. 2006; Burger et al. 2004; Ebert 1992; Wandsnider and Camilli 1992), with representative units of analysis distributed in geomorphological, topographical and ecological spaces. This subdivision enabled us to address the different geographical areas into which we considered that the length and breadth of the archipelago could be divided. In general terms, these operational areas included inshore channels, the mainland coast, the Pacific border, and the northern and southern extremes of the archipelago—one isolated by large expanses of water and the other enclosed by the Taitao Peninsula. This does not mean to say that the predefined prospection areas

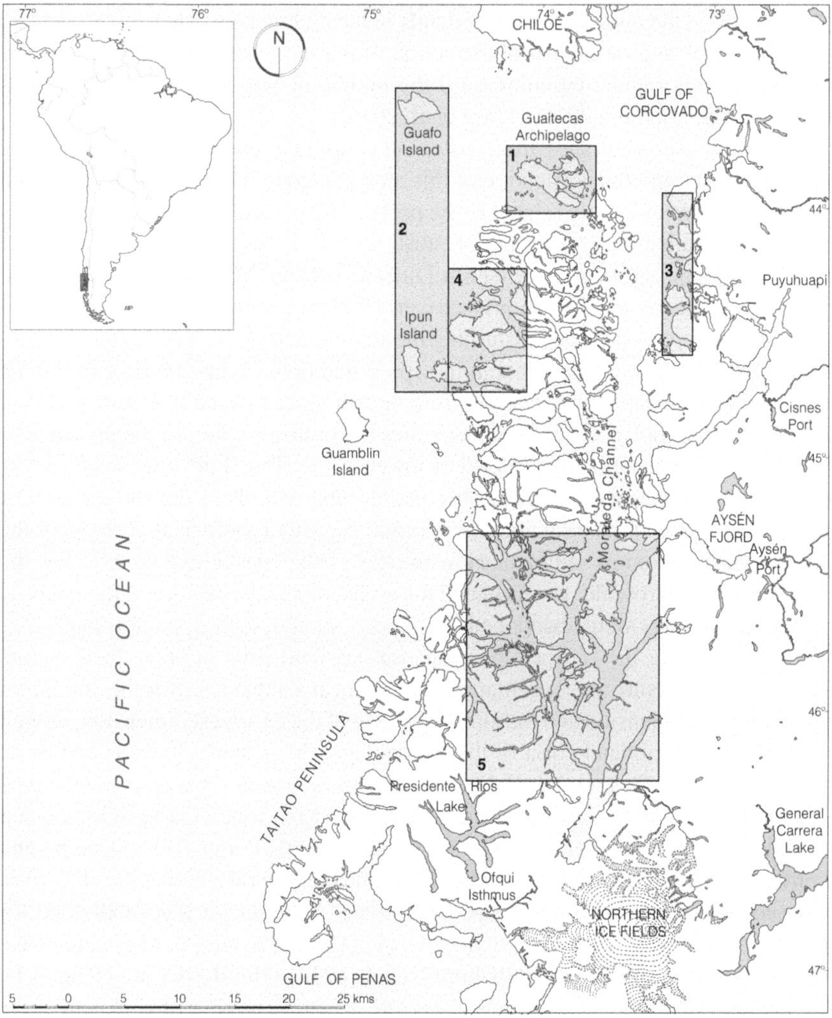

Fig. 3.1 Map of the archipelago showing the five predefined areas for archaeological prospection: 1.- Northern area of the Chonos Archipelago; i.e. the Guaitecas Archipelago; 2.- Offshore Islands and Pacific Border area; 3.- Mainland coast area; 4.- Central area; and 5.- South-central area

differed from one another per se. The whole archipelago is a mosaic, depending on the orientation of the channels, the drainage of the rivers, the types of shoreline, depths, etc. (Viviani 2004); if one starts to look for differences, every island or sector may present particularities. Our idea was to understand, in a wider context of distribution across the region and with broader geographical variables, how the archaeological record could be represented and what aspects of that representation could be discussed.

This prospection design, based on a concept of the 'continuous use' of space (Belardi et al. 1998; Belardi and Campan 1999), was founded on our previous experiences on the mainland and the transitional/steppe zone of the Aysén Region, where forest cover and accessibility were determining factors in the design of prospection and the evaluation of the later distributional record (Mena 1988–9, 1991, 1999, Mena and Lucero 2007; Mena and Ocampo 1993; Reyes et al. 2006a). Furthermore, we adjusted our plans in the light of our previous experience of searching and recording in situations when we prospected on the forested shorelines of the Chonos Archipelago by boat (Reyes et al. 2006b, 2007, 2011).

Thus the Northern coastal prospection area was the sector of the Guaitecas Archipelago, the northernmost part of the Chonos Archipelago between 43°46′ and 44°S. This group of islands is cut off on all sides by expanses of water: to the north the Gulf of Corcovado, 40 km wide, separates it from the southern end of Chiloé Island, while to the east the Moraleda Channel, 30 km wide, separates it from the mainland coast; to the west lies the Pacific Ocean and to the south the Tuamapu and Leucayec Channels cut it off from the rest of the archipelago—which extends for another 300 km. Guaitecas forms the head of the Chonos Archipelago, and one of the portions most remote from *terra firma*.

The Mainland coast area was defined as the zone on the east shore of the Moraleda Channel, fenced in by high mountains. This channel separates the mainland coast from the islands and channels of the archipelago by a distance which narrows from 30 km in the north to 10 km in the south. The climate and vegetation of the mainland coast are practically identical to those of the rest of the archipelago, however, there are differences: the fauna, which includes larger mammals; the large rivers and watercourses that descend from the Andes to the sea; and the presence of volcanic areas which were sources of raw materials. We explored part of this coast opposite the Guaitecas Archipelago, between 43°46′ and 44°S, on the coast of the Melimoyu Peninsula.

The Central area was a sector of the inshore islands and channels between 44°23′ and 44°46′S. Basically, it included Level, Izaza and Benjamín Islands, between the Simpson Channel in the north, Byron in the south, Memory in the west and Chaffers in the east.

The South-central area was the sector of inshore islands and channels between Aysén Fjord in the north, Barros Arana Channel in the east, the outlet of Presidente Ríos Lake on the east coast of the Taitao Peninsula in the south, and the Pulluche Channel in the west, lying between 44°15′ and 46°15′S.

Finally, the Offshore Islands and Pacific Border area formed the most westerly sector, subject to the influence of the open ocean with no protective barrier or obstacle. Offshore Islands lie further away from the main archipelago: they *"are those close to continents but separated by sea deep enough for them to remain islands during ice ages when sea-level is lower… The limiting depth is usually about 120 m, but may be more or less than this because of local emergence or subsidence"* (Rackham 2012:88). The limits were set as 43°31′ to 44°55′S. The most northerly is Guafo Island in the Pacific Ocean, equidistant from Chiloé Island and the Guaitecas Archipelago (40 km). Further south lies a series of islands whose west coasts are exposed to the ocean, of which we prospected parts of Johnson, Forsyth, Level, Stokes and Ipún Islands. Guamblín Island (the most exposed and isolated together with Guafo) is equidistant from the nearest groups of islands in the archipelago (30 km). However, it could not be prospected for practical reasons, and this work remains pending.

In this context, in recognition of the work carried out and as a word of advice to other researchers, it must be said that plans and designs made in the office are easily altered in the archipelago. Any archaeological prospection has its problems (e.g. working time, climate, obtaining permission from landowners, distance from access points and a long list of *et ceteras*). In this case, coastal prospection of the archipelago, by definition only possible by boat, is sensitive to a number of factors which determine the effective working time, the intensity of prospection and the specific archaeological work to be carried out. These include the weather, the sea state (which may vary from one channel to the next), accessibility for landing on the coast, sailing controls imposed by the Chilean Navy, distance from supply centres (especially fuel) and our own capacity to carry out the archaeological work and sail at the same time. We are not, and will never be, seamen—much less canoe people. There were moments when our plans were modified or changed completely due to the problems of sailing in an unstable climate; and there were field trips which failed because the weather never improved and much of the time available was spent vainly in a port—even considering the margin allowed in planning for lost time which is a *sine qua non* for work in this area. Mike Tyson says wisely: *"Everybody has a plan until they get punched in the face"*. Carrying out archaeological work in the channels requires the same philosophy; boat travel requires time and patience.

3.2 Fieldwork

The archaeological prospections were mainly coastal, using shallow-draft inflatable dinghies (Zodiac), sailing from the larger boat used for longer voyages and as a base (Fig. 3.2). This was the only way of getting to the various pre-established sectors of the shoreline (offshore islands, inshore channels and more inaccessible parts of the mainland coast). No large-scale charts of the area are available (they are mostly 1:250,000). The images on Google Earth and the aerial photographs of the Aerophotogrametric Service of the Chilean Air Force are of limited use since they do not provide details of the local geomorphology (e.g. rocky coasts, currents, steep coasts and intertidal zones); they serve only as a reference for times and distances when navigating and planning which sectors to prospect. Some shorelines change dramatically on a daily basis depending on the state of the tide, and whether it is at springs or neaps. Thus the intensity and cover of the prospection design were repeatedly adjusted to the local circumstances of the moment (e.g. weather, tide, discovery of areas with a wider intertidal zone rather than steep coast and coastal relief of the islands).

It must be borne in mind that throughout the Chonos Archipelago, the only areas clear of forest are those cleared by anthropic intervention (felled, burnt and occupied

Fig. 3.2 Coastal archaeological prospection in the Guafo island

by fishermen) and the peat bogs present on waterlogged land. All the rest is covered by evergreen forest (Álvarez et al. 2010; Gajardo 1994; SERPLAC 2005) down to the tideline (Camus 2001; Viviani 2004).

This is also the place for a word about the intertidal zone, the coastal strip between the lines of the highest and lowest tides. It is the only marine space regularly exposed to the air, which affects strongly the physical and biological processes of the organisms that live there (Camus 2001; Díaz-Naveas and Frutos 2010). Intertidal zones are classified as having soft (sand, mud, swampy), hard (rock, shingle, platforms) and mixed bottoms. They may be virtually absent in near-vertical coasts, as in steep-sided fjords and islands, or hundreds of metres wide in areas with terrace formations (Camus 2001; Díaz-Naveas and Frutos 2010). The systematic prospection of these areas depends on access exclusively when the tide is out (Fig. 3.3).

Considering the data cited above, much of the search for sites followed geomorphological criteria (Butzer 2008), i.e. looking for and identifying landscape features (e.g. marine or fluvial terraces located above the current coastline and profiles with tidal erosion or intertidal zones) which would provide information on past coastline changes possibly related to fluctuations in the sea level or local tectonics (Lomnitz 1970; Plafker and Savage 1970; Vargas et al. 2013). Accordingly, all the post-contact archaeological sites (*cholgueros*) detected and reported in this work were geo-referenced in order to estimate their geomorphological situation and their relation with the current sea level.

Fig. 3.3 Prospecting the shoreline at low tide, north shore of Jacaf channel

Once on terra firma, the archaeological prospection included superficial reconnaissance of the area and its particularities (e.g. bays, terraces, stream mouths and exposed profiles), and subsuperficial prospection in zones where visibility was reduced by forest cover. This was done to detect and characterise sites, i.e. to determine the volume of the anthropic deposits discovered (stratigraphic potential and area), and to recover deep organic samples for radiocarbon dating (e.g. shells, charcoal and soils). In this way, we could establish sedimentation rates and the accumulation rates of some of the larger waste deposits (shell-middens) (Reyes et al. 2015). The search strategy implied the use of bores, using AMS equipment with 'French' and 'Dutch' bits of 60 mm diameter, to allow extraction of sediment cores more than 8 m deep. The cores enabled us to detect highly obtrusive deposits (Lightfoot 1986; Ponce 2004; Stein 1986), such as shell-middens and palaeo-beaches, in zones with dense plant cover and little or no archaeological visibility. Detection of coastal sites of this kind was based on finds mainly of mollusc shells or the periostracum of *Mytilus spp.*, which remain as evidence of archaeological camps, or else dark organic soils differentiated from the reddish-brown, acidic organic matter produced by the forest, or the levels of sandy sediments and pebbles which characterise palaeo-beaches. Nonetheless, biases exist for the discovery of small, low-density sites with marked concentrations of cultural remains. Furthermore, subsuperficial prospection techniques (cores, digging, test pits) present an operational disadvantage in that they demand more time and work for each area prospected. Despite this, the use of cores in conjunction with other systematic subsuperficial methods is an efficient way of discovering cultural lenses below the surface, and recognising geo-archaeological patterns which help in assessing potential locations of anthropic deposits (Stein 1986).

To record the cultural evidence, the variables considered when characterising and contextualising an archaeological site were transferred to previously prepared record cards (sensu Reyes et al. 2006a; San Román and Morello 2000). These have been used by other authors in the context of coastal prospections with the emphasis on distribution (e.g. Borrero and Lanata 1992), where not only the presence and type of sites are important, but also their absence (Bintliff 2000; Foley 1981). Thus, we incorporated reference variables for the geomorphology and local placing of the sites which might have had an impact on both the location and the preservation of the site. The necessary basic data were included, such as UTM coordinates, altitude and distance from the current coastline, current height above sea level, distance to the nearest fresh water, estimated area and depth of the deposit, type of site based on its main characteristics, surroundings, presence of artefacts, presence of ecofacts and traces.

The archaeological visibility was related to the percentage of cover (high visibility 0–30% cover, medium 31–60%, low 61–90% and nil 91–100%). All these variables were organised to evaluate the cultural deposit and later archaeological intervention in situ. This allowed us to project qualitative inferences (e.g. function of the site, cultural attribution and potential for intervention). Emphasis was placed on observation of the formation of the contexts, distinguishing their cultural and natural processes (Harris 1991; Schiffer 1996) into 'taphonomically active zones' (sensu Borrero 2014). Taphonomic variables in archaeofaunal remains (Lyman 1994), reassemblies (Cahen et al. 1979), differential erosion (Borrazzo 2007; Hiscock 1985) and micro-alterations in stone instruments lead to a certain degree of control over some of these processes (Borrero 2001). Finally, the distributional and spatial information was transferred to satellite records of the area geo-referenced in UTM WGS84 (Google Earth images) for data processing. They were studied as geographical entities, in which a particular location was associated with a set of significant attributes (e.g. type of site, location, chronology obtained and presence of raw materials).

The stratigraphic record, and the cultural and/or bio-anthropological and/or ecofactual materials were obtained through a network of cores, test pits, characterisation of exposed profiles, survey of surface material and extended excavations of superficial and subsuperficial sites, to allow reconstruction and cultural inferences of the area occupied. Priority was given to recovering contextual evidence by natural/cultural stratigraphic layers, rather than artificial units of measurement (Legoupil 2003). In particular, the excavations were undertaken from a geo-archaeological perspective (Dincauze 2000; Goldberg and Macphail 2006), using some basic analytical techniques to characterise the lithostratigraphic, archaeostratigraphic (e.g. malacofauna) and pedostratigraphic deposits (characterisation of sediments, pH, organic matter). Special care was taken in shell-middens—formed by anthropic accumulation and deposit of mollusc valves—since the shells appear as sedimentary particles with their own deposit processes (Stein 1992; Stein et al. 2003).

We also made geomorphological records of the surroundings of the archaeological sites in which we intervened. This is a basic tool for understanding stratigraphic units and site formation processes, and for obtaining evidence from which to project the geographical and environmental conditions at the time of occupation (Borrero 2013; Morales et al. 2009; Sandweiss et al. 2004).

In the first instance, we used broad spatial scales (sensu Dincauze 2000). This was justified by the small number of archaeological sites in the study area in terms of archaeological frequency (Curry 1997; Reyes et al. 2015) and the intensity of our investigations, since the occupations identified can be represented as isolated points in space. Site identification was treated on the mesoscale (10^2 to 10^4 km^2; e.g. predefined prospection areas, resource capture areas, sources of raw materials and mainland resources) and compared and contextualised on the macroscale (10^4 to 10^7 km^2 physiographical province, Aysén Region, Western Patagonia, and biogeographical barriers like the Moraleda Channel, the Gulf of Corcovado, and the mountain chains and volcanic areas adjacent to the mainland coast). Finally, to reflect particular cultural processes in the general framework, site micro- and mesoscales were used. We also worked on a mesoscale in time, covering part of the Mid and Late Holocene, and on microscales, for example, for the chronology of site occupation, considering the particular variables in each case (start and end of occupations, sedimentation rate, introduction of allochthonous strata, etc.).

To support our chronologies, we obtained samples for radiocarbon dating from most of the archaeological deposits identified (including cholguero sites) in order to associate the start of occupation with the geoform and substrate on which the first occupants of the area settled (Erlandson and Moss 1999). In some cases, samples were taken to establish a stratigraphic sequence of occupation and the formation of the cultural deposit, and/or to date precisely a particular feature at the site (e.g. hearths, burials, intrusion of sediments or hiati). In some ossuaries, we dated 100% of the individuals recovered and identified, in order to define these funerary contexts in a limited area and a known time sequence. Correction factors (calibration) were considered when local and regional chronostratigraphic comparisons were made (Dincauze 2000) (Local Reservoir Effect; Albero et al. 1986, 1988; Fabier Dubois 2009; Merino et al. 2019; Reyes et al. 2013). The cultural sequences used in the Mid- and Late Holocene follow the temporal references defined by Walker and others (2012). Series of ^{14}C datings is a widely used method of statistical analysis for understanding sets of data and their magnitudes over time, and thus relations between the environment and cultural processes (Gamble et al. 2004; Steele 2010; Williams 2011). It serves to understand the varying intensity of the use of spaces, especially periods of abandonment (Araujo et al. 2006), and analytic scales (Cameron and Tomka 1993; Reyes et al. 2019).

The radiocarbon analyses were carried out by Beta Analytic Inc., DirectAMS Lab and the Center for Applied Isotope Studies (CAIS), University of Georgia. The AMS technique was used (except for BETA 34,404—conventional, Porter 1993). Stable isotope analyses were carried out by CAIS. All the radiocarbon dates presented here ($n = 86$ dates, 85 from our investigation and one from Porter 1993) were corrected by calibration at 2σ with Calib 7.0.4 (Stuiver et al. 2013) and are expressed in calculated years before the present (cal. BP). Samples of materials belonging to terrestrial trophic chains were matched with the SHCal13 curve (Hogg et al. 2013). For the samples of individuals with a marine or mixed diet, a local reservoir effect was applied of 141 ± 43 years if they originated in the Chiloé Archipelago, or 52 ± 47 years if they came from the Chonos Archipelago, according to the closest available values (Merino et al. 2019) and using the Mixed Marine SoHem curve (Reimer et al. 2013). The calibration was based on the % of marine resources in the diet of each individual. This was estimated from the $d^{13}C$ of the individuals, and by establishing estimated 100% terrestrial and 100% marine diets (Coltrain et al. 2016), based on the isotope ecology compiled for the area from bibliographic and own data (Reyes et al. 2019).

3.3 Laboratory Work

The subsequent laboratory and office work included the complete assessment of the results. To this was added analysis of the geomorphological placements identified and their relation with the distribution and type of archaeological sites discovered, together with their stratigraphic potential (defined by cores and/or test pits). The morphofunctional characteristics of the cultural materials recovered in prospection and tests were also analysed, as well as the origin of raw materials and their link with the chronological series obtained.

Study of lithic technology in the collections included morphofunctional and typological evaluation (Aschero 1983) to recognise techno-stylistic indicators for the populations studied (see Orquera and Piana 1999; Orquera et al. 1977; Pelegrin 1990; Schidlowsky 2004). Basic analyses of the technologies used (Andrefsky 1994, 1998) and the origins of the raw materials (geochemical analysis) were carried out in the Laboratory for Environmental Geochemistry at the University of Colorado (Boulder), directed by C. Stern; these gave indications of the direction and flows of population movements (Legoupil 1997; Morello et al. 2002; Stern and Porter 1991) based on the raw materials identified and their possible origins. The purpose of the technological characterization was to elucidate the 'technology organisation' (Nelson 1991) of the groups studied and their design strategies (Hayden et al. 1996).

Study of the archaeofaunal remains, which helped us to analyse the subsistence strategies of the past, included taxonomic analysis of the remains of terrestrial and marine mammals and birds (Lyman 1982; San Román et al. 2016), fish (Colley 1990; Meléndez et al. 1993; Torres 2016) and malacological fauna (Cárcamo 2015; Claassen 1998, 2000). The taxonomic diversity and abundance (NISP, MNE, MNI and MAU) were calculated for all the species identified, and the representation of anatomical parts was calculated where possible (sensu Claassen 1998, 2000; Colley 1990; Grayson 1984; King 1983; L'Heureux and Borella 2011; Waselkov 1987). The analyses included evaluation of taphonomic variables and a record of stigmas of natural origin (Pearce 2008; Reynard 2014) and anthropic traces (Behrensmeyer 1978; Lyman 1994). Special attention was paid to the record of cultural modes of processing different taxa, in order to establish the existence of culturally important patterns (Binford 1981; Haynes 1983; Mengoni 1999).

Analysis of the human bone remains recovered from ossuaries and archaeological excavations focused first on differentiating individuals rather than analysing elements in the case of ossuaries (minimum number of individuals—MNI), since the burials excavated presented more clarity in the definition of anatomical components for each individual. The bone remains which could not be assigned to an individual were analysed by ossuary sets, separating them by bone or type of bone represented (Galimany 2016, 2017). The MNI in each ossuary was determined using the bone with the highest representation, and bone sizes. For the basic bio-anthropological record of adult individuals, sex was estimated by observation of the morphology of the pelvic girdle (Phenice 1969) and the cranium (Buikstra and Ubelaker 1994; Walrath et al. 2004). Age was estimated by the method of the morphological changes present in the pubic symphysis (Brooks and Suchey 1990), the auricular surface of the ilium (Lovejoy et al. 1985; Osborne et al. 2004) and the sternal border of the fourth rib (Iscan et al. 1984, 1985), as well as other indicators. Palaeo-pathological conditions were also recorded (Capaso et al. 1999; Hillson 1996; Lewis and Roberts 1997; Ortner 2003; Waldron 2009) and stature could be estimated in some cases (Genovés 1967; Trotter and Gleser 1958). The age of subadult individuals was estimated based on the development and eruption of the deciduous and permanent teeth, and the fusion and development of epiphyses and long bones (Al Qahtani et al. 2010; Buikstra and Ubelaker 1994; Schaefer et al. 2009).

Although the majority of the human bone remains had been plundered and moved, a basic bio-anthropological analysis could be drawn up (sex, age, MNI) with a contextual report of the site and synchronic associations. Absolute chronologies and isotope analyses (δ^{13}C collagen, δ^{15}N collagen) were also obtained, and in the majority of cases the carbon/nitrogen (C/N) ratio, which is an elemental indicator of diagenesis, in the samples analysed (Ambrose 1990; De Niro 1985). In a later stage, this allowed us to establish relations with the various trophic chains present (Barberena 2008). For this, we used unpublished human bone remains obtained in this study (n = 18, Reyes et al. 2019, integrating them with information obtained from samples collected previously (n = 20, Reyes et al. 2011, 2013, 2015), both from the Chonos Osteological Collection of the University of Chile and the Research Centre in Patagonian Ecosystems in Coyhaique, recovered in judicial investigations carried out by the anthropologist Mauricio Osorio and the archaeologist Francisco Mena (Osorio 2011), together with the Chilean Navy and Carabineros (police).

To obtain a basic local isotope ecology (Ambrose and Katzenberg 2000; Schoeninger 1995) for the archaeofaunal resources identified, we obtained samples of terrestrial mammals from the archaeological sites and current samples for comparison with archaeological samples obtained previously in the neighbouring areas of the Aysén Region (Barberena et al. 2011; Méndez et al. 2014; Reyes et al. 2007). The isotope values obtained from the bone collagen of the archaeofaunal samples were projected to muscle (Fernándes 2015), in order to homologate them and compare them with the isotope results obtained for a series of modern species of fish, crustaceans, molluscs, marine mammals and birds from adjacent geographical zones both on the Pacific coast between 37° and 43°S (Huckstadt et al. 2007; Mayrs et al. 2011; Muñoz et al. 2013) and the Atlantic coast between 40° and 46°S (Forero et al. 2004). This enabled us to establish reference values for the resources consumed. In this way, we projected a regional isotope ecology, assessing its economic importance for this group of hunter-gatherers in the Patagonian Archipelago (sensu Alfonso et al. 2015; Barberena 2002; Borrero et al. 2001; Borrero and Barberena 2006; Guichón et al. 2001; Kochi 2017; Panarello et al. 2006; Reyes et al. 2019; Tafuri et al. 2017; Tessone et al. 2003, 2009a, b; Tykot et al. 2009; Yesner et al. 2003).

References

Alberó M, Angiolini F, Piana E (1986) Discordant ages related to reservoir effect of associated archaeological remains from Tunel site (Beagle Channel, Argentine). Radiocarbon 28(2A):748–753

Alfonso M, Giles B, Misarti N, San Román M, Morello F (2015) Antiquity and geographic distribution of cranial modification among the prehistoric groups of Fuego- Patagonia. Am J Phys Anthropol 158:607–623

Al Qahtani S, Héctor M, Liversidge H (2010) Brief communication: the London atlas of human tooth development and eruption. Am J Phys Anthropol 142(3):481–490.4

Álvarez M, San Martin C, Novoa C, Toledo G, Ramírez C (2010) Diversidad florística, Vegetacional y de hábitats en el Archipiélago de los Chonos (Región de Aisén, Chile). Anales del Instituto de la Patagonia 38:35–56

Ambrose S (1990) Preparation and characterization of bone and tooth collagen for isotopic analysis. J Archaeol Sci 17:431–451

Ambrose S, Katzenberg A (2000) Biogeochemical Approaches to Paleodietary Analysis. University of Illinois, Urbana-Champaign, Urbana

Ammerman A (1981) Surveys and archaeological research. Annual Rev Anthropol 10(1):63–88

Andrefsky W (1994) Raw-material availability and the organization of technology. Am Antiq 59(1):21–34

Andrefsky W (1998) Lithics. Macroscopic approaches to analysis. Cambridge University Press, Cambridge

Aragón E, Franco N (1997) Características de rocas para la talla por percusión y propiedades petrográficas. *Anales del Instituto de la Patagonia* . Sr Cs Hs 25:187–199

Aschero C (1975) 1983. Manuscrito. Ensayo para una clasificación morfológica de los artefactos líticos, Buenos Aires

Aspillaga E, Castro M, Rodríguez M, Ocampo C (2006) Paleopatología y estilo de vida: El ejemplo de los Chonos. Magallania (Punta Arenas) 34(1):77–85

Bailey G (2004) World prehistory from the margins: The role of coastlines in human evolution. J Int Stud Hist Archaeol 1:39–50

Bailey G, Flemming N (2008) Archaeology of the continental shelf: marine resources, submerged landscapes and underwater archaeology. Quatern Sci Rev 27(23):2153–2165

Banning E, Hawkins A, Stewart S (2006) Detection functions for archaeological survey. American Antiquity, 723–742

Barberena R (2002) Los límites del mar: Isótopos estables en Patagonia meridional. Sociedad Argentina de Antropología. Buenos Aires

Barberena R (2008) Arqueología y biogeografía humana en Patagonia Meridional. Sociedad Argentina de Antropología. Buenos Aires

Barberena R, Méndez C, Mena F, Reyes O (2011) Endangered species, archaeology, and stable isotopes: Huemul (Hippocamelus bisulcus) isotopic ecology in Central-Western Patagonia (South America). J Archaeol Sci 38(9):2313–2323

Behrensmeyer A (1978) Taphonomic and ecologic information from bone weathering. Paleobiology 4(02):150–162

Belardi J, García M, Campán P (1998) Las distribuciones artefactuales y las poblaciones humanas. In: Borrero L (ed.), Arqueología de la Patagonia Meridional, pp 53–71. Búsqueda de Ayllu, Buenos Aires

Belardi, J, Campan, P (1999) Estepa y bosque: La utilización de lagos y lagunas en la región de Lago Argentino, Provincia de Santa Cruz. Actas de las III Jornadas de Arqueología de la Patagonia. Bariloche, Argentina

Bennett K, Haberle S, Lumley S (2000) The last glacial-holocene transition in Southern Chile. Science 290:325–328

Binford L (1981) Bones: ancient men and modern Myths. Academic Press, New York

Binford L (2001) Constructing frames of reference. An analytical method for archaeological theory building using ethnographic and environmental data sets. University of California Press, California

Bintliff J (2000) The concepts of 'site'and 'offsite'archaeology in surface artefact survey. Non-destructive techniques applied to landscape archaeology 4

Borrazzo K (2007) Aporte de la tafonomía lítica al estudio de distribuciones artefactuales en ambientes lacustres: el caso del sistema lacustre al sur del Lago Argentino. Comeching Virtual 1(3):132–153

Borrero L (1991) Los "modelos de situaciones excepcionales" y el estudio de las sociedades de cazadores y recolectores. *Comechingonia.* Revista de Antropología e Historia. Año 8, N°7

Borrero L (1997) The origins of ethnographic subsistence patterns in Fuego-Patagonia. In: McEwan C, Borrero LA, Prieto A (eds.), Patagonia. Natural history, prehistory and ethnography at the uttermost end of the Earth, pp 60–81. British Musem Press, London

Borrero L (2001) Regional taphonomy: the scales of application to the archaeological record. In: Buitenhuis H, Prummel W (eds.), Animals and man in the past, pp 17–20. ARC-Publicatie 41. Groningen. Netherlands

Borrero L (2013) Paisajes desconocidos, geografía cultural y tafonomía total. Anuario de Arqueología, Rosario, 5:17–30

Borrero L, Guichon R, Tykot R, Kelly J, Prieto A, Cárdenas P (2001) Dieta a partir de isótopos estables en restos óseos humanos de Patagonia Austral. Estado actual y perspectivas. Anales del Instituto de la Patagonia. 29:119–127

Borrero L, Barberena R (2006) Hunter-gatherer home range and marine resources. Current Anthropol 47:855–867

Borrero L, Lanata J (1992) Análisis Espacial en la Arqueología Patagónica. (compiladores) Ediciones Ayllu, Buenos Aires

Buikstra J, Ubelaker D (1994) Standards for data collection from human skeletal remains: proceedings of a seminar at the field museum of natural history. Fayetteville: Arkansas Archeological Report Research Series

Butzer K (2008) Challenges for a cross-disciplinary geoarchaeology: the intersection between environmental history and geomorphology. Geomorphology 101(1):402–411

Burger O, Todd P, Burnett T, Stohlgren T, Stephens D (2004) Multi-scale and nested-intensity sampling techniques for archaeological survey. J Field Archaeol 29(3–4):409–423

Cahen D, Keeley L, Van Noten F (1979) Stone tools, toolkits, and human behavior in prehistory. Current Anthropol 20(4):661–672

Camus P (2001) Biogeografía marina de Chile Continental. Revista Chilena de Historia Natural. 74:587–617

Capasso L, Kennedy, K, Wilczak C (1999). Atlas of occupational markers on human remains 1st ed., Teramo, Italy: Edigrafial

Cárcamo J (2015) El consumo de moluscos en el conchal GUA-10, Archipiélago de las Guaitecas, Patagonia septentrional: *Un Estudio Arqueomalacológico*. Universidad de Tarapacá, Memoria de Título

Chapman A (2012) Yaganes del Cabo de Hornos. Encuentros con los europeos antes y después de Darwin. Ediciones Pehuén. Santiago, Chile

Church T (1994) Lithic resource studies: a sourcebook for archaeologists. Special publications 3. Lithic Technology. University of Tulsa, Oklahoma

Claassen C (1998) Shells. Cambridge Manuals in Archaeology. Cambridge University Press

Claassen C (2000) Quantifying shell: comments on Mason, Peterson, and Tiffany. Am Anti 65:415–418

Colley S (1990) The analysis and interpretation of archaeological fish remains. In: Schiffer MB (ed.), Advances in archaeological method and theory, vol 2, pp 207–253. Academic Press, San Diego

Coltrain JB, Tackney J, O'Rourke DH (2016) Thule Whaling at Point Barrow, Alaska: The Nuvuk Cemetery Stable Isotope and Radiocarbon Record. J Archaeol Sci Reports 9:681–694

Curry P (1991) Distribución de sitios e implicaciones para la movilidad de los canoeros en el canal Messier. En Anales del Instituto de la Patagonia. 20:145–154

Curry P (1997) Manuscrito. Archaeological exploration of the central Chonos Archipelago. Informe Arqueológico Consejo Monumentos Nacionales. Santiago

Darwin Ch (1845) The Voyage of the Beagle (Journal of researches) into the natural history and geology of the countries visited during the voyage of H.M.S. Beagle round the world, under the Command of Capt. Fitz Roy, R.N. 2ª edición, Londres: John Murray, Cap 13:273–290

De Niro MJ (1985) Post-mortem preservation and alteration of in-vivo bone collagen isotope ratios in relation to paleodietary reconstruction. Nature 317:806–809

de Porras ME, Maldonado A, Abarzúa A, Cárdenas M, Francois JP, Martel A, Stern C, Méndez C, Reyes O (2012) Postglacial vegetation, fire and climate dynamics at Central Chilean Patagonia (Lake Shaman, 44°S). Quatern Sci Rev 50:71–85

Díaz-Naveas J, Frutos J (2010) Geología Marina de Chile. Comité Oceanográfico Nacional de Chile—Pontificia Universidad Católica de Valparaíso—Servicio Nacional de Geología y Minería de Chile

Dincauze D (2000) Environmental archaeology. Principles and practice. Cambridge University Press

Ebert J (1992) Distributional archaeology. University of Utah Press

Emperaire J (1963) Los nómades del mar. Ediciones de la Universidad de Chile, Santiago

Erlandson J (2001) The archaeology of aquatic adaptations: paradigms for a new millennium. J Archaeol Res 9(4):287–350

Erlandson J, Moss M (1999) The systematic use of radiocarbon dating archaeological survey in coastal and other erosional environments. Am Antiq 64(3):431–443

Favier Dubois C (2009) Valores de efecto reservorio marino para los últimos 5.000 años obtenidos en concheros de la costa atlántica norpatagónica (golfo San Matías, Argentina). Magallania 37(2):139–147

Fernandes R (2015) A simple (r) model to predict the source of dietary carbon in individual consumers. Archaeometry. https://doi.org/10.1111/arcm.12193

Foley R (1981a) A model of regional archaeological structure. In: Proceedings of the Prehistoric Society, vol 47, pp 1–17. Cambridge University Press

Forero M, Bortolotti G, Hobson K, Donazar J, Bertelloti M, Blanco G (2004) High trophic overlap within the seabird community of Argentinean Patagonia: a multiscale approach. J Anim Ecol 73(4):789–801

Franco N, Borrero L (1999) Metodología de análisis de la estructura regional de recursos líticos. Prácticas de recolección en el cono Sur de América, En Los tres Reinos, pp 27–37

Gajardo R (1994) La vegetación natural de Chile. Editorial Universitaria, Santiago

Galimany J (2016) Manuscrito. Informe bioantropológico. Informe Manuscrito, Proyecto FONDECYT 1130151. Santiago

Galimany J (2017) Manuscrito. Informe bioantropológico del sitio Alero Low1, Melimoyu 1 y Yalac 1. Informe Proyecto FONDECYT 1130151. Santiago

Gamble C, Davies W, Pettitt P, Richards M (2004) Climate change and evolving human diversity in Europe during the last glacial. Philosoph Trans Royal Soc B 359:243–254

García J, Maldonado A, de Porras ME, Nuevo A, Ebensperger C, Reyes O, Binnie S, Luethgens C, Méndez C (2019) Early deglaciation and paleolake history of the Río Cisnes Glacier, Patagonian Ice Sheet 44°S. Quat Res 91(1):194–217

Genovés S (1967) Proportionality of the Long Bones and their Relation to Stature among Mesoamericans. Am J Phys Anthropol 26:67–78

Goldberg P, Macphail R (2006) Practical and theoretical geoarchaeology. Blackwell Publishing

Grayson D (1984) Quantitative zooarchaeology: topics in the analysis of archaeological faunas. Academic Press, Orlando, Florida

Guichón R, Borrero L, Prieto L, Cárdenas P, Tykot R (2001) Nuevas determinaciones de isótopos estables para Tierra del Fuego. Revista Argentina de Antropología Biológica, 3

Gusick A, Faught M (2011) Prehistoric archaeology underwater: anascent subdiscipline critical to understanding early coastal occupations and migration routes. In: Bicho N, Haws J, Davis L (eds.), Trekking the shore: changing coastlines and the antiquity of coastal settlement, pp 27–50. Springer, New York

Gusinde M (1951) Hombres primitivos en la Tierra del Fuego: de investigador a compañero de tribu, vol. 63. Escuela de Estudios Hispano-Americanos, Consejo de Superior de Investigaciones Científicas

Haberle S, Bennett K (2004) Postglacial formation and dynamics of North Patagonian Rainforest in the Chonos Archipelago Southern Chile. Quat Sci Rev 23(23):2433–2452

Harris E (1991) Principios de estratigrafía arqueológica. Editorial Crítica, Barcelona

Hayden B, Franco N, Spafford J (1996) Evaluating lithic strategies and design criteria. In: Odell G (ed.), Stone tools: theoretical insights into human prehistory, pp 9–45. Plenum Press

Haynes G (1983) Frecuencies of spiral and green-bones fractures on ungulate limb bone in modern surface assemblages. Am Antiq 48:102–114

Heusser C (1960) Late-Pleistocene Environments of the Laguna de San Rafael Area Chile. Geograph Rev 50(4):555–577

Heusser C (2002) On glaciation of the southern Andes with special reference to the Peninsula de Taitao and adjacent Andean cordillera (46°30'S). J South Am Earth Sci 15:577–589

Hillson S (1996) Dental anthropology. Cambridge University Press

Hiscock P (1985) The need for a taphonomic perspective in stone artefact analysis. Queensland Archaeol Res 2:82–95

Hückstädt L, Rojas C, Antezana T (2007) Stable isotope analysis reveals pelagic foraging by the Southern sea lion in central Chile. J Exp Mar Biol Ecol 347(1):123–133

Iscan M, Loth S, Wright R (1984) Age estimation from the rib by phase analysis: white males. J Forensic Sci 29(4):1094–1104

Iscan M, Loth S, Wright R (1985) Age estimation from the rib by phase analysis: white females. J Forensic Sci 30(3):853–863

Kempf P, Moernaut J, Van Daele M, Vandoorne W, Pino M, Urrutia R, De Batist M (2017) Coastal lake sediments reveal 5500 years of tsunami history in south central Chile. Quatern Sci Rev 161:99–116

King J (1983) Seals of the World. Cornell University Press, Ithacan New York

Kochi S (2017) Paleodietas en cazadores-recolectores del Canal Beagle durante el Holoceno tardío. Intersecciones en Antropología 18:329–339

Legoupil D (1997) Bahía Colorada (île d'Englefield): Les Premiers Chasseurs de Mammifères Marines de Patagonie Australe. Éditions Recherche sur les Civilisations, Paris

Legoupil D (2003) Cazadores-recolectores de Ponsonby (Patagonia austral) y su paleoambiente desde VI al III milenio A. C. Magallania 31, tirada especial

Legoupil D, Bearez P, Lepetz S, San Román M, Salas K (2007) De la pesca a la caza: evolución económica del sitio Stuven 1, al sur del Golfo de Penas. In: Morello F, Martinic M, Prieto

A, Bahamonde G (eds.), Arqueología de Fuego-Patagonia. Levantando piedras, desenterrando huesos… y develando arcanos, pp 279–294. Punta Arenas: Ediciones CEQUA
L'Heureux G, Borella F (2011) Guía osteométrica para el estudio de elementos óseos de Otaria flavescens. Dunken, Buenos Aires, Argentina
Lewis M, Roberts C (1997) Growing pains: the interpretation of stress indicators. Int J Osteoarchaeol 7:581–586
Lightfoot K (1986) Regional surveys in the Eastern United States: the strengths and weaknesses of implementing subsurface testing programs. Am Antiq 51:484–504
Lomnitz C (1970) Major earthquakes and tsunamis in Chile during the period 1535 to 1955. Mexico: Geologische Rundschau 59(3):938–960
Lovejoy C, Meindl R, Pryzbeck T, Mensforth R (1985) Chronological metamorphosis of the auricular surface of the ilium: a new method for the determination of adult skeletal age at death. Am J Phys Anthropol 68(1):15–28
Lyman L (1982) Archaeofaunas and subsistence studies. In: Schiffer MB (ed.), Advances in archaeological method and theory 5, pp 331–93. Academic Press, New York
Lyman L (1994) Vertebrate taphonomy. Cambridge University Press, Cambridge
Marín A (2014) Palena, Quitralco y Melimoyu: Intentos fallidos de colonización en el litoral de la Región de Aysén (1889–1983). Rev Austral Cienc Soc 27:137–156
Martial L, Hyades P, Deniker J (2007) Etnografía de los indios Yaghan en la misión científica del Cabo de Hornos, pp 1882–1883. Universidad de Magallanes
Mayorga M (2016) Antecedentes históricos referidos a la caza de Lobos Marinos y su interacción con el medio geográfico y humano en el extremo Austral Americano: El caso del lobero escocés William Low. Magallania (Punta Arenas) 44(2):37–64
Mayrs C, Försterra G, Häussermann V, Wunderlich A, Grau J, Zieringer M, Altenbach A (2011) Stable isotope variability in a Chilean fjord food web: implications for N-and C-cycles. Mar Ecol Prog Ser 428:89–104
Meléndez R, Falabella F, Vargas L (1993) Manuscrito. Osteometría e identificación de restos ictioarqueológicos de Chile Central. Informe final Proyecto FONDECYT 0089-91. Santiago
Mena F (1988–9) Hacia un panorama del registro arqueológico regional: promesas y frustraciones. Arqueología Contemporánea 2(2). Buenos Aires, Argentina
Mena F (1991) Prehistoric resource space and settlement at the Río Ibáñez Valley (Central Patagonian Andes) Ph.D. dissertation; UCLA. Los Angeles
Mena F (1999) La ocupación prehistórica de los valles andinos centro-patagónicos (XI Región, Chile): Generalidades y localismos. Actas de las Terceras Jornadas de Arqueología de la Patagonia, pp 57–64. Bariloche
Mena F, Lucero V (2007) Manuscrito. Informe Línea de Base Proyecto Hidroeléctrico Aisén, XI Región, Chile. SEIA, ENDESA, Chile
Mena F, Ocampo C (1993) Distribución, localización y caracterización de sitios arqueológicos en el Río Ibáñez (XI Región). Boletín del Museo Regional de la Araucanía. Temuco 4(1):33–58
Méndez C, Velásquez H, Reyes O, Trejo V (2006) Tras los moradores del bosque Conjuntos arqueológicos de Alero El Toro (Valle del río Cisnes, XI Región de Aisén). Revista Werkén 8:101–115. Santiago
Méndez C, Reyes O (2008) Late Holocene human occupation of Patagonian forests: a case study at Cisnes River basin (44° S, Chile). Antiquity. 82:560–570
Méndez C, Reyes O (2015) Archaeology near the southern ice-end. Current advances in human interdisciplinary research in central western Patagonia. SAA Archaeol Record 15(3):21–26
Méndez C, Stern C, Reyes O (2008/9) Transporte de obsidianas a lo largo de los Andes de Patagonia Central (Aisén, Chile). Cazadores-Recolectores del Cono Sur. Revista de Arqueología 3:51–68. Mar del Plata
Méndez C, Stern C, Reyes O, Mena F (2012) Early Holocene long-distance obsidian transport in Central-South Patagonia. Chungara Revista de Antropología Chilena 44(3):363–375
Mengoni G (1999) Cazadores de guanacos de la estepa patagónica. Colección Tesis Doctorales, Sociedad Argentina de Antropología, Buenos Aires

Merino V, De Pol-Holz R, Southon J, Latorre C, Collado S (2019) Marine radiocarbon reservoir age along the chilean continental margin. Radiocarbon 61(1):195–210

Morello F, San Román M, Prieto A (2002) Puntas de proyectil lanceoladas en Patagonia meridional y Tierra del Fuego. Anales del Instituto de la Patagonia, Serie Ciencias Humanas 30:155–166

Muñoz L, Pavez G, Quiñones R, Oliva D, Santos M, Sepúlveda M (2013) Diet plasticity of the South American sea lion in Chile: stable isotope evidence. Revista de Biología Marina y Oceanografía 48(3):613–622

Nelson M (1991) The study of technological organization. In: Schiffer M (ed.), Archaeological method and theory, vol 3, pp 57–100. University of Arizona Press, Tucson

Núñez A, Molina R, Aliste E, Bello A (2016) Silencios geográficos de Patagonia-Aysén: Territorio, nomadismo y perspectivas para re-pensar los márgenes de la nación en el siglo XIX. Magallania (Punta Arenas) 44(2):107–130

Ocampo C, Aspillaga E (1984) Breves notas sobre una prospección arqueológica en los archipiélagos de las Guaitecas y de los Chonos. Revista Chilena de Antropología 4:155

Orquera L, Piana E (1999) La vida material y social de los Yámana. Eudeba-IFIC, Buenos Aires

Orquera L, Sala A, Piana E, Tapia A (1977) Lancha Packewaia. Editorial Huemul, Buenos Aires

Ortner D (2003) Identification of pathological conditions in Human skeletal remains, 2nd ed. Elsevier, USA

Osborne D, Simmons L, Nawrocki S (2004) Reconsidering the auricular surface as an indicator of age at death. J Forensic Sci 49(5)

Osorio M (2011) Manuscrito. Informe Peritaje. Hallazgo osamentas en Isla Victoria, canal Carrera del Chivato. CIEP, Coyhaique

Panarello H, Zangrando F, Tessone A, Kozameh L, Testa N (2006) Análisis comparativo de paleodietas humanas entre la región del canal Beagle y península Mitre: Perspectivas desde los isótopos estables. Magallania 34(2):37–46

Pearce T (2008) When a snail dies in the forest, how long will the shell persist? Effect of dissolution and micro-bioerosion. Am Malacol Bull 26(1–2):111–117

Pelegrin J (1990) Prehistoric lithic technology. Archaeol Rev Cambridge 9(1):116–125

Phenice T (1969) An analysis of the human skeletal material from burial mounds in north central Kansas. In: Standards for Data Collection from Human Skeletal Remains. Proceedings of a Seminar at the Field Museum of Natural History, Fayetteville: Arkansas Archeological Report Research Series

Plafker G, Savage J (1970) Mechanism of the Chilean Earthquakes of May 21 and 22, 1960. Geol Soc Am Bull 81(4):1001–1030

Porter C (1993) GUA-010, un sitio costero erosionado en una zona sísmica activa. En: Actas del XII Congreso Nacional de Arqueología Chilena. Boletín del Museo Regional de la Araucanía 4:81–88

Punke M, Davis L (2006) Problems and prospects in the preservation of late Pleistocene cultural sites in southern Oregon coastal river valleys: implications for evaluating coastal migration routes. Geoarchaeol 21(4):333–350

Rackham O (2012) Island landscapes: some preliminary questions. J Marine Island Cult 1(2):87–90

Reeder-Myers L, Erlandson J, Muhs D, Rick T (2015) Sea level, paleogeography, and archeology on California's Northern channel islands. Quatern Res 83(2):263–272

Reimer PJ, Bard E, Bayliss A, Beck WJ, Blackwell PG, Ramsey CB, Buck CE, Cheng H, Edwards L, Friedrich M, Grootes PM, Guilderson TP, Haflidason H, Hajdas I, Hatté C, Heaton TJ, Hoffmann DL, Hogg AG, Hughen KA, Kaiser FK, Kromer B, Manning SW, Niu M, Reimer RW, Marian DAR, Scott E, Southon JR, Staff RA, Turney CSM, van der Plicht J (2013) IntCal13 and Marine13 Radiocarbon Age Calibration Curves, 0-50,000 years cal BP. Radiocarbon 55:1869–1887

Reyes O, Méndez C, Velásquez H, Trejo V (2006a) Distribuciones espaciales y contextos arqueológicos de cazadores recolectores esteparios en alto río Cisnes: (XI Región de Aisén). Magallania. (Punta Arenas) 34(2):75–90

Reyes O, Méndez C, Trejo V, Velásquez H (2006b) De las estepas a los canales septentrionales: uso humano de distintos ambientes. Bosques del valle del río Cisnes (Holoceno tardío, XI Región de

Aisén). Actas del XVII Congreso Nacional de Arqueología Chilena [2010]. Tomo 2:1487–1495. Valdivia
Reyes O, Méndez C, San Román M, Cárdenas P, Velásquez H, Trejo V, Morello F, Stern C (2007) Seno Gala 1; Nuevos Resultados En La Arqueología De Los Canales Septentrionales (44°S, XI Región de Aisén, Chile). Magallania. 35(2):105–119
Reyes O, Méndez C, Maldonado A, Velásquez H, Trejo V, Cárdenas M, Abarzúa A (2009) Uso del espacio de cazadores recolectores y paleoambiente Holoceno en el valle del río Cisnes, Región de Aisén Chile. Magallania 37(1):91–107
Reyes O, San Román M, Moraga M (2011) Archipiélago de los Chonos: Nuevos registros arqueológicos y bioantropológicos en los canales septentrionales. Isla Traiguén, XI Región de Aisén. Magallania 39(2):295–303
Reyes O, Moraga M, Aspillaga E (2013) El registro bioantropológico y las evidencias de ocupación en el Archipiélago de los Chonos (XI Región de Aisén, Chile). Avances en la arqueología de los canales septentrionales del extremo sur. In: Tendencias Teórico-Metodológicas y Casos de Estudio en la Arqueología de la Patagonia, pp 227–232. Argentina
Reyes O, Moraga M, Mendez C, Cherkinsky A (2015) Maritime Hunter-Gatherers in the Chonos Archipelago (43°50'-46°50' S), Western Patagonian Channels. J Island Coast Archaeol 10(2):207–231
Reyes O, Kuzminsky S, Méndez C, San Román M (2016a) Examining biological variation among the maritime hunter-gatherers of the Chonos Archipelago, Western Patagonian Channels, Chile. Libro Resúmenes 81 The Annual Meeting of the Society for American Archaeology. Orlando
Reyes O, San Román M, Morello F (2016b) Search for maritime hunter-gatherer archaeological record in the shifting shorelines of the South Pacific Coast (Chonos and Guaitecas Archipelago, Chile). In: Marine ventures: archaeological perspectives on human-sea relations, pp 141–155. Equinox Publishing, United Kingdom. https://doi.org/10.1558/equinox.24563
Reyes O, Méndez C, San Román M, Francoise J (2018) Earthquakes and coastal archaeology: assessing shoreline shifts on the southernmost Pacific coast (Chonos Archipelago 43°50′–46°50′ S, Chile, South America). Quatern Int 463:161–175
Reyes O, Tessonne A, San Román M, Méndez C (2019a) Dieta e isótopos estables de cazadores recolectores marinos en los canales occidentales de Patagonia Chile. Latin Am Anti 30(3):550–568
Reynard J (2014) Trampling in coastal sites: an experimental study on the effects of shell on bone in coastal sediment. Quatern Int 330:156–170
Richardson JBIII (1981) Modeling the development of sedentary maritime economies on the coast of Peru: a preliminary statement. Annals Carnegie Museum 50:139–150
Sandweiss D, Maasch K, Chai F, Andrus C, Reitz E (2004) Geoarchaeological evidence for multidecadal natural climatic variability and ancient Peruvian fisheries. Quatern Res 61(3):330–334
San Román M, Morello F (2000) Manuscrito. Catastro Georeferenciado de sitios arqueológicos de Magallanes. CONADI XII Región, Punta Arenas
San Román M, Reyes O, Méndez C, Cárcamo J, Torres J (2016b) Western patagonia subsistence strategies: zooarchaeological studies of marine hunter-fisher-gatherers of the chonos archipelago, chile. Libro Resúmenes 81. The Annual Meeting of the Society for American Archaeology. Orlando
Schaefer M, Black S, Scheuer L (2009) Juvenile Osteology: a laboratory field manual. Academic Press, USA
Scheinsohn V (2001) Odisea del espacio: paisajes y distribuciones artefactuales en arqueología. Relaciones de la Sociedad Argentina de Antropología, 26
Schidlowsky V (2004) Primeros cazadores marítimos y terrestres de patagonia austral y tierra del fuego: tecnología lítica e identidad cultural. Chungara. Revista de Antropología Chilena. Volumen Especial, tomo 1:89–94
Schiffer M (1996) 1987 Formation processes of the archaeological record. University of Utah Press
Schiffer M, Sullivan A, Klinger T (1978) The design of archaeological surveys. World Archaeol 10(1):1–28

Schoeninger M (1995) Stable isotope studies in human evolution. Evol Anthropol 4:83–98
SERNAGEOMIN (2003) Mapa Geológico de Chile: versión digital. Servicio Nacional de Geología y Minería, Publicación Geológica Digital, N°4, Santiago
SERPLAC (2005) Atlas de Aysén. Ediciones LOM Ltda
SHOA (2000) El maremoto del 22 de mayo de 1960 en las costas de chile. Ediciones Servicio Hidrográfico y Oceanográfico de la Armada de Chile, Valparaíso
Steele J (2010) Radiocarbon dates as data: quantitative strategies for estimating colonization front speeds and event densities. J Archaeol Sci 37:2017–2030
Stein J (1986) Coring archaeological sites. Am Antiq 51(3):505–527
Stein J (1992) Sediment analysis of the British Camp shell midden. In: Stein J (ed.), Deciphering a shell midden, pp 135–162. Academic Press, San Diego
Stein J, Deo J, Phillips L (2003) Big sites-short time: accumulation rates in archaeological sites. J Archaeol Sci 30(3):297–316
Stern C (2004) Obsidian in Southern Patagonia: review of the current information. Contra viento y marea. Arqueología de Patagonia, pp 167–176
Stern C, Porter C (1991) Obsidiana en yacimientos arqueológicos de Chiloé y las islas Guaitecas. Anales del Instituto de la Patagonia. Serie Ciencias Humanas 20:205–209
Stuiver M, Reimer P, Reimer R (2013) CALIB 7.0.2 www program and documentation. https://calib.qub.ac.uk/calib/
Tafuri M, Zangrando F, Tessone A, Kochi S, Cecchi J, Di Vincenzo F, Profico A, Manzi G (2017) Dietary resilience among hunter-gatherers of Tierra del Fuego: Isotopic evidence in a diachronic perspective. PLoS ONE 12(4):e0175594
Tessone A, Zangrando F, Valencio S, Panarello H (2003) Isótopos estables del carbono en restos óseos humanos de la región del canal Beagle, Isla Grande de Tierra del Fuego. Revista Argentina de Antropología Biológica. 2:33–43
Tessone A, Zangrando F, Barrientos G, Goñi R, Panarello H, Cagnoni M (2009a) Stable isotope studies in the Salitroso Lake basin (Southern Patagonia, Argentina): assessing diet of Late Holocene hunter-gatherers. Int J Osteoarchaeol 19:297–308
Tessone A, García Guraieb S, Panarello H, Goñi R (2009b) Variaciones en el $\delta^{15}N$ en relación con la edad de muerte de restos óseos humanos de cazadores-recolectores de Patagonia. Actas de las Novenas Jornadas Nacionales de Antropología: 102. Puerto Madryn, Chubut
Torres J (2016) Manuscrito. Informe análisis ictioarqueológico de seis sitios del archipiélago de las Guaitecas y los Chonos, XI Región de Aisén. FONDECYT 1130151. Santiago
Trotter M, Gleser G (1958) A re-evaluation of estimation of stature based on measurements of stature taken during life and of long bones after death. AJPA 16:79–123
Tykot RH, Falabella F, Planella MT, Aspillaga E, Sanhueza L, Becker C (2009) Stable isotopes and archaeology in central Chile: methodological insights and interpretive problems for dietary reconstruction. Int J Osteoarchaeol 19:156–170
Vargas G, Rebolledo S, Sepúlveda S, Lahsen A, Thiele R, Townley B, Padilla C, Rauld R, Herrera MJ, Lara M (2013) Submarine earthquake rupture, active faulting and volcanism along the major Liquiñe-Ofqui Fault Zone and implications for seismic hazard assessment in the Patagonian Andes. Andean Geol 40:141–171
Vidal Gormaz F (1901) Hundimientos y solevantamientos verificados en las costas chilenas. Revista Chilena de Historia Natural 10:213–224
Viviani C (2004) Manuscrito. *Caracterización del litoral de la Región de Aisén*. Universidad de Magallanes, Punta Arenas
Waldron T (2009) Palaeopathology. Cambridge University Press, Cambridge, UK
Walker M, Berkelhammer M, Björck S, Cwynar L, Fisher D, Long A, Lowe J, Newnham R, Rasmussen O, Weiss H (2012) Formal subdivision of the Holocene Series/Epoch: a discussion paper by a working group of INTIMATE (Integration of ice-core, marine and terrestrial records) and the Subcommission on Quaternary Stratigraphy (International Commission on Stratigraphy). J Quat Sci 27(7):649–659

Walrath D, Turner P, Bruzek J (2004) Reliability test of the visual assessment of cranial traits for sex determination. AJPA 125:132–137

Wandsnider L, Camilli E (1992) The character of surface archaeological deposits and its influence on survey accuracy. J Field Archaeol 19(2):169–188

Waselkov G (1987) Shellfish gathering and shell midden archaeology. Adv Archaeol Method Theory 10:93–210

Waters M, Kuehn D (1996) The geoarchaeology of place: the effect of geological processes on the preservation and interpretation of the archaeological record. American Antiquity, pp. 483–497

Williams A (2011) The use of summed radiocarbon probability distributions in archaeology: a review of methods. J Archaeol Sci

Yesner D, Torres M, Guichon R, Borrero L (2003) Stable isotope analysis of human bone and ethnohistoric subsistence patterns in Tierra del Fuego. J Anthropol Archaeol 22(3):279–291

Chapter 4
The Archaeological Record in the Chonos Archipelago (43°30′–47° SOUTH)—Western Patagonia

Abstract This chapter describes all the archaeological sites discovered and registered in this study in the different areas of coastal prospecting representative of the Chonos Archipelago: Northern area, Mainland coast, Central, South-central, and Offshore islands and Pacific Border. We describe the main characteristics of each site, position, geographic location, work carried out, chronology and material, ecofactual and bio-anthropological characterisation, in addition to briefly inferring its role based on the information collected both in the field and in laboratory analyses.

Keywords Western patagonia · Maritime hunter-gatherers · Coastal survey · Holocene chronology · Archaeological sites · Archipelagic settlements · Human remains · Ossuaries · Shell-middens · Historical sites

4.1 Northern Area—Guaitecas Archipelago at the Northern End of the Chonos Archipelago (43°46′–44°S)

Below, we describe the archaeological sites recorded and studied in the Guaitecas Archipelago, located at the northern end of the Chonos Archipelago (Fig. 4.1).

4.1.1 GUA-010 Terraza

The GUA-010 Terraza (terrace) site is located on the south coast of the central isthmus of Gran Guaiteca Island, in a small, south-facing bay. The Cucuquén Stream, flowing generally from north to south and with water all year round, runs out at the northernmost point of the bay. The archaeological occupation identified lies on a marine terrace on the eastern bank of the stream. The terrace has emerged to a height of 5.5 m above the present sea level and is 50 m from the current shoreline, up a steep slope of 70°. Its archaeological visibility is nil, and it was only found—and subsequently rediscovered and recorded (Porter 1993)—by the use of test pits and

O. Reyes, *The Settlement of the Chonos Archipelago, Western Patagonia, Chile*, The Latin American Studies Book Series,
https://doi.org/10.1007/978-3-030-54326-6_4

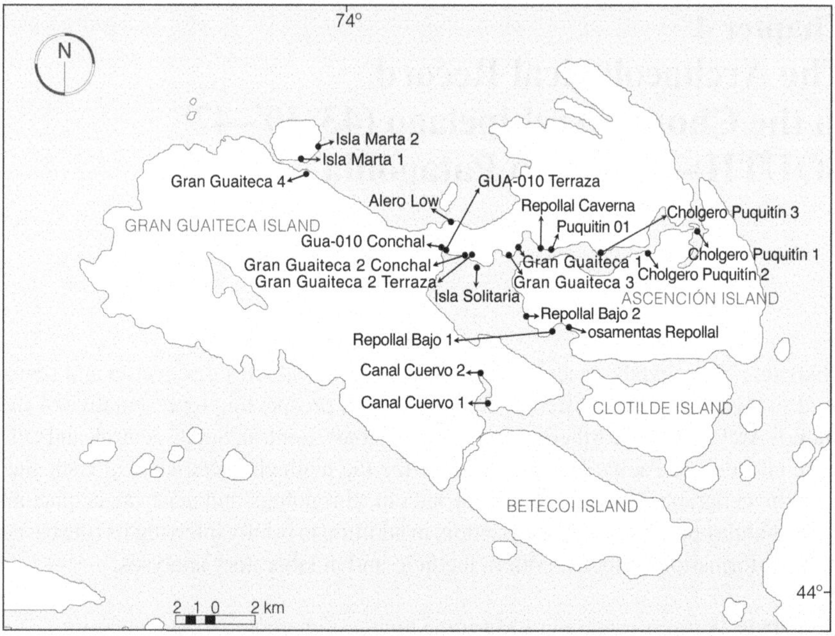

Fig. 4.1 Distribution of sites recorded and studied in the northern area, the Guaitecas Archipelago, which is the northern end of the Chonos Archipelago (43°S)

sediment cores. It is covered by thick forest, shrubby vegetation and an abundant layer of plant material formed by centuries of forest formation. The vegetation in the bay where it is sited is evergreen forest with peat bogs, coming right down to sea level.

The total area of the archaeological site has not been estimated accurately. By taking test cores, we determined the presence of a dark organic sediment containing cultural materials (recorded previously in a test pit, Porter 1993); however, as no lithic material appeared in the cores, the area was estimated following the presence of the cultural layer with sediments of silt and dark sand. The site is at least 6 m from north to south and 8 m from east to west, totalling at least 48 m^2 with 20 cm depth of cultural deposit recorded. In general terms, the site is an archaeological occupation with unclear presence of lithic material and dating based on a speck of carbon from 5020 ± 90 BP (Table 4.1). This makes it the oldest archaeological occupation of the Chonos Archipelago (Reyes et al. 2016b).

The first characterisation of sediments was carried out in an earlier field trip to identify and evaluate the site. The old 1 × 1 m test pit dug by Porter (1993) was cleared and a small column 20 × 20 cm was excavated in the NW corner of the pit to define the stratigraphy of the site and take samples (^{14}C, sediments).

Our stratigraphic record established a thick layer of organic humus (decomposing leaves, roots and twigs) between 0 and 20 cm. Between 20 and 30 cm (28 cm), we recovered an anvil made of a beach pebble in a matrix composed of silt and sand, grey in colour and with small clasts. Carbon specks were recovered between 40 and 50 cm, and also between 50 and 60 cm. The layer of sandy silt 30 cm thick is the archaeological occupation defined by Porter (1993) for the fifth millennium BP. At 65 cm, a yellow sand matrix with gravel appeared, the sterile palaeo-beach on which the first occupants settled. This matrix continues down to 80 cm. Below it is a layer of finer, gravel-free sand down to one metre, where the excavation was closed due to phreatic flooding.

After identifying the site and carrying out stratigraphic characterisation by finding Porter's old pit and clearing its profiles, we planned an excavation of 12 m^2, made up of 1 × 1 m units in a grid 3 m north–south and 4 m east–west, in what was taken to be the centre of the site defined by the cores. However, this grid was reduced to 6 m^2 after we had excavated the first organic layer and exposed the dark cultural sediment defined previously in the stratigraphic analysis. We excavated in artificial 5 cm levels. The template was set adjacent to and south of Porter's 1 × 1 m unit (Fig. 4.2).

Our excavation allowed us to define different stratigraphic units consistent with our observations when clearing the old test pit. The archaeological site is buried

Fig. 4.2 Archaeological excavation on the terrace 5 masl in the evergreen forest at the GUA-010 Terraza site

under a layer of organic plant material composed of pale brown silt–clay sediments between 20 and 40 cm thick. Below this thick layer of plant material there is a dark grey sandy silt matrix, 3 to 10 cm thick, at the base of which the cultural deposit starts, no more than 20 cm thick. The dark grey layer then turns into compact, black, organic clay silt, approximately 10 to 15 cm thick, from which most of the lithic waste was recovered. In this layer, we recorded lenses of burnt material and ash, carbon specks, knapping waste and bifacial instruments. This matrix gives place to a 5 cm layer of clay silt, also black but with inclusions of gravel. The amount of cultural material diminishes greatly in this level. Finally, the sterile component on which the occupation rests is a transition layer of dark, sandy clay, average thickness 5 cm, and below it a layer of dark brown sand more than 30 cm thick (measured by cores) which formed the palaeo-beach on this marine terrace.

Note that no organic materials (e.g. bones and shells) were found in the cultural deposit (interface of dark grey layer and black layer). This may be explained by the acidity of the soil (pH between 2.9 and 4.5) which may have degraded the organic materials and also some lithic materials made of local raw material (degraded whitish kaolinite). We believe that as a result many pieces of lithic waste disintegrated during excavation when touched with the brush. This acidity is stimulated by the large component of plant matter in degradation which forms part of the matrix and results from the abundant layer of forest material throughout the area. This makes it a *podzol* soil.

Interpretation of the sediments suggests that the area was located close to sea level, to judge by macroscopic comparisons of the types and variations of the sediments excavated (e.g. texture, size and shape of clasts, proportion and roundness, plant cover, presence of natural molluscs) with those currently observed and recorded along a 200 m transect in the adjacent intertidal which presented ten sets of distinctive sediments. The proportion and types of clasts located in the dark cultural matrix indicate that it is analogous to the dark organic sediments now recorded at the mouth of the Cucuquén Stream, projecting the original position of the site during the formation of the terrace before it was raised to 5.5 m above current sea level. The site was surveyed and a plan was made of its position and relation to existing sea level.

The assemblage excavated and recovered consists mainly of degraded lithic material and carbon speck samples. We think that the high acidity of the soil (podzol), which explains the degradation of the lithic material recovered, would also explain in part the absence of osseous material. Very little cultural material was recovered. What there was totalled 55 pieces of lithic material including flakes, preforms of stone projectile heads and thick-section lanceolate stone projectile heads, in 20 cm of cultural deposit over 6 m^2 of excavation. We also obtained specks of carbon for radiocarbon dating.

The excavation in the GUA-010 Terraza site exposed an occupation surface with infrequent deposits of knapping waste, cores and whole or fragmented stone projectile heads and preforms in early stages of manufacture ($n = 55$ in 6 m^2). No other type of stone instrument was detected. The knapping waste presented no cortex and accounted for almost half of the sample ($n = 42$). The assemblage presented a high level of fragmentation, with only three complete pieces. The whole and fragmented projectile heads (n = 5) and the preforms (n = 7) account for 21.8% of the total assemblage recovered. Microsplintering due to use was found in most pieces, however, the traces were less clear in the case of softer stone.

The raw material rock was of very good quality in four cases (black rhyolithic obsidian and siliceous tuff) and middling quality in six (soft sedimentary rock containing talc). The rhyolithic obsidian was obtained from Chaitén Volcano on the mainland coast of the archipelago, 150 km north-east of the site. The geochemical analyses carried out in the Laboratory for Environmental Geochemistry at the University of Colorado (Boulder) by C. Stern corroborated our macroscopic observation. Flakes obtained in earlier excavation of the area of GUA-010 without detailed references (shell-midden or terrace) also indicated that the rhyolithic obsidian present came from the Chaitén Volcano source (Stern and Porter 1991).

The carbon sample obtained between 50 and 60 cm, at the base of the cultural deposit, gave a date of 5370 ± 30 years BP (6210–5994, 2-sigma years cal. BP) (Reyes et al. 2016b). This date differs statistically from the calibration applied to the dating obtained by Porter of 5020 ± 90 BP (5918–5583, 2-sigma years cal. BP), establishing different chronological events at the start of the occupations (Table 4.1). More datings will allow us to specify whether the chronology of the area of this occupation can be extended for a few more centuries, or whether, as appears so far, the occupation events were discrete and chronologically limited.

A varied assemblage of lithic material was recovered from the excavated area of the site ($6m^2$) (knapping waste, thick-section lanceolate stone projectile heads and preforms, Fig. 4.3), reflecting specific activities related to the use and/or discarding of stone projectile heads around ~6000 years cal. BP. This was a knapping area where the raw materials were limited to a talc-type rock, degraded by the high soil acidity, and two other very good quality types, namely the tuff and the obsidian from the Chaitén Volcano source. It may well be that we excavated one specific area of the site, and not others where other activities may have been carried out, or that preservation in such an acid podzol soil prevented the persistence of materials indicating other activities (e.g. no archaeofaunal record). With the evidence obtained to date, we can propose that this was a site where specific activities were carried out, and where the operating chain was linked with the manufacture and maintenance of medium-sized stone projectile heads, lanceolate in shape with a thick cross-section (almost 1 cm on average), for hunting. The presence of obsidian from the Chaitén Volcano source also associates the groups who settled in the area with a wide network of circulation and/or exchange of raw materials.

4.1.2 GUA-010 Conchal and Intertidal Zone

The GUA-010 Conchal (shell-midden) site, with the adjacent intertidal zone, lies on the south-facing shoreline of Gran Guaiteca Island, in the same bay formed by the mouth of the Cucuquén Stream as the GUA-010 Terraza site. It lies below the marine terrace of GUA-010 Terraza and adjacent to the beach and intertidal zone. The base

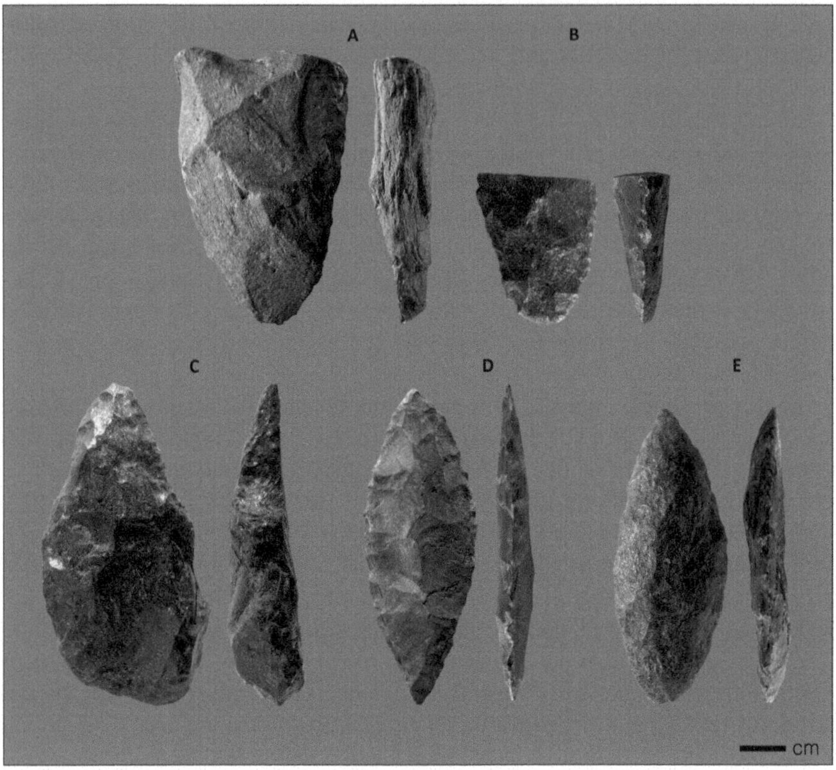

Fig. 4.3 Lithic material obtained from the excavation of the GUA-010 Terraza site and collection in the intertidal. GUA-010 Terraza site: **a** Fragment of bifacial piece; **b** Base fragment of lanceolate stone projectile head; **c** Fragment of bifacial piece and **d** Lanceolate stone projectile head. Lithic material recovered from the intertidal: **e** Lanceolate stone projectile head

of the deposit is 1 m above current sea level, and its southern edge is 10 m from the present shoreline. The site is a shell-midden slightly eroded on the south-west side due to subsidence caused by the May 1960 earthquake. Roots of the evergreen forest with peat bogs which grew there and were flooded at that time can still be seen. The area of the Guaitecas Archipelago subsided by almost 2 m from its previous level in the earthquake (Plafker and Savage 1970; Reyes et al. 2016b).

The shell-midden is located on the east bank of the Cucuquén Stream, dominating the centre of the bay. The slope at the site is 10°, falling southwards towards the shoreline. The archaeological visibility is medium, since part of the surface has suffered anthropic intervention and grass has grown due to the activity of fishermen, exposing the shell-midden in some areas. The shell-midden can also be observed in the profiles exposed by the stream and high tides; shells, bones and carbon specks form part of the sediments that characterise the archaeological shell-midden.

The site, measured by drilling cores, measures 20 m from north to south and 30 m from east to west, totalling at least 600 m^2 with a maximum depth of 1 m of cultural deposit. There is a filled-in test pit of 1 × 1 m (Porter 1993) in the highest part of the shell-midden.

The intertidal zone below the shell-midden measures 160 m from east to west and 100 m from north to south, covering approximately 16,000 m^2. These dimensions vary according to the high and low tidelines. The intertidal zone in this bay connects the GUA-010 Conchal and Gran Guaiteca 2 Conchal sites along the adjacent coastline, as well as GUA-010 Terraza and Gran Guaiteca 2 Terraza, in similar locations but between 4 and 6 m above the actual high tideline.

The intertidal zone slopes down gently at 10° (so it is easily flooded by the rise and fall of the tide and by sudden changes in sea levels, e.g. in earthquakes or marine transgressions). It consists of sediments—sand, silt, mud and swamp—with the inclusion of pebbles interspersed with the rocky outcrops that border the intertidal in places and surround the whole bay of the Cucuquén Stream. This habitat supports a variety of malacological fauna on both the sandy and the rocky substrates. In this extensive area, which has sunk below sea level and is subject to the tides, Porter (1993) recovered a large quantity of lithic material (n = 421), with no more detailed records than those given in his publication. The lithic material recovered consists mainly of medium or large, thick cross-section, lanceolate stone projectile heads, either whole or fragmented (n = 12), net weights (n = 108), flakes with and without cortex (n = 112), hammerstones (n = 24), cores (n = 26) and polished axes (n = 4).

The test pit dug by Porter (1993) was reopened to characterise the profiles of the shell-midden. Then a pit 1 × 1 m was dug close to the western edge of the shell-midden to extract a column of malacological fauna and obtain samples for radiocarbon dating. A compact matrix of shells was recorded with no significant hiati (e.g. lenses of sand) down to the base of the shell-midden which rests on the palaeo-beach.

The matrix of this shell-midden consists of three general stratigraphic units. The first is between 0 and 4 cm deep and consists of a layer of humus and organic earth covering the archaeological site. It contains no archaeological evidence. Next is the shell-midden matrix, between 4 and 60 cm. This matrix consists of the remains of malacological fauna, either whole or fragmented, in different levels and consisting mainly of picorocos (*Austromegabalanus psittacus*), cholgas (*Aulacomya ater*), choro mussels (*Choromytilus chorus*), Chilean mussels (*Mytilus edulis chilensis*), oysters (*Ostrea chilensis*), clams (*Protothaca thaca, Venus antiqua*), locos (*Concholepas concholepas*), keyhole limpets (*Fissurella nigra*), snails (*Tegula atra*), chitons (*Chiton sp.*) and sea urchins (*Loxechinus albus*). A dark, sandy silt sediment is interspersed with the remains of malacological fauna. Very fragmented remains of human bones (teeth, bone fragments) were recorded between 0 and 10 cm. We believe that this represents an ancient anthropic disturbance of a subsuperficial burial in the shell-midden.

This shell matrix contains different degrees of compaction and the presence of lenses of ash, hearths and carbon specks. The interfaces between strata, with the exception of the hearths, are diffuse because the malacological fauna presents different intensities of fragmentation (e.g. comparing mussel valves with loco or picoroco shells). Stratigraphic resolution will certainly require more extensive excavations and exposure of a larger area of the exposed profiles. Finally, a third stratigraphic unit was recognised below 60 cm, namely a compact, inorganic sand and gravel matrix including small pebbles which made up the palaeo-beach where the first occupants of the site settled.

A new superficial collection of materials was carried out on the extensive area of the intertidal zone adjacent to GUA-010 Conchal. A total of 13 pieces of lithic material were recovered in the sectors exposed only at low tide (Fig. 4.3). The small quantity of lithic material collected is due to the collection by Porter (1993), who practically 'emptied' the intertidal. He recovered 421 lithic pieces, the present whereabouts of which are unknown.

The superficial collection from the intertidal zone produced 4 choppers, 2 stone projectile heads (one complete and one fragmented) and 8 flakes. A hammerstone was recovered from the superficial area of the GUA-010 Conchal site. During clearing of the test pit dug by Porter (1993) and the exposed profiles, samples of carbon and shells were recovered for dating; no lithic material was obtained. No cultural material was recovered from the test pit dug by us; however, fragmented, incomplete remains of human bones were found.

Macroscopic examination of the lithic waste showed that the flakes, the hammerstone and the choppers were all made of local raw material obtained from the abundant pebbles on the beach. The lithic projectile heads were made of a better raw material, veined grey tuff. They present a thick cross-section, almost 10 mm.

Of the large quantity of lithic material in the intertidal, Porter says: *"A superficial search recorded 421 lithic artifacts from the intertidal zone, 4 from the shell-midden* (GUA-010 Conchal) *and none from the terrace* (GUA-010 Terraza). *The superficial artifacts include pebbles with chipped out grooves classified as net weights (n = 108) 25%, secondary flakes (n = 82) 19%, flakes from initial cortex removal (n = 30) 7%, side-scrapers (n = 31) 7%, cores (n = 26) 6%, choppers (n = 24) 6%, hammerstones (n = 24) 6%, stone points (n = 12) 3%, notches (n = 11) 3%, perforators (n = 6) 1%, and polished axes (n = 4) 1%."* (Porter 1993:84). Samples of lithic pieces obtained in this collection showed early on that rhyolithic obsidian from Chaitén Volcano was the source of some of the raw materials detected (Stern and Porter 1991).

The 778 remains of malacological fauna recovered from the shell-midden column represented a total of 84 individuals of 18 species of molluscs and crustaceans (Cárcamo 2015). The species with the highest MNI were *Venus antiqua,* with 17.86% (n = 15) and *Nacella magellanica chiloensis,* with 13.10% (n = 11). The lowest frequencies were found for *Choromytilus chorus,* 1.19% (n = 1)*; Concholepas concholepas,* 1.19% (n = 1)*; Crepidula philippiana,* 1.19% (n = 1)*; Fissurella crassa,* 1.19% (n = 1)*;* and *Trophon sp.*, 1.19% (n = 1). The habitats of the species collected vary widely: rocky intertidal, rocky subtidal, parasitic and mixed intertidal (sandy and/or rocky), reflecting provisioning strategies which include various molluscs. There were also 26 remains of taxonomically identifiable fish; the majority (n = 25) were snoek (*Thyrsites atun*) and only one specimen of blenny (*Eleginops maclovinus*) (Torres 2016). Remains of a canid were also found (San Román et al. 2016b), probably transported to the site.

Of the human bone remains recovered, we recorded fragments of a thoracic vertebra, 4 proximal phalanges of the hand, 2 medial phalanges of the hand, one scaphoid, one fourth metacarpal, fragments of clavicle and mandibular condyle, and diaphyses of two long bones, as well as two molars. These portions of bone can be assigned to a single adult individual, of indeterminate age and sex. Given the depth at which the remains were recovered (between 0 and 10 cm only), it can be said that this was a subsuperficial burial disturbed by sub-present anthropic activity on the surface of the site (tree-felling and seasonal camping by fishermen).

The carbon sample obtained at the base of the deposit at a depth of 60 cm gave a date of 2170 ± 30 years BP (1840–1554, 2-sigma cal. BP) for the start of the formation of the shell-midden (Table 4.1) (Reyes et al. 2016b).

The GUA-010 Conchal site represents a camp where many activities were carried out and there were successive reoccupations from almost 1800 years cal. BP. The discarding of waste from mollusc gathering is the most evident activity, and reflects locally obtained resources in which the different types of coastline and sea-floor are represented in a generalised gathering strategy; however, the presence of hearths and the lithic material from the intertidal zone (e.g. net weights, stone projectile heads, stone axes and choppers) also indicates other camping and food-collecting activities (gathering, hunting and fishing). On the other hand, the human remains found, while scarce, reveal that the site was also used to bury individuals at the camp periphery.

It must be added that the area presents occupations three millennia older than the first occupation from the base of the shell-midden, supported by the date of the GUA-010 Terraza site; the lithic material present in the intertidal may, therefore, represent different situations in time which are indistinguishable to us. It is reasonable to suppose that the material recovered in the intertidal zone bears a relation with what was happening on the adjacent coast in terms of typology, artefacts and similarity of the raw materials. Porter's description (1993) of the collection of the pieces, their number and nature is vitally important, since it enables us to deduce various activities such as maintenance of lanceolate stone projectile heads for hunting (similar types of thick-section lanceolate projectile heads, from the same raw material, were recovered in the excavation of the terrace); the stone net weights are directly related to fishing with nets and thus the presence of snoek and blenny among the fauna in the shell-midden, while the stone axes are related to exploiting and working wood, perhaps as wedges for stripping the bark, or in the building/maintenance of canoes. On the other hand, flakes and choppers are involved in immediate, efficient use of stone objects, taking advantage of the abundant pebbles on the beach. Canoe transport suggests the possibility of replacing, discarding and/or losing stone tools when embarking or disembarking (whether at the time of GUA-010 Terraza or later from the time of GUA-010 Conchal). Finally, the presence of a canid and of an obsidian from the Chaitén Volcano source also associates the groups who settled in the area with a wide network of circulation and/or exchange associated with obtaining raw materials.

4.1.3 Gran Guaiteca 2 Terraza

The Gran Guaiteca 2 Terraza site lies on the south-facing shoreline of Gran Guaiteca Island, in the same bay formed by the mouth of the Cucuquén Stream, and 250 m east of the GUA-010 Terraza and GUA-010 Conchal sites and associated intertidal area. It is on the same marine terrace as GUA-010 Terraza, 5–6 m above current sea level and 30 m from the shoreline. Its archaeological visibility is nil, and it was only found by drilling cores and archaeological test pits (Fig. 4.4). It is covered by thick forest, shrubby vegetation and an abundant layer of plant material. The site lies on the east side of the bay, facing west, with a slope of no more than 10°.

The Gran Guaiteca 2 Conchal (shell-midden) site lies on the same transect, but 2 m above current sea level and 10 m from the shoreline. As we will see, its dates are later, but the test pits dug in the forest on the terrace above were suggested by its location.

The total area of the archaeological site has not been estimated accurately. By taking test cores, we determined the presence of a dark organic soil containing scarce cultural materials; however, as no lithic material appeared in the test pits, the area was estimated following the presence of the cultural layer with sediments of silt and dark sand. The site is at least 6 m from north to south and 3 m from east to west, totalling at least 18 m^2 with 15 cm depth of cultural deposit recorded.

Fig. 4.4 View of the Gran Guaiteca 2 Conchal site (adjacent to the beach) and Gran Guaiteca 2 Terraza site in the evergreen forest 4 m above sea level. Topographical survey to measure the height of the sites above current sea level

Three test pits of 50 × 50 cm were dug. We found knapping waste, carbon specks and a bifacial fragment in a matrix of very dark organic soil, clearly defined between the upper layer of organic humus (podzol) and the sand layer (palaeo-beach) on which the occupants of the site settled. The site was also surveyed and a plan was made of its position and relation to existing sea level.

The composition of the sediments and the stratigraphy were similar in all the test pits. First, we found 20 cm of a reddish-brown organic matrix, with fine sand, humus, rootlets and large roots from the forest trees on the site, with a high proportion of composted plant matter and an acid pH (podzol with pH 4.5). Next, between 20 and 25 cm appeared the transition to a grey sand-silt layer containing gravel. At 25 cm, the grey sandy silt layer with gravel and a scattering of small pebbles, schists and fragments of quartz was clearly defined. This layer became darker, more compact and silty between 40 and 50 cm. It is in this layer that carbon specks are recorded in all the pits; a flake was recovered from pit 2 and a bifacial fragment from pit 3. Below 60 cm, the matrix changed to light brown sandy sediments which form part of the palaeo-beach.

The scarce assemblage recovered consists mainly of degraded lithic material and carbon specks. We believe that the high soil acidity (pH around 4.5) explains the degradation of the lithic material recovered from the levels excavated, just as we observed in the GUA-010 Terraza site. A total of three pieces of knapping waste were recovered and a fragment of a medium-sized, thick-section lanceolate stone projectile head.

Macroscopic examination of the lithic waste and the bifacial fragment showed that the raw material was grey obsidian from Chaitén Volcano on the mainland coast of the archipelago, 150 km north-east of the site. This observation was subsequently confirmed by geochemical analyses carried out in the Laboratory for Environmental Geochemistry at the University of Colorado (Boulder) by C. Stern. The mid portion of the stone projectile head is 4 cm wide, while the cross-section is 7 mm thick. The projected shape is lanceolate.

The carbon sample obtained between 50 and 60 cm, at the base of the occupation floor, gave a date of 3958 ± 24 years BP (4428–4243, 2-sigma cal. BP, Table 4.1).

With the evidence obtained to date (area excavated less than 1 m^2), we can comment that this was a site where specific activities were carried out more than ~4200 years cal. BP, and where the operating chain was related to the manufacture and maintenance of medium-sized stone projectile heads, lanceolate in shape with a thick section (almost 1 cm on average), related to hunting. The presence of obsidian from the Chaitén Volcano source also associates the groups who settled in the area with a wide network of circulation and/or exchange of raw materials.

4.1.4 Gran Guaiteca 2 Conchal

The Gran Guaiteca 2 Conchal (shell-midden) site lies on the south-facing shoreline of Gran Guaiteca Island, in the same bay formed by the mouth of the Cucuquén Stream

Fig. 4.5 Interior view of the Alero Low site (7 masl). Disturbed human remains can be observed on the surface

and 250 m east of the GUA-010 Terraza and GUA-010 Conchal sites and associated intertidal area. This archaeological shell-midden is located on a small sandy beach surrounded by rocky coast and backed by the marine terrace which has emerged to a height of 5 m. The base of the deposit is 2 m above the present sea level and the site is 10 m from the shoreline (Fig. 4.5). It presents low archaeological visibility since the surface is covered by forest; however, being adjacent to the beach, it is at the mercy of the high tides and storms which have exposed its western profile. The exposed shell matrix, 1.7 m high, was observed during prospection of the shoreline. It is covered by an abundant layer of plant material formed by the evergreen forest and peat bogs. The site lies on the east side of the bay, facing west, with a pronounced slope of no more than 30°. The Gran Guaiteca 2 Terraza site lies on the same transect towards the forest, on the forested marine terrace 5–6 m above current sea level and 25 m from the shoreline.

The site is a shell-midden of anthropic origin, 25 m from east to west and 8 m from north to south. The current area of this site, which we rediscovered, is 200 m^2 with a cultural deposit 2 m deep. We found that a test pit 1 m × 1 m had been made in the edge of the exposed profile and a small excavation of 1 m × 50 cm also in the exposed profile of the shell-midden. On investigating, we discovered that three skeletons had been excavated from the site profile and that the site was originally called REP-02 (Repollal 02, Aspillaga personal communication 2014). Individual

N°1 was excavated in 1985 and consisted of the incomplete cranium of an adult male and remains of the upper limbs (due to collapse of the profile); individual N°2 was a complete infant, possibly with congenital syphilis or a similar condition. Only scarce remains existed of individual N°3 (excavated in 1987, like individual N°2, presumably in the edge of the exposed profile).

The burials were found at depths of between 5 and 20 cm, similar to the remains recovered at the GUA-010 Conchal site, 200 m away. It is also known that part of the skeleton was found falling from the exposed profile due to erosion. There is no further information on the characteristics of the burials excavated or the tests done, except that necklace beads were found. We dated the rib of individual N°1 (excavated and removed from the profile) when we started the characterisation and archaeological contextualisation of the Chonos Osteological Collection held in the University of Chile (Reyes et al. 2013).

We reopened the test pit discovered to characterise the profiles of the shell-midden and obtain samples for radiocarbon dating. We also obtained a column of fauna extracted from the north profile (10 × 10 cm), obtaining the samples in artificial levels of 10 cm. We obtained 6 levels for the whole depth of the shell-midden. The site was surveyed and a plan was made of its position and relation to existing sea level.

The exposed profile presents three stratigraphic units. The first is a layer of organic earth with light brown silt–clay, 10 cm thick, which covers the shell-midden. Next is the shell-midden matrix, between 10 and 100 cm. This matrix consists of the remains of malacological fauna, either whole or fragmented, in different levels and consisting mainly of picorocos (*Austromegabalanus psittacus*), cholgas (*Aulacomya ater*), choro mussels (*Choromytilus chorus*), Chilean mussels (*Mytilus edulis chilensis*), oysters (*Ostrea chilensis*), clams (*Protothaca thaca, Venus antiqua*), locos (*Concholepas concholepas*), keyhole limpets (*Fissurella nigra*), snails (*Tegula atra*) and chitons (*Chiton sp.*). This matrix includes dark sandy silt sediments interspersed with the remains of malacological fauna. Different degrees of compaction are observed and the presence of a hearth lens between 43 and 50 cm. Finally, a third stratigraphic unit was recognised below the depth of 100 cm, namely a compact, inorganic sand and gravel matrix including small pebbles which made up the palaeo-beach where the first occupants of the site settled. No archaeological material was recovered either on the surface or in the adjacent intertidal. We obtained shell samples from the base of the deposit for radiocarbon dating. No lithic material was recovered from the clearing of the pit.

The 143 remains of malacological fauna recovered from the shell-midden column represented a total of 60 individuals of 12 species of molluscs and crustaceans (Cárcamo 2015). The species which presented the highest MNI was *Venus antiqua,* with 31.37% (MNI = 19). The species with the lowest frequency were *Acanthina sp., Austromegabalanus psittacus, Concholepas concholepas, Crepidula unguiformis, Fissurella nigra, Fissurella sp.* and *Protothaca thaca,* all with the same frequency of 1.67% (MNI = 1). No fragments of sea urchins (*Loxechinus sp.*) were found. The habitats of the species collected vary widely: rocky intertidal, rocky subtidal, parasitic and mixed intertidal (sandy and/or rocky), reflecting provisioning strategies which include various malacological fauna. We identified the presence of three species of fish: tollo (*Mustelus mento*), jurel (*Thrachurus murphyi*) and snoek (*Thyrsites atun*) (Torres 2016).

The shell sample obtained at the base of the deposit at a depth of 100 cm gave a date of 1963 ± 25 years BP (1585–1329, 2-sigma cal. BP). The sample of rib from buried individual N°1 (approx. 0–30 cm) gave a dating of 830 ± 30 BP (624–491, 2-sigma cal. BP, Reyes et al. 2013, 2016, Table 4.1).

The Gran Guaiteca 2 Conchal site was a camp where various activities are recorded from around ~1400 years cal. BP. The most evident is the discarding of waste from the gathering and consumption of molluscs, reflecting local malacological fauna obtained from the different types of coast and sea-floor. At least three burials took place on the edge of the shell-midden in the highest levels of occupation, around ~600 years cal. BP. The site was reused over at least 800 years, unlike the area of the bay in general which presents a much longer period of occupations. Although no lithic material was recovered, this site is adjacent to the same intertidal environment which connects it with the GUA-010 Conchal site, where the lithic material recovered allowed us to distinguish various subsistence activities related to hunting, fishing and woodworking at least 200 years earlier, according to the basic data of the site. The isotope values obtained in the dated individual allowed us to establish the consumption of an exclusively marine diet (see Chapter 6, Evaluation and Discussion of the Evidence).

4.1.5 Gran Guaiteca 1

The site is located on the south coast of the central isthmus of Gran Guaiteca Island, opposite Ascensión Island across the Puquitín Channel. It lies at the end of a small peninsula between two bays. The two bays form the mouths of small watercourses that flow all year round. The area is heavily forested with evergreen forest and peat bogs, and the shoreline is formed by a wide beach with small pebbles, ground shells and fine sand.

The site is 1 m above the current sea level and 3 m from the shoreline, facing south. It has medium archaeological visibility, thanks to anthropic deforestation which has thinned the forest that covered part of the site. The profiles along the southern edge have been exposed by tidal action. On the upper terrace, however, where there is no shell-midden matrix, the cultural deposit could only be recorded by drilling cores and digging test pits. There are no materials on the surface. There is a narrow open area which rises gently but continuously at 20° from the shoreline to the start of the trees and peat bogs, 3 m above the current sea level.

The site is an archaeological shell-midden covered by a layer of plant matter 10 cm thick on average, and an adjacent area where cultural material was recovered on the marine terrace 3 m high. According to estimates based on drilling cores and digging test pits, the area is 50 m from east to west and 80 m from north to south, covering approximately 4,200 m^2. There is an area of shell-midden adjacent to the coast 1 m above the current high tideline, measuring 10 m from north to south and 20 m from east to west; it covers 200 m^2 with a shell deposit 110 cm deep. Carbon and cultural materials are also recorded in the sandy silt sediment, with no shell-midden matrix, 20 m further inland and 3 m above the present shoreline.

The exposed profiles were searched for cultural materials and four test pits were dug, two 30 × 30 cm and two 1 × 1 m; they yielded lithic material and carbon specks, and enabled us to define the stratigraphic units which make up this cultural deposit, as well as its dimensions. The cores showed a deposit of shell-midden matrix with maximum thickness 110 cm underneath the intertidal zone, proving that the shell-midden has been flooded due to changes in sea level. The contact of the base of the shell-midden with the palaeo-beach is 30 cm below current sea level. No soil or hiati were found in the formation of the deposit. Clearing of the exposed southern profile of the shell-midden shows the predominant presence of fragments of malacological fauna like cholgas (*Aulacomya ater*), choro mussels (*Choromytilus chorus*), Chilean mussels (*Mytilus edulis chilensis*), oysters (*Ostrea chilensis*), clams (*Protothaca thaca, Venus antiqua*), locos (*Concholepas concholepas*), keyhole limpets (*Fissurella nigra*), snails (*Tegula atra*), chitons (*Chiton sp.*) and sea urchins (*Loxechinus albus*). We also found carbon specks and knapping waste (flakes and microflakes).

The stratigraphy outside the area of the shell-midden matrix is similar in all the test pits dug. Between the surface and a depth of 10 cm is dark brown organic sediment, mainly humus. Below this, between 10 and 20 cm, the humus becomes more compact and includes clay-silt sediments. A preform of a projectile head and a flake were recovered from this sediment in test pit 1, at a depth of 20 cm. Between 20 and 30 cm, we observed the transition to the beach, with pale brown gravel.

The other test pits produced similar stratigraphies. In pit 2, however, 40 m from the shoreline and at a depth of 30 cm, there was a small lens of ground clam shells. A sample was taken for radiocarbon dating. It was decided to date material in this area with samples obtained from the pits on the upper terrace above the base of the shell-midden adjacent to the coast, in order to determine the possibility of an older occupation, following the criteria which enabled us to locate and record the Gran Guaiteca 2 Terraza and Isla Solitaria sites, 4 and 3 km to the west, respectively.

In pit 1, we recovered a fragment of a preform of a stone projectile head and a flake in the layer between 10 and 20 cm. In pit 2, we took a sample of shells at a depth of 30 cm for radiocarbon dating, and found two flakes between 10 and 20 cm. Macroscopic examination of the lithic material showed that the stone projectile head preform was made of local raw material obtained from the abundant pebbles on the beach. The fragment of preform was 5 cm long, with a section almost 1 cm thick.

The geochemical analyses of the two flakes obtained from pit 2, carried out in the Laboratory for Environmental Geochemistry at the University of Colorado (Boulder) by C. Stern, showed that the raw material used was grey obsidian from Chaitén Volcano on the mainland coast of the archipelago, 150 km north-east of the site. The shell sample obtained at the base of the deposit at a depth of 35 cm gave a date of 1120 ± 30 years BP (723–531, 2-sigma cal. BP) (Reyes et al. 2016b, Table 4.1).

The Gran Guaiteca 1 site is an extensive camp formed by discards of gathered material and consumption of malacological fauna obtained from the shoreline; it shows successive reoccupations. In this shell-midden, we observed activities related to the manufacture and maintenance of stone instruments (record of knapping waste in the profile). There were activities around the area, in the terrace above the shell-midden, where evidence was found of the manufacture of stone projectile heads for hunting. The late date of this site, around 700–500 years cal. BP, may make it one of the last pre-Hispanic occupation events in the area. It still remains to date the base of the shell-midden to determine for how long and from what start date this site was used. The presence of obsidian from the Chaitén Volcano source, recorded in the later levels, associates the groups who settled in the area with a wide network of circulation and/or exchange of raw materials.

4.1.6 Gran Guaiteca 3

This site is located in the central isthmus of Gran Guaiteca Island on the west side of a small bay with a permanently running stream. It lies 1 km south-east of the archaeological sites recorded in the bay of the Cucuquén Stream. The area is heavily forested with evergreen forest and peat bogs, and the shoreline is formed by a wide beach with medium and large pebbles. The site is 1 m above the current sea level and 3 m from the shoreline. It stands right on the shoreline facing south. It has medium archaeological visibility due to anthropic deforestation which has thinned the forest that covered part of the site. The profiles along the south-western edge have been exposed by tidal action. There was no archaeological material either on the surface or in the adjacent intertidal zone. The slope is between 10° and 20°.

The site is an archaeological shell-midden covered by a layer of plant matter 10 cm thick. According to estimates based on drilling cores, the area is 40 m from east to west and 20 m from north to south, covering approximately 800 m^2. The shell deposit is 1.7 m deep.

The cores show an uninterrupted deposit of shell-midden matrix with no soil or hiati in its formation. This matrix consists mainly of picorocos (*Austromegabalanus psittacus*), cholgas (*Aulacomya ater*), choro mussels (*Choromytilus chorus*), Chilean mussels (*Mytilus edulis chilensis*), oysters (*Ostrea chilensis*), clams (*Protothaca thaca, Venus antiqua*), locos (*Concholepas concholepas*), keyhole limpets (*Fissurella nigra*), snails (*Tegula atra*), chitons (*Chiton sp.*) and sea urchins (*Loxechinus albus*). A sample of shells was taken from the centre of the shell-midden at a depth of 1.6 m for radiocarbon dating. The date assigned was 1929 ± 28 BP (1540–1300, 2-sigma cal. BP) (Reyes et al. 2016b, Table 4.1).

The Gran Guaiteca 3 site is an extensive camp formed by discards from the gathering and consumption of malacological fauna; successive reoccupations of the area are recorded from ~1400 years cal. BP. According to the record and the archaeological work carried out, no other activities are inferred for the site.

4.1.7 Isla Solitaria

The site lies on the northern tip of a small island, 500 m by 100 m, off the south coast of the central isthmus of Gran Guaiteca Island, separated from it by a narrow channel 300 m across. It lies 2 km south-east of the archaeological sites recorded in the bay of the Cucuquén Stream. The island presents a distinctive geoform due to the deforestation of its original vegetation (characterised as evergreen forest with peat bogs). At the northern end is a sharply cut terrace 6 m above current sea level, which falls away gradually to the south of the island. The forest has been thinned (*Persea lingue, Luma apiculata, Fuchsia magellanica*) and the landscape is marked by an old farm with meadows and animal tracks (centre of the island).

The site is 3 m above current sea level and 10 m from the shoreline, on the marine terrace in the north of the island. The archaeological visibility is nil. There were no materials on the surface; all the finds came from the test pits dug. The ground slopes down to the south from the site (10°). There is a cultural deposit restricted to 10 × 10 m with scarce lithic material and carbon specks distributed in a matrix 20 cm thick. It lies directly on the marine terrace with no remains of malacological fauna. There are no watercourses, either seasonal or permanent. The beach all round the island consists of medium and large pebbles.

Three test pits, each 1 × 1 m, were dug; they yielded lithic material and carbon specks, and allowed us to determine the stratigraphic units of the deposit. The site was also surveyed and a plan was made of its position and relation to existing sea level. The profile exposed in the test pits presents two stratigraphic units. The first is a layer of semi-compact organic earth with light brown silt–clay, 10 cm thick; the same matrix, with inclusion of gravel, is found between 10 and 30 cm. The matrix then turns to a very compact silt–clay with more gravel and larger round stones forming part of the terrace down to at least than 50 cm. No lenses of ash or defined hearths were recorded, only isolated specks of carbon.

A quartz flake and a distal fragment of a thick-section, lanceolate stone projectile were recovered between 10 and 30 cm. Samples of carbon specks were taken from both levels; the deepest sample found, at 35 cm (limit of the base of the deposit), was taken to avoid the possibility of using more recent carbon associated with farming activities on the island. An assemblage of 5 flakes was recovered from pits 2 and 3, at depths between 10 and 30 cm.

Macroscopic examination of the lithic waste showed that the raw material was grey obsidian from Chaitén Volcano on the mainland coast of the archipelago, 150 km north-east of the site. This observation was subsequently confirmed by geochemical analyses carried out in the Laboratory for Environmental Geochemistry at the University of Colorado (Boulder) by C. Stern. The carbon sample obtained at the base of the deposit at a depth of 35 cm gave a date of 607 ± 22 years BP (629–529, 2-sigma cal. BP, Reyes et al. 2016b, Table 4.1).

According to the lithic material recovered, the Isla Solitaria site appears to be a logistical site or lookout point, with very low occupation density. Based on its position close to the coast on a small island with no water but with good visibility and connections with other islands, we infer that the principal activity was knapping and discards of bifacial instruments around ~600 years cal. BP. No ecofactual evidence or associated fauna were observed in the area tested, which would permit the inference of prolonged camp activities. The presence of obsidian from the Chaitén Volcano source also associates the groups who settled in the area with a wide network of circulation and/or exchange of raw materials.

4.1.8 Alero Low

The Alero (shelter) Low site is located in Low Bay on the north coast of the central isthmus of Gran Guaiteca Island, facing the open sea. The site is a small rock shelter facing east-north-east. It is hidden in the forest almost 7 m above the current sea level and 20 m from the shoreline. The only access is from the beach up a steep cliff with a slope of 80°. The rock shelter is as it were 'hanging' like an overhung ledge. It is covered by thick forest, shrubby vegetation and an abundant layer of plant material formed by centuries of forest formation. In front of the site, there is a small island which is connected to Gran Guaiteca by a wide sand bar at low tide. The shore of Low Bay is formed by a wide sand beach, however, the shoreline close to the rock shelter consists entirely of large boulders and pebbles which make access and movement difficult. The area around the site is covered with thick evergreen forest vegetation and peat bogs, which come down to sea level and hide the rock shelter. The archaeological visibility from the shore is nil. This site is mentioned by the researcher Eugenio Aspillaga; we searched for it without success in a previous season.

The rock shelter is formed of compact, agglomerated sediments containing schists, quartz, beach pebbles and some rock, on an eroded marine terrace exposed to the sea and subsequently uplifted. The east-north-east orientation protects it from the prevailing south and south-west winds, and with the thick surrounding vegetation it is a well-protected refuge. The sediments in the shelter are dry. The site is 7 m long from north-west to south-east, and almost 4 m deep from north-east (the entrance) to south-west, with total area 28 m^2. The height varies between 1 and 3 m. Immediately below the drip-line, the bank falls abruptly (80°) to the rocky border of the adjacent shoreline.

On the surface are human postcranial bone remains, weathered and fragmented, which have been disturbed and moved in the sub-present, and finally redeposited against the wall of the shelter (Fig. 4.5). In the centre and towards the back of the shelter, there is an excavation which we understand to be the work of fishermen. During this anthropic intervention, a fire was lit and other domestic activities carried out, taking advantage of the special refuge provided by the site. On the surface, we observed fragmented and entire remains of malacological fauna (*Mytilus*, locos, keyhole limpets, clams, sea urchins) and two bone punches made from a bird's humerus and carpometacarpus.

Considering that this was a small rock shelter, and from the superficial material observed, we decided on an excavation 2 × 1 m close to the bank, in two contiguous pits in the centre of the shelter protected from the drip-line. The site was also surveyed and a plan was made of its position and relation to current sea level.

The excavation showed the following stratigraphy: the shell-midden matrix lies immediately down from the surface. It is mixed with dark grey, fine sand-silt sediments which have accumulated in the rock shelter, together with clasts and schists which have dropped from the wall and roof. This very dense shell-midden matrix is defined from 0 to 80 cm depth. It presents carbon specks, bones of sea lions and birds, and remnants of fish vertebrae and bones. The shell fragments consist mainly of cholgas (*Aulacomya ater*), choro mussels (*Choromytilus chorus*), Chilean mussels (*Mytilus edulis chilensis*), oysters (*Ostrea chilensis*), clams (*Protothaca thaca, Venus antiqua*), locos (*Concholepas concholepas*), keyhole limpets (*Fissurella nigra*), snails (*Tegula atra*), chitons (*Chiton sp.*) and sea urchins (*Loxechinus albus*). Allochthonous pebbles were observed in the shell-midden matrix, which might have been brought into the site and/or fallen off the shelter wall.

Fragments of human bones were recovered between 0 and 10 cm. They were weathered but not exposed to sunlight, reflecting the sub-present disturbance of these remains found on the surface. Fragmented, incomplete human bone remains were found between 10 and 40 cm, some of which were semi-articulated (portions of vertebrae and lower extremities); this shows that the individuals were buried in the shell-midden matrix and not laid on top of it. Two small trenches or hollows could be observed in the stratigraphic record, 40 cm deep, corresponding to the excavation of the shell-midden to deposit the bodies. These trenches at a depth of 50 cm form the principal feature by which the first 80 cm of shell-midden can be segregated: one anthropic disturbance of the matrix and one archaeological disturbance to effect at least two burials. The bone remains recovered at the site show that individuals from neonates to adults were buried there. As we will discuss below, burial in flexed lateral decubitus position can be inferred from the arrangement of some anatomical segments which survived the sub-present disturbances. Below 50 cm, the remains of malacological fauna start to appear more complete, and there is an increase in specks and pieces of carbon.

From 70 cm, the matrix starts to assume a reddish brown colour, with greater presence of ashy sediments, down to 80 cm. It then turns completely ashy, red in colour, with inclusion of fragments of shell, fish bones and bird bones down to 90–100 cm. At this depth, a large lens of ash was found, of mean thickness 30 cm, after which there is again a dense matrix of dark grey shell-midden, with sand-silt sediments, down to 140 cm. In excavation unit 2, this grey shell-midden matrix (90 to 140 cm) is interrupted again by another large lens of reddish ash 20 cm thick. In this unit, the shell-midden terminates abruptly where it lies on the base rock at a depth of almost 150 cm. In unit 1, the shell-midden is mixed with abundant medium and large schists. It becomes smaller as the base rock advances from unit 2. In unit 1, the archaeological shell-midden continues below 150 cm of shell-midden mixed with schist to a depth of 210 cm, where the base rock appears. No diagnostic lithic material was recovered. Small fragments of quartz, schists and some tiny rounded pebbles were recorded; however, all can be explained as having fallen from the support which makes up the shelter. There was no record of any other kind of raw material.

The 859 remains of malacological fauna recovered from the shell-midden column represented a total of 94 individuals of 11 species of molluscs and crustaceans (Cárcamo 2015). The species which presented the highest MNI was *Tegula atra,* with 45.74% (MNI = 43). The least-represented taxon was *Chiton sp.*, with a relative frequency of 1.06% (MNI = 1). Unlike the other shell-middens recorded, no remains of picorocos (*Austromegabalanus psittacus*) were recorded. The habitats of the species collected vary widely: rocky intertidal, rocky subtidal, parasitic and mixed intertidal (sandy and/or rocky), reflecting varied provisioning strategies, none of which prevailed over the others.

The fish remains consisted of 723 pieces, of which 181 could be identified (Torres 2016). They appeared in all the levels excavated down to 180 cm. Blenny is the predominant species in the assembly (69.6% NISP—number of identified specimens), followed by jurel with 10.9% NISP of the total assemblage. The frequency of these species is observed to be constant in the stratigraphic sequence, with blenny always predominant. This indicates its importance in the occupation and the constant presence of an estuarine environment over the whole period of the different occupations. It is also remarked that snoek appears in the stratigraphy when the presence of jurel starts to diminish.

Although the fish remains are not abundant, this site presents the greatest taxonomic diversity and evidence of exploitation of diverse environments (Torres 2016): fishing for neritic-pelagic species like jurel and snoek; rocky bottom fish like Patagonian redfish (*Sebastes oculatus*), tomollo (*Auchenionchus sp.*) and congrio (*Genypterus sp.*); and finally fish from sandy bottom environments like small-eyed flounder (*Paralichthys microps*). Judging by the size of a maxilla and a vertebra, one individual of this last taxon would have weighed around 7 kg. Remains of a canid were also found (San Román et al. 2016b), probably transported to the site.

The MNI of the individuals recovered at the surface and in the excavation was 14 (Galimany 2017), with 9 infants (4 foetuses/perinates; and 5 children aged 6–18 months, 1–2 years, 2–5 years, 4–8 years and 7–11 years), a juvenile aged between 14 and 18 years and 4 adults of indeterminate age, two of whom were identifiable as female (Schaefer et al. 2009). Almost all the levels between 0 and 40 cm in both trenches contained remains of adults and subadults, associated both within and between excavation levels and the surface. All age ranges are represented at this site. The majority of the bone portions represented, from which the number of individuals could be distinguished, were hand and foot bones (phalanges, metacarpals and metatarsals). They were much more numerous than crania and long bones, probably absent due to anthropic disturbance because they were more visible.

A series of samples was taken from different stratigraphic levels. One human bone sample was obtained between 0 and 10 cm from the disturbed superficial level (ALOW1-1) and a second was recovered between 30 and 40 cm (ALOW1-2); these represented two types of funerary deposition (burial and surface deposition), giving different moments for these samples with consistent stratigraphy. Carbon samples were taken for dating the lower levels at 150–160 cm (ALOW1-5) and samples of fauna for dating the levels 160–170 cm (ALOW1-3) and 180–190 cm (ALOW 1–4); these also showed stratigraphic consistency with occupations of the shelter between ~2000 and 350 years cal. BP (Reyes et al. 2019b, Table 4.1).

The area excavated at this site (2 m^2) enables us to infer that the shelter was a camp where various activities were carried out and that it was formed by discards from the gathering and consumption of malacological fauna; successive reoccupations of the site are recorded from ~2000 years cal. BP. The fact that the space was small and sheltered allowed us to observe big hearths around which diverse malacological fauna were consumed and discarded, however, there was no record of stone tool maintenance and/or manufacture. Based on the faunal record, other activities can be inferred such as fishing, hunting for birds and smaller species like coypu and marine otter, and hunting for larger marine fauna like adult and juvenile sea lions. The presence of a canid projects a wide network of mobility and/or exchange to obtain certain resources not available in the islands.

This shelter was used as a camp for almost 1,000 years, and at the end of the sequence it was also used as a funerary space to bury and deposit a series of individuals. The bio-anthropological results related the superficial and excavated bone remains, showing the disturbance suffered by some burials. The dates obtained in two of the individuals who were deposited in the shelter give a mean range between 650 and 350 years cal. BP; the extreme calibrated ranges of the dates do not overlap, so this ossuary may have been formed by various events over a relatively short period of time in which the same place was used repeatedly to deposit the dead. The isotope values obtained in the dated individuals allowed us to establish the consumption of an exclusively marine diet (see Chapter 6, Evaluation and Discussion of the Evidence) (Reyes et al. 2019a).

4.1.9 Isla Marta 1

The Isla Marta 1 site is located on the south coast of Marta Island, separated from the north coast of Gran Guaiteca Island by a narrow channel 140 m across. Twenty metres from the eastern end of the site is a stream which runs all year round, from north to south. The trees in the area around the site have been felled by fishermen who settled here. Today, there is a fishermen's hut on the site which is still visited, and a large hearth for smoking seafood. The evergreen forest and peat bog begin immediately beyond the perimeter. The site is 2 m above the current sea level and 50 m from the shoreline. It stands on the shore facing south over a beach covered with small pebbles. The archaeological visibility is nil, and detection depended on a series of cores and test pits in the area. The site is a large archaeological shell-midden covered by a layer of plant matter 3 cm thick. According to estimates based on drilling cores, the area is 12 m from east to west and 10 m from north to south, covering approximately 120 m^2. The shell deposit is 1.75 m deep.

At least three levels of marine terraces can be observed on the south and east coasts of Marta Island. We drilled a series of cores in the highest terrace, located between

10 and 11 masl, then in the second terrace, between 5 and 6 m, and finally in the third terrace, between 2 and 4 m. The archaeological shell-midden was discovered in the latter.

This core test showed an uninterrupted deposit consisting mainly of picorocos (*Austromegabalanus psittacus*), cholgas (*Aulacomya ater*), choro mussels (*Choromytilus chorus*), Chilean mussels (*Mytilus edulis chilensis*), oysters (*Ostrea chilensis*), clams (*Protothaca thaca, Venus antiqua*), locos (*Concholepas concholepas*), keyhole limpets (*Fissurella nigra*), snails (*Tegula atra*), chitons (*Chiton sp.*) and sea urchins (*Loxechinus albus*). Throughout the stratigraphic record, we observed bones of sea lions, birds (cormorants), fish, coypu and pudú, as well as carbon specks.

On a subsequent visit, a 1 × 1 m test pit was dug to obtain a better characterisation of the stratigraphy, and a column to check for faunal remains. The test pit had to be enlarged by 30 cm in the SE corner because a burial was found in situ at a depth of 20 cm. The excavation matrix produced the following stratigraphic record: 0–5 cm a covering layer of grass and roots, leading immediately to a very dark organic sandy silt, 5 cm thick. At a depth of 10 cm, a shell-midden matrix becomes defined with remains of fragmented malacological fauna. At 20 cm, an adult human molar and fragments of the cranium of a subadult appeared, later better defined between 20 and 30 cm. It should be added that down to 20 cm, we found sub-present rubbish (nails, plastic items) resulting from the activity of the fishermen settled immediately on top of the site. This activity disturbed the first 20 cm but did not alter the burial recorded between 20 and 30 cm. The shell-midden continues present with the inclusion of fragments of bones of common sea lion, coypu, marine otter, birds and fish, as well as carbon specks. At 170 cm, the shell-midden matrix becomes less dense and starts to present inclusion of more sandy silt sediments; the amount of faunal remains in the matrix also diminishes. Between 170 and 185 cm, fragments of shells are observed at the interface with a pale brown sandy matrix, with inclusion of small pebbles. This matrix belongs to the palaeo-beach of the third terrace where the first occupants of the area settled. The excavation was continued to 215 cm, where only sandy sediments were found with increasing numbers of medium and small pebbles from the palaeo-beach.

A molar and a proximal phalange of one hand were recovered from the burial found at 20 cm (Individual N°1). The phalange belongs to an individual aged 16 years or less (the epiphysis is not fused, Schaefer et al. 2009). The molar, on the other hand, could belong to an adult individual. It presents stage 5 wear (Smith 1984) with elimination of the cusp and exposure of dentin. The other bone remains belong to an infant recovered in situ, deposited in hyperflexed right lateral decubitus with its head to the north and feet and hips to the south (Individual N°2). Its age was estimated to be between 4 and 6 years from the maxillary and mandibular tooth development and eruption (Buikstra and Ubelaker 1994), as well as the state of fusion and the size of the cranial and postcranial bones (Schaefer et al. 2009). No cultural or ecofactual elements were recovered that could be interpreted as funerary offerings and/or trappings.

The data on fish remains recovered from the column to check for fauna (10 × 10 cm) produced taxonomical identification of 498 remains (Torres 2016). The abundance fluctuates down to 150 cm and then diminishes towards the lowest levels excavated (185 cm). The whole assemblage shows a predominance of snoek (*Thyrsites atun*), with 38.1% NISP; this species remains prevalent throughout the stratigraphy, except between 60 and 100 cm where a different assemblage is observed.

Three different moments or sets of taxonomic diversity are observed in the stratigraphic sequence of the site: in the first 50 cm there is greater diversity, although snoek predominates. Jurel (*Thrachurus murphyi*) only appears in this range, and is absent lower down. Between 50 and 120 cm, there are more bilagay (*Cheilodactylus variegatus*), Patagonian redfish (*Sebastes oculatus*) and blenny (*Eleginops maclovinus*). The two former species inhabit rocky subtidal environments, while the latter frequent estuaries and other environments with a freshwater component. Finally, snoek and blenny predominate between 120 and 170 cm (Torres 2016).

The samples obtained for radiocarbon dating at 125 cm (pudú) and at 175 cm at the base of the deposit (shell) gave dates of 1520 ± 25 (1310–1406, 2-sigma cal. BP) and 2074 ± 30 BP (1428–1733, 2-sigma cal. BP) respectively, representing a difference of approximately ~300 years in 50 cm of shell-midden deposit (Table 4.1).

The Isla Marta 1 site is a camp where multiple activities were carried out. It was formed mainly by discards from gathering and consumption of malacological fauna; successive reoccupations of the site are recorded from ~1700 years cal. BP. Based on the faunal record, other activities occurred such as fishing for a diversity of species, hunting for birds and smaller species like coypu (*Myocastor coypus*) and marine otter (*Lontra provocax*), and hunting for larger marine fauna like adult and juvenile sea lions. The presence of a pudú bone (*Pudu pudu*) (around ~1300 years cal. BP) proves at least exchange with the habitat of this small cervid on larger islands (e.g. Chiloé) or the mainland coast (Jiménez 2010). The camp also witnessed at least two burials on the southern edge of the shell-midden: a juvenile aged approximately 16 years and an infant aged between 4 and 6 years; no funerary offering or trappings were observed in these final moments of pre-Hispanic occupation in the area.

4.1.10 Isla Marta 2

The site is located on the south-east coast of Marta Island, 1 km from Isla Marta 1 and Gran Guaiteca 4. It lies at the south-westerly entrance of a small channel formed between Marta and Guacanec Islands. The area is heavily forested with evergreen forest and peat bogs, however, the vegetation around the site has been cleared by fishermen, leaving a grassy area. The site therefore enjoys medium archaeological visibility, as the side profiles of the site have been exposed by tidal action and can be observed. The shoreline consists of a pebble and sand beach. The site is 2 m above the current sea level and 1 m from the shoreline. It faces east and lies between the third marine terrace and the adjacent beach on the shoreline. There was no archaeological material either on the surface or in the adjacent intertidal zone. The site presents a slope of 30°.

It is a large archaeological shell-midden covered by a layer of plant matter 10 cm thick on average. According to estimates based on drilling cores, the area is 25 m from east to west and 65 m from north to south, covering approximately 1625 m^2. The shell deposit is 2 m deep. Immediately to the north of the shell-midden runs a small stream which flows all year round. We also drilled a series of cores in the highest terrace, between 10 and 11 m, then in the second terrace, between 5 and 6 m, and finally in the third terrace, between 2 and 4 m, to discard the presence of anthropic sediments. The archaeological shell-midden is located on the lowest terrace.

Clearing of the east profile revealed an uninterrupted deposit of shell-midden matrix from 10 to 200 cm. The base of the deposit lies on the palaeo-beach, composed of grey sandy silt sediments with inclusion of clasts and medium-sized pebbles. The shell-midden matrix consists mainly of picorocos (*Austromegabalanus psittacus*), cholgas (*Aulacomya ater*), choro mussels (*Choromytilus chorus*), Chilean mussels (*Mytilus edulis chilensis*), oysters (*Ostrea chilensis*), clams (*Protothaca thaca, Venus antiqua*), locos (*Concholepas concholepas*), keyhole limpets (*Fissurella nigra*), snails (*Tegula atra*), chitons (*Chiton sp.*) and sea urchins (*Loxechinus albus*).

The Isla Marta 2 site is an extensive camp formed by discards from gathering and consumption of malacological fauna; successive reoccupations of the area are recorded. According to the record and the archaeological work carried out, no other activities are inferred for the site.

4.1.11 Gran Guaiteca 4

The site is located on the north coast of Gran Guaiteca Island, between a 140 m-wide channel which separates the island from the south coast of Marta Island, and a permanent stream which flows from the north into the sea south-west of the site. The area around the site has been cleared by fishermen who settled here; however, the evergreen and peat bog forest starts immediately beyond the perimeter. The site lies 1 m above the present sea level and 1 m from the shoreline, facing north-west. The beach immediately below is covered with small pebbles. It has medium archaeological visibility due to anthropic deforestation which has thinned the forest covering part of the site, allowing observation of the profiles along the south-western edge which have been exposed by tidal action.

A stone net weight and a retouch flake were recovered in the intertidal, both made of pebbles from the same zone. The flat area has a gentle slope of 10°, but the whole northern edge is cut off by an abrupt, almost vertical drop.

The site is a large archaeological shell-midden covered by a layer of plant matter 3 to 10 cm thick. According to estimates based on drilling cores, the area is 20 m from east to west and 30 m from north to south, covering approximately 600 m^2. The shell deposit is 2.3 m deep. The lithic material was recovered from the intertidal

Fig. 4.6 Clearing the profile and obtaining samples for radiocarbon dating in Gran Guaiteca 4

adjacent to the south-west profile; an exposed profile was cleared and checked, and samples were obtained for radiocarbon dating (Fig. 4.6).

The core tests showed an uninterrupted deposit of shell-midden matrix consisting mainly of picorocos (*Austromegabalanus psittacus*), cholgas (*Aulacomya ater*), choro mussels (*Choromytilus chorus*), Chilean mussels (*Mytilus edulis chilensis*), oysters (*Ostrea chilensis*), clams (*Protothaca thaca, Venus antiqua*), locos (*Concholepas concholepas*), keyhole limpets (*Fissurella nigra*), snails (*Tegula atra*), chitons (*Chiton sp.*) and sea urchins (*Loxechinus albus*). Throughout the stratigraphic record, we observed sea lion bones as well as carbon specks. Fine sand hiati were observed in the stratigraphy, attributed to fluvial events (high river levels and sandy backwaters) in the nearby stream. The base of the shell-midden deposit lies at a depth of 2.3 m, on a stratum of grey sand with inclusions of gravel that continues for more than 30 cm below the shell-midden. Shell samples were taken from the contact layer of the beach with the archaeological shell-midden.

Only a stone net weight and a retouch flake were recovered in the intertidal, both made of pebbles from the same zone. No cultural material was obtained during clearing of the profile. The shell sample obtained at the base of the deposit, at a depth of 230 cm, gave a date of 2259 ± 26 years BP (1664–1946, 2-sigma cal. BP, Reyes et al. 2016b, Table 4.1).

The Gran Guaiteca 4 site is an extensive camp formed by discards of gathered material and consumption of malacological fauna; successive reoccupations of the area are recorded from ~1800 years cal. BP. The archaeofaunal record shows consumption of marine fauna and fishing, the latter also confirmed by the lithic material (net weight) recovered in the intertidal.

4.1.12 Repollal Bajo 1

The site is on the west coast of Ascensión Island, on the western side of a small, south-facing bay protected by a series of small islands and rocks. The area is heavily forested with evergreen forest and peat bogs, and the shoreline is formed by a small pebble and sand beach.

The site is 2 m above the current sea level and 2 m from the shoreline. It stands on the shore facing south. It has medium archaeological visibility due to anthropic deforestation which has thinned the forest covering a large part of the site. The profiles along the south-western edge have been exposed by tidal action. Furthermore, a family has settled on the site, building a house, sheds and a wharf. There was no archaeological material either on the surface or in the adjacent intertidal zone. The slope is between 10° and 20°. There is an extensive archaeological shell-midden covered by a layer of plant matter 10 cm thick where the house and sheds have been built. According to estimates based on drilling cores, the area is 10 m from east to west and 20 m from north to south, covering approximately 200 m^2. The shell deposit is 3.1 m deep. On the west side, the shell-midden is cut by a small, permanent stream (Fig. 4.7).

The test cores indicate an uninterrupted deposit of shell-midden matrix from 10 to 310 cm, where it lies on the palaeo-beach. This matrix consists mainly of picorocos (*Austromegabalanus psittacus*), cholgas (*Aulacomya ater*), choro mussels (*Choromytilus chorus*), Chilean mussels (*Mytilus edulis chilensis*), oysters (*Ostrea chilensis*), clams (*Protothaca thaca, Venus antiqua*), locos (*Concholepas concholepas*), keyhole limpets (*Fissurella nigra*), snails (*Tegula atra*), chitons (*Chiton sp.*) and sea urchins (*Loxechinus albus*).

The Repollal Bajo 1 site is an extensive camp formed by discards from gathering and consumption of malacological fauna; successive reoccupations of the area are recorded. According to the record and the work carried out, no other activities are inferred for the site.

4.1.13 Repollal Bajo 2

The site is on the west coast of Ascensión Island. The area is forested with evergreen forest and peat bogs, and lies close to the shoreline formed by a beach with sand and pebbles interspersed with rocky outcrops. The site is 2 m above the current sea level and 2 m from the shoreline. It stands on the shore facing south. It has been affected by the rural road joining the hamlet of Repollal Bajo with Melinka. The site was found

Fig. 4.7 Repollal Bajo 1 site. Extensive shell-midden mound on which a house has been built. Note the grassed over side of the site, more than 3 m high, which looks like a natural geoform of the landscape

due to the exposure of its west profile, giving it archaeological visibility. There was no archaeological material either on the surface or in the adjacent intertidal zone. The site presents a slope of 40°.

It is an archaeological shell-midden covered by a layer of plant matter 20 cm thick. The western end has been cut by a rural road exposing part of the profile. According to estimates based on drilling cores, the area is 20 m from east to west and 80 m from north to south, covering approximately 1600 m^2. The shell deposit is 1.9 m deep. Test cores were drilled to define the dimensions and obtain samples of the base of the deposit, and 2.6 m of profile were cleared to record the stratigraphy. The site was also surveyed and a plan was made of its position and relation to existing sea level.

The test cores indicate an uninterrupted deposit of shell-midden matrix from 20 to 190 cm, where it lies on the palaeo-beach. This matrix consists mainly of picorocos (*Austromegabalanus psittacus*), cholgas (*Aulacomya ater*), choro mussels (*Choromytilus chorus*), Chilean mussels (*Mytilus edulis chilensis*), oysters (*Ostrea chilensis*), clams (*Protothaca thaca, Venus antiqua*), locos (*Concholepas concholepas*), keyhole limpets (*Fissurella nigra*), snails (*Tegula atra*), chitons (*Chiton sp.*) and sea urchins (*Loxechinus albus*). The shell sample obtained at the base of the deposit when clearing the profile to a depth of 70 cm gave a date of 1960 ± 30 years BP (1322–1586, 2-sigma cal. BP, Reyes et al. 2016b, Table 4.1).

The Repollal Bajo 2 site is a camp formed by discards from gathering and consumption of malacological fauna; successive reoccupations of the area are recorded since at least ~1400 years cal. BP. According to the record and the archaeological work carried out, no other activities are inferred for the site.

4.1.14 *Osamentas Repollal*

The site is located on the west coast of Ascensión Island, on the northern edge of a small, south-facing bay closed by a series of small islands and rocks. It is close to Repollal Bajo 1 and 2, respectively 700 m and 1400 m further west. The area is heavily forested with evergreen forest and peat bogs, and the shoreline is formed by a pebble beach. The site is a small rock shelter just over 2 m wide by 1 m deep, total area approximately 3 m^2. In this restricted shelter, there is a matrix of silt-sand sediments, with remains of shells, mainly *Mytilus* and keyhole limpets. At least two individuals were buried in this sediment at a depth of 10 cm.

The site was diagnosed by the fortuitous discovery of bones by settlers in the area. Subsequently, the remains were assigned to us by a judicial investigation for bio-anthropological analysis (Reyes and Trejo 2010). Finally, in our field trips to prospect and identify archaeological sites, we visited the island to record the site. The site is 10 m above the current sea level and 50 m from the shoreline. The archaeological visibility is nil, as the site is covered and hidden by vegetation.

The principal characteristics were recorded. Inspection of the sediments and the surrounding area produced no cultural or bio-anthropological material. From our bio-anthropological analysis, we concluded that the bone remains recovered belong to at least two individuals, both adults of indeterminate age and sex.

The Osamentas Repollal site is a small rock shelter set back from the shoreline in thick forest. The activity at the site included discarding of malacological fauna, and at least two individuals were buried in the sediment, presumably (to judge by the dimensions of the shelter) in hyperflexed lateral decubitus. No evidence was found of funerary offerings and/or trappings.

4.1.15 *Cholguero Puquitín 1*

The site is located on the north-east coast of Ascensión Island, in a narrow bay enclosed by a peninsula. The cholguero port is on the south coast of the peninsula, on the pebble and sand beach just above the intertidal. The area is forested with evergreen forest and peat bogs, and lies close to the shoreline formed by a beach with sand and pebbles interspersed with rocky outcrops. The site is 2 m above the current sea level and 20 m from the shoreline. It stands on the shore facing south. The forest has been thinned and cleared, so there is now a grassy clearing with a few trees, with an old apple tree and the humps of vegetable beds, covered over with weeds and grass. It has medium archaeological visibility, being distinguishable by the wooden structure and rags of plastic used as a shelter, and the hearth area (very overgrown) inside. Some materials can be observed on the surface such as metal, wood and plastic. The site presents a slope of 20°.

It is an abandoned historical cholguero, covered by a layer of plant material 10 cm thick in some places, and beneath it a shell-midden deposit consisting exclusively of species of mytilidae (mussels and cholgas) no more than 20 cm thick. Below this shell matrix is a layer of inorganic earth mixed with fine sand 10 cm thick, and then the palaeo-beach. According to estimates based on drilling cores and the location of the wooden structure, the apple tree and the old vegetable beds, the area is 50 m from east to west and 80 m from north to south, covering approximately 4200 m^2.

No occupation or archaeological material was found below the cholguero, nor on the surface or in the nearby intertidal. The test cores indicated a discrete deposit of shell-midden matrix 20 cm thick, consisting mainly of cholgas (*Aulacomya ater*), choro mussels (*Choromytilus chorus*) and Chilean mussels (*Mytilus edulis chilensis*).

Cholguero Puquitín 1 is a settlement site developed by seasonal visits, possibly reiterated, related to intensive, selective gathering of mytilids for the canning industry. This activity probably implied other associated practices, such as vegetable-growing and very probably hunting and fishing for the fishermen's and mussel gatherers' subsistence. We also recorded the plantation of fruit trees and old potato beds, organised in the expectation of recurrent visits to the area. The start of occupation may be dated to the beginning or middle of the twentieth century.

4.1.16 Cholguero Puquitín 2

The site is located on the north-east coast of Ascensión Island, on the south shore of the Puquitín Channel which separates Gran Guaiteca from Ascensión Island. It is at the head of a small, north-facing bay. The cholguero is on the sand and small pebble beach just above the intertidal. The area is forested with evergreen forest and peat bogs, and lies close to the shoreline formed by a beach with sand and pebbles interspersed with rocky outcrops. It presents nil archaeological visibility due to the shrub cover in the area. The site is 1 m above the current sea level and 5 m from the shoreline. It stands on the shore facing north. The forest in the area has been felled and cleared, such that now there is a grassy clearing covered with scrub. No cultural material was observed on the surface. The site presents a slope of 40°.

It is an abandoned historical cholguero, covered by a layer of plant material 10 cm thick in some places, and beneath it a shell-midden deposit consisting exclusively of species of mytilids (mussels and cholgas), approximately 20 cm thick. Below this shell matrix is the base rock, part of the rocky promontory which limits the site to the west. According to estimates based on drilling cores, the area is 10 m from east to west and 20 m from north to south, covering approximately 200 m^2. No occupation or archaeological material was found beneath the cholguero, nor on the surface or in the nearby intertidal.

The test cores indicated a deposit of shell-midden matrix 100 cm thick, consisting mainly of cholgas (*Aulacomya ater*), choro mussels (*Choromytilus chorus*) and Chilean mussels (*Mytilus edulis chilensis*). Cholguero Puquitín 2 is a campsite developed by seasonal visits, possibly reiterated, related to intensive, selective gathering of mytilids for the canning industry. This activity probably implied other associated practices, such as hunting and fishing for the fishermen's and mussel gatherers' subsistence. The start of occupation may be dated to the beginning or middle of the twentieth century.

4.1.17 Cholguero Puquitín 3

The site is located on the south-east coast of Gran Guaiteca Island, on the north shore of the Puquitín Channel which separates Gran Guaiteca from Ascensión Island. It stands on a small rocky isthmus with accessible beaches on either side. The cholguero is between the two beaches in the partly cleared forest, above the intertidal zone which consists of sand and small pebbles. The area is forested with evergreen forest and peat bogs, and lies close to the shoreline formed by a beach with sand and pebbles interspersed with rocky outcrops. The site is 1 m above the current sea level and 10 m from the shoreline, facing south. The forest in the area has been felled and cleared, so now there are grassy clearings covered with scrub. The archaeological visibility is medium. Cultural material was recorded on the surface, such as metal objects and glass bottles, as well as a wooden structure at the centre of which is a hearth ringed with rocks. The slope of the site is 20° (Figs. 4.8).

It is an abandoned historical cholguero, covered by a layer of plant material 10–20 cm thick in some places, and beneath it a shell-midden deposit consisting exclusively of species of mytilids (mussels and cholgas), approximately 20 cm thick. Below this shell matrix is the base rock of the isthmus. According to estimates based on drilling cores, the area is 20 m from east to west and 20 m from north to south, covering approximately 400 m^2. No occupation or archaeological material was found below the cholguero, nor on the surface or in the nearby intertidal.

The test cores indicated a deposit of shell-midden matrix 100 cm thick, consisting mainly of cholgas (*Aulacomya ater*), choro mussels (*Choromytilus chorus*) and Chilean mussels (*Mytilus edulis chilensis*). Cholguero Puquitín 3 is a campsite developed by seasonal visits, possibly reiterated, related to intensive, selective gathering of mytilids for the canning industry. This activity probably implied other associated practices, like hunting and fishing for the fishermen's and mussel gatherers' subsistence. The start of occupation may be dated to the beginning or middle of the twentieth century.

Fig. 4.8 View of Cholguero Puquitín 3 and part of the remains of a shelter constructed of branches and stakes in the evergreen forest

4.1.18 Cholguero Canal Cuervo 1

The site is located on the south coast of Gran Guaiteca Island, on the north shore of the Cuervo Channel which separates Gran Guaiteca from Betecoi Island. It is at the eastern end of the channel, at the mouth of a south-facing bay. The cholguero is situated in the partly cleared forest, above the intertidal zone which consists of small pebbles. The area is forested with evergreen forest and peat bogs, and lies close to the shoreline formed by a beach with sand and pebbles interspersed with rocky outcrops. The site is 3 m above the current sea level and 1 m from the shoreline, facing south. The forest in the area has been felled and cleared, so now there are grassy clearings covered with scrub. The archaeological visibility is nil. There is a hearth overgrown with grass, as well as sticks, stakes and a stove chimney. The site presents a slope of 40°.

It is an abandoned historical cholguero, covered by a layer of plant material 10–20 cm thick in some places, and beneath it a shell-midden deposit consisting exclusively of species of mytilids (mussels and cholgas), approximately 20 cm thick. Below this shell matrix is the base rock which forms the foundation of the cholguero. According to estimates based on drilling cores, the area is 10 m from east to west and 8 m from north to south, covering approximately 80 m^2. No occupation or archaeological material was found below the cholguero, nor on the surface or in the nearby intertidal.

The test cores indicated a deposit of shell-midden matrix 100 cm thick, consisting mainly of cholgas (*Aulacomya ater*), choro mussels (*Choromytilus chorus*) and Chilean mussels (*Mytilus edulis chilensis*). Cholguero Canal Cuervo 1 is a campsite developed by seasonal visits, possibly reiterated, related to intensive, selective collection of mytilids for the canning industry. This activity probably implied other associated practices, like hunting and fishing for the fishermen's and mussel gatherers' subsistence. The start of occupation may be dated to the beginning or middle of the twentieth century.

4.1.19 Cholguero Canal Cuervo 2

The site is located on the south coast of Gran Guaiteca Island, on the north shore of the Cuervo Channel which separates Gran Guaiteca from Betecoi Island. It is at the head of a south-facing bay. The cholguero is situated in the partly cleared forest, above the intertidal zone which consists of small pebbles and large rocks. The area is forested with evergreen forest and peat bogs, and lies close to the shoreline formed by a beach with sand and pebbles interspersed with rocky outcrops. The site is 1 m above the current sea level and 1 m from the shoreline, facing south. The forest in the area has been felled and cleared, so now there are grassy clearings covered with scrub. The archaeological visibility is medium. Cultural material was found such as metal objects, bottles, remains of a hearth ringed by rocks and fragments of cement. The site presents a slope of 20°.

It is an abandoned historical cholguero, covered by a layer of plant material 10–20 cm thick in some places, and beneath it a shell-midden deposit consisting exclusively of species of mytilids, approximately 50 cm thick. Below this shell matrix is the base rock which forms the foundation of the cholguero. According to estimates based on drilling cores, the area is 10 m from east to west and 20 m from north to south, covering approximately 200 m^2. No occupation or archaeological material was found below the cholguero, nor on the surface or in the nearby intertidal.

The test cores indicated a deposit of shell-midden matrix 100 cm thick, consisting mainly of cholgas (*Aulacomya ater*), choro mussels (*Choromytilus chorus*) and Chilean mussels (*Mytilus edulis chilensis*). Cholguero Canal Cuervo 1 is a campsite developed by seasonal visits, possibly reiterated, related to intensive, selective gathering of mytilids for the canning industry. This activity probably implied other associated practices, like hunting and fishing for the fishermen's and mussel gatherers' subsistence. The start of occupation may be dated to the beginning or middle of the twentieth century.

4.2 Mainland Coast area of the Chonos Archipelago (43°46′–44°S)

Below, we describe the archaeological sites recorded and studied on the mainland coast of the Chonos Archipelago, east of the Moraleda Channel (Fig. 4.9).

4.2.1 Yalac 1

The Yalac 1 site is on the south-east coast of Yalac Island, opposite the Pedregoso Channel which is formed by this island, Refugio Island and the Melimoyu Peninsula. The site is a small cave in a rocky outcrop, located on a small, north-east facing bay. The coast is mainly rocky and the intertidal zone has medium-sized pebbles and rocks. The area around the site is covered with thick evergreen forest vegetation and peat bogs, which come down to sea level and hide the rock shelter, making access difficult. (Fig. 4.10). The site offers high archaeological visibility from the coast,

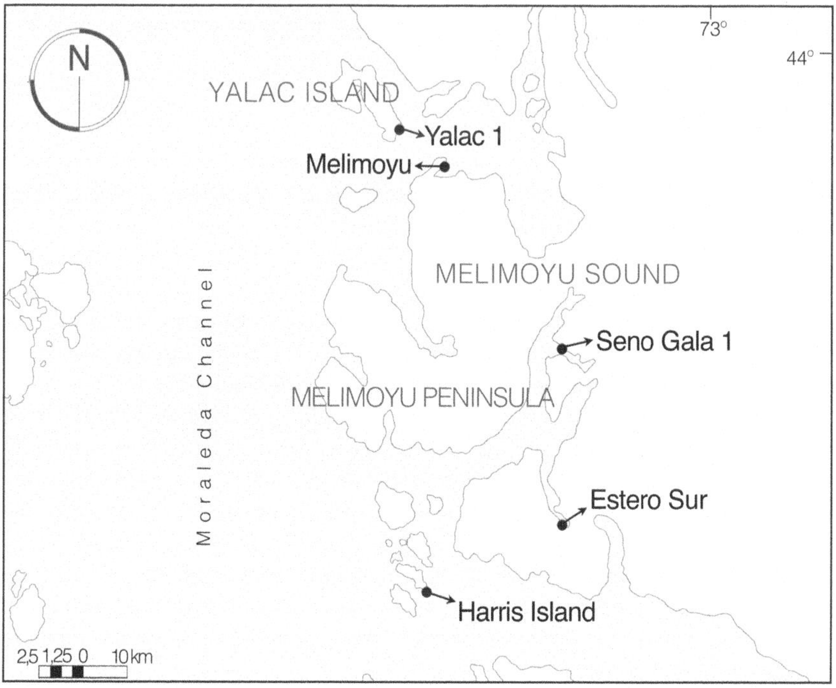

Fig. 4.9 Distribution of sites recorded and studied to the east of the Moraleda Channel, on the mainland coast of the Chonos Archipelago

Fig. 4.10 Entrance to Yalac 1 site. Note the thick forest cover hiding the entrance to the cave. The exposed rocks indicate the high tideline

with the upper part of the small cave visible among the vegetation. It is 3 m above the current sea level and 10 m from the shoreline.

The Yalac 1 cave was used as an ossuary. The cave entrance faces east and is raised 1 m above the general ground level. Immediately below the drip-line, the bank falls sharply (60°) to the rocky shore (1 m). It measures 4.8 m from north to south and 4.8 m from east to west, total approximately 23 m^2. The height fluctuates between 2.5 m and the 5.5 m. There are several watercourses close to the site.

A series of human remains were found on the surface, grouped or scattered in an area 3 m from north to south by 2.3 m from east to west, on a dry floor consisting of fine sand-silt sediments, with medium-sized angular clasts. It is clear that some bones have been moved repeatedly by visitors, who put the decontextualised crania together and placed a wooden cross in the area where the bones are concentrated, a typical fishermen's practice which has been documented in other sites with human remains in the archipelago. We decided on a superficial collection of all the human remains present on the cave floor and checked the surrounding area in search of more bones and/or cultural materials.

Analysis of the bone remains from the Yalac 1 site produced a minimum number of 13 individuals, consisting of 4 infants and 9 adults, all with fragments of crania and/or mandibulas as the most represented portions (Galimany 2017). The four infants presented different ages at death: between 2 and 4 years, between 4 and 7 years, between 7 and 9 years and finally between 9 and 12 years. Among the adults, there was a young male adult (minimum age 20 years), a young female adult (minimum age 20 years), a middle-aged adult, probably female (minimum age 35 years), a young adult, probably female (minimum age 20 years), a middle-aged male adult (minimum age 35 years), a female adult (approximate age 20 years), an adult, probably a young male (minimum age 20 years), an adult of indeterminate age, probably female, and an adult of indeterminate sex and age (Galimany 2017).

The Yalac 1 site is a cave used exclusively for funerary purposes. This type of site is known as an 'ossuary', given the scale of the accumulation of human remains. The cave is a rock shelter adjacent to the shore, protected from the rain and surrounded by forest. The site has been altered by later historical/sub-present anthropic activities. No artefactual or ecofactual material was recorded to indicate whether the site was occupied apart from the funerary ritual of depositing bodies. There is no evidence of rock art on the walls of the cave, and no remains of funerary offerings or trappings.

The results of the datings obtained from two adult individuals 1 (YAL1-1) and 6 (YAL1-6) of the 13 recorded indicate deposition on the cave floor between ~500 and 300 years cal. BP (Reyes et al. 2019b, Table 4.1). The extremes of the calibrated age ranges overlap, possibly indicating the formation of this ossuary in a synchronic event, or several events over a short lapse of time which cannot be differentiated chronologically, and clearly show reiterated use of the same site to deposit the dead. The isotope values obtained in the dated individuals allowed us to establish the consumption of an exclusively marine diet (see Chapter 6, Evaluation and Discussion of the Evidence).

4.2.2 Osario Melimoyu

The Osario Melimoyu (ossuary) site is located on the north coast of the Melimoyu Peninsula, on the mainland coast of the archipelago. This site was reported due to a judicial investigation on an assumed accident when the bones were found. The bones were collected by the Chilean Navy (Mena 2015) and sent to the CIEP laboratories in Coyhaique, where we were given access to them to allow general description and to take samples for radiocarbon dating. The site is located on the north-west coast of a small peninsula overlooking the broad channel separating the Melimoyu Peninsula from Yalac and Refugio Islands. There is no intertidal zone, and the shore drops vertically. The area around the site is covered with thick evergreen forest vegetation and peat bogs, which come down to sea level and hide the rock shelter, making access difficult. The archaeological visibility is nil and the site was found during clearing work carried out in the forest by a salmon-farming company. The site is a small overhang or shelter formed of rocky outcrops. It is 5 m above the current sea level and 20 m from the shoreline, in the forest.

The Osario Melimoyu site is a shelter where a large number of human bone remains were found on the surface, some grouped, some scattered. It is 2 m long and 1 m wide, total approximately 2 m^2. The height fluctuates between 0.5 and 1.5 m. It is clear that the bones have been moved repeatedly by locals who have visited the site, gathering the crania together and decontextualising them. The Navy took away 100% of the human bone remains deposited on the surface (n = 133 remains).

Analysis of the bone remains found at the Osario Melimoyu site produced a minimum number of 11 individuals: 5 subadults and 6 adults (Galimany 2017). Among the subadults, we found different ages at death: one unweaned baby aged between 3 and 24 months; and four infants, aged between 2 and 5 years, between 4 and 8 years, between 7 and 9 years and finally between 7 and 11 years. The adult individuals include one male adult (between 35 and 56 years), one male adult (older than 30 years), one male adult (between 33 and 44 years), one female adult (between 40 and 50 years), one male adult (between 35 and 49 years) and one male and one female adult (both older than 19 years).

The Osario Melimoyu site is a small shelter selected and used exclusively for funerary purposes, known as an ossuary. It is close to the shore, protected from the rain and surrounded by forest. The site has been altered by later anthropic activities. No artefactual or ecofactual material was recorded to indicate whether the site was occupied apart from funerary rituals; the small size of the space would preclude habitation. There is no evidence of rock art on the walls of the cave, and no remains were reported of funerary offerings or trappings (Mena 2015).

The results of the datings obtained from adult individuals 1 (MEL1-1) and 3 (MEL1-3) of the 11 individuals deposited on the cave floor give a mean range between ~428 and 667 years cal. BP (Reyes et al. 2019b Table 4.1). The extremes of the calibrated age ranges overlap, possibly indicating the formation of this ossuary in a synchronic event, or several events over a short lapse of time which cannot be differentiated chronologically, and clearly show reiterated use of the same site to deposit the dead. The isotope values obtained in the dated individuals allowed us to establish the consumption of an exclusively marine diet (see Chapter 6, Evaluation and Discussion of the Evidence).

4.2.3 Estero Sur, Gala Sound

The Estero Sur site lies at the head of an inlet 4 km long running generally from north-west to south-east. It is on the mainland coast of the Chonos Archipelago, separated from the islands by the Moraleda Channel.

The site is in one of the many basalt outcrops typical of the area, which forms a deep shelter. It is 30 m above the current sea level and 40 m from the shoreline, completely hidden in the forest. The shoreline of the surrounding area is mixed, with vertical rock outcrops down to an intertidal with sandy sediments and small and medium pebbles; the topography therefore changes with the state of the tide. The area is covered with thick evergreen forest vegetation and peat bogs, which come down to sea level and hide the rock shelter, making access difficult. The archaeological visibility is nil. The site was discovered by chance, by fishermen who were cutting firewood and reported the site to the police. When the judicial investigation began, we were advised on the strength of our investigations, and we were authorised to carry out an archaeological study.

The Estero Sur site is a rock shelter used as an ossuary where a large number of human bone remains were found on the surface. The shelter lies below a high rock wall, approximately 50 m high; the upper part is visible above the forest. It faces north-east but receives little sunlight, due to the protecting forest cover and because the trees growing on the bank, and the roots and creepers which hang down from above, create a natural screen which covers almost the whole shelter. There is a pronounced slope of approximately 30° from the south-west to the north-east end of the shelter, with a fall of at least 2 m.

The shelter is around 40 m long from south-west to north-east, with a depth varying between 2 and 10 m. The total area is approximately 30 m^2. The height also fluctuates between 5 and 10 m; it is fairly irregular and presents large blocks of fallen rock which define the topography for access to deposit the bodies. Small spaces and openings allowed entry to deposit or redeposit bodies. There is also a sort of protected passage where various assemblages of bones were found, defined by the longitudinal wall of the shelter and some huge fallen blocks approximately 7 m long × 3 m wide. Immediately below the drip-line, the bank falls sharply (60°) to the rocky shore. Several watercourses were observed close to the site.

The site contained a series of human remains, either grouped together or scattered. In some places, there was no superficial evidence, such as the access point at the sloping north-east end where scattered vertebrae and rotulas were found under the plant material (1 cm), some actually on the bank. It is clear that the bones have been moved repeatedly by locals who have visited the site, gathering the crania together and decontextualising them (e.g. assemblage 1 consisting of just three crania). There are also three crosses that have been placed in the shelter by fishermen. The local who reported the site declared that he had counted 19 crania; however, we recorded only 13 in situ (although this number was increased to 16 by reconstruction of fragments in the laboratory).

In view of the topography of the site and the superficial material observed, we decided to collect all the superficial human remains present. For the purposes of making a record and a drawing of the shelter, and surveying the bones distributed there, we decided to define assemblages of bones as we found them, separated by variable distances, whether or not they had been redeposited in the sub-present (Fig. 4.11). To do this, we drew a longitudinal axis parallel to the lie of the shelter to define the locations of the bone assemblages. Nine assemblages were defined, varying both in the number of bone remains and in the presence/absence of cranial and postcranial skeletal remains. The remains found in the passage were then collected, as they were highly exposed to anthropic agents.

Fig. 4.11 View of the interior of the Estero Sur site showing the narrow passage in which the bodies were deposited

It should be noted that we did not collect 100% of the bones from the rock shelter. Between the assemblages defined as 6 and 9, there is a long crack formed by the fallen block which helps to form the passage; this contains scattered bones, some partly buried. In assemblage 9, we identified a partly buried skeleton (right lateral decubitus), some parts of which were obtained for recording. Bones were therefore left in place precisely in order to be able to carry out future excavations to complement the information obtained from this general survey. The bone remains were positioned, drawn, photographed and then collected in the assemblages defined.

The majority of the human bone remains on the surface were surveyed by 'assemblages' which could later be rearranged in the laboratory. Samples were taken from the individuals identified as 1, 2, 7, 9 and 11. The distribution and completeness of the remains were decisive for choosing the individuals for dating, so as to obtain information associated with sex, age and any pathologies noted. In the field we thought that, whether or not they had been moved in the sub-present, the individuals present in the passage were deposited later than the individuals found in the various cracks between the rocks. The latter, or perhaps some of them, may have been redeposited in order to clear space in the passage to deposit new remains. Our choice of dating samples also took into account whether the individuals identified were found in cracks or in the passage.

Analysis of the bone remains found at the Estero Sur site produced a minimum number of 16 individuals: 7 subadults and 9 adults (Galimany 2017). The minimum number of subadults was based on the size and degree of development of the bones found and the repetition of elements in the same range. The number of adults was based on the number and characteristics of the crania, hip bones and femurs described. The most represented bones were portions of crania and remains of lower limbs (tibias and femurs). An MNI of 16 individuals was identified in the laboratory. Among the subadults, we found the remains of a perinate (less than 6 months), two infants aged between 3 and 6 years, two infants aged between 5 and 8 years, a juvenile, probably female, aged between 14 and 21 years and a male juvenile aged between 17 and 22 years. Among the adult individuals, we found the remains of three young male adults (between 20 and 35 years), one young female adult (between 20 and 35 years), three middle-aged male adults (between 30 and 50 years) and two middle-aged female adults (between 30 and 50 years).

Samples obtained from cranial fragments assignable to individuals were taken from 5 of the 16 individuals recorded: 1(ES1), 2 (ES2), 7 (ES7), 9 (ES9) and 11 (ES11). The results of the dates obtained from all the individuals deposited on the surface of the shelter gave a mean range between ~1700 and 1400 years cal. BP (Reyes et al. 2019b, Table 4.1).

The Estero Sur site in Gala Sound is a shelter used exclusively for funerary purposes known as an ossuary. It is a shelter in the forest 50 m above the shoreline, protected from the rain and located at the head of an inlet. The site has been altered by later anthropic activities. No artefactual or ecofactual material was recorded to indicate whether the site was occupied apart from funerary rituals. There is no evidence of rock art on the walls of the shelter, and no remains of funerary offerings or trappings.

The dates obtained in all of the individuals who were deposited on the surface of the shelter give a mean range between ~1700 and 1400 years cal. BP; the extreme calibrated ranges of all the dates overlap, so this ossuary may have been formed by a synchronic event or several events over a short lapse of time which cannot be differentiated chronologically, and clearly show reiterated use of the same site to deposit the dead. The isotope values obtained in the dated individuals allowed us to establish the consumption of an exclusively marine diet (see Chapter 6, Evaluation and Discussion of the Evidence) (Reyes et al. 2019a).

4.3 Central Area of the Chonos Archipelago (44°23′–44°46′S)

Below, we describe the archaeological sites recorded and studied in the central area of the Chonos Archipelago (Fig. 4.12).

4.3.1 Isla Benjamín 1

The Isla Benjamín 1 site is located in the central part of Benjamín Island, on the south coast of an inlet 20 km long and 2 km wide running east–west, which almost cuts the island into two. The site lies on a small, hidden bay formed by a river mouth and well protected from the elements. Access is by a narrow channel 50 m long and 8 m wide. The intertidal in this very sheltered area consists of sandy silt, flanked by a big wall of basaltic rock forming a very protected cave. It is almost 3 m above the current sea level and 10 m from the shoreline. It is covered by thick forest, shrubby vegetation and an abundant stratum of plant material, formed by centuries of forest formation, which hinders movement and access. The area around the site is covered with thick evergreen forest vegetation and peat bogs, which come down to sea level and hide the cave. The archaeological visibility from the shore is nil. The site was reported by the investigator Eugenio Aspillaga, who carried out a field trip (Ocampo and Aspillaga 1984) when unreported excavations were made and human remains from a minimum of 12 individuals were recovered from the surface. We were allowed access to the osteological collection of the University of Chile for our research to take samples of the individuals collected in that work for the purposes of radiocarbon dating (Reyes et al. 2013), from which it was established that the bodies were deposited in this ossuary approximately between ~300 and 500 years cal. BP (Table 4.1).

The cave consists of a space hollowed out under the rock wall, facing north. The site is 19 m long from east to west, and 8–10 m deep from north to south, total area 190 m^2. The height varies between 8 and 11 m. Immediately below the drip-line, the bank falls abruptly (70°) to the adjacent shoreline.

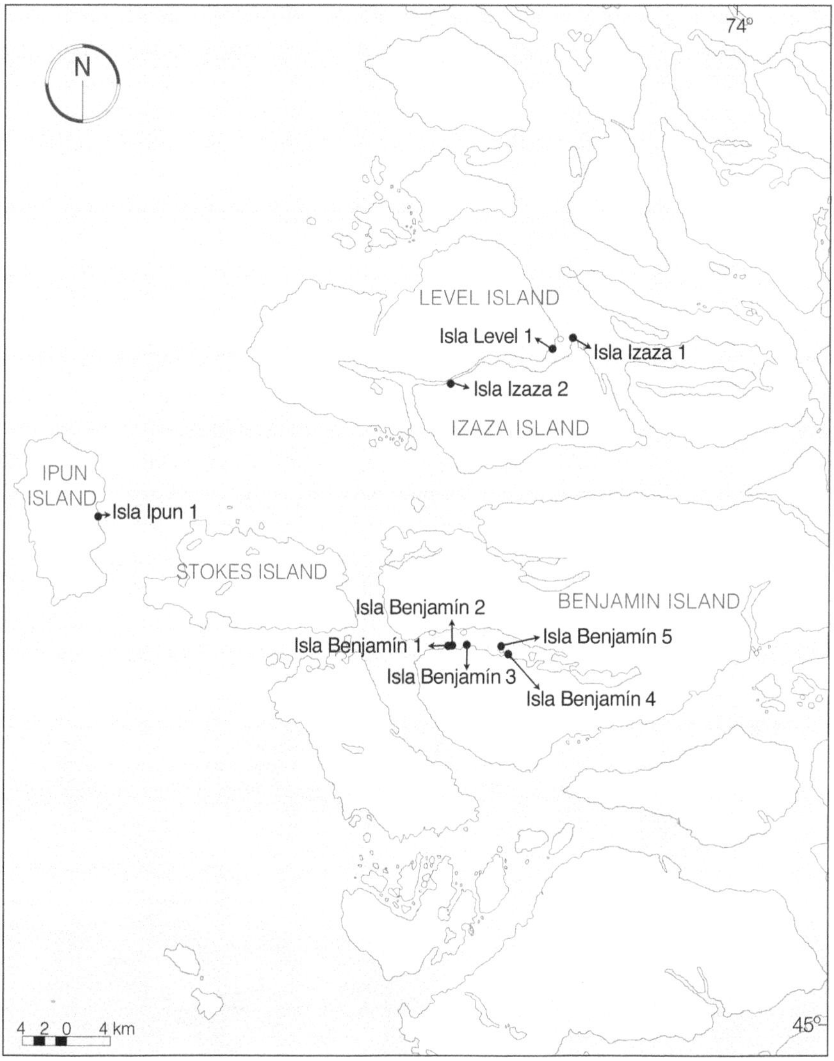

Fig. 4.12 Detail of the central portion of the Chonos Archipelago and the distribution of the sites recorded and studied in the area

The interior of the cave is very dry, with very friable, pale brown sand-silt sediments and low-density gravel inclusions. On the surface, there are fragmented and weathered human bone remains, both cranial and postcranial. Earth movements are observed in the centre, probably from the excavations carried out, as well as scattered heaps which are probably where the sediments were sieved. On the south wall of the cave, there is writing in black paint (*'national cultural heritage'*, *'No'*, *'CH 011'*, etc.), which were left when the archaeological work was carried out. We also observed fragmented and entire remains of malacological fauna (*Mytilus*, locos, limpets, clams, *Fissurella*, sea urchins, picorocos); some of these remains were piled up as a result of the sieving. To summarise, the site is a very well-protected cave, containing a shell-midden matrix with weathered human bone remains on the surface belonging to a minimum (at the time of our investigation) of 5 individuals (diagnosis in situ).

In view of the state of the site and the superficial material observed, we decided on an excavation of 1 × 1 m close to the bank, adjacent to the drip line, the spot which we estimated would be best preserved after the excavations and earth-moving described above. The site was also surveyed and a plan was made of its position and relation to existing sea level (Fig. 4.13).

The stratigraphy revealed by the excavation was as follows: the shell-midden matrix lies immediately at the surface; however, from our observations of the surface, we think that part of the first layers of the matrix derive from sieving by the earlier, unreported archaeological work. We therefore excavated in controlled levels of 10 cm, finding that the shell-midden had been disturbed down to 40 cm. We observed very broken remains of malacological fauna, loose and friable, with little sediment. After 40 cm the matrix becomes more compact, with fragmented and incomplete malacological remains included in fine, grey sandy silt sediments—not in separate pockets. The malacological remains recorded consist mainly of picorocos (*Austromegabalanus psittacus*), cholgas (*Aulacomya ater*), choro mussels

Fig. 4.13 Test pit and search in Isla Benjamín 1 site

(*Choromytilus chorus),* Chilean mussels (*Mytilus edulis chilensis*), oysters (*Ostrea chilensis*), clams (*Protothaca thaca, Venus antiqua*), locos (*Concholepas concholepas*), keyhole limpets (*Fissurella nigra*), snails (*Tegula atra*) and chitons (*Chiton sp.*). There were also records of sea lions, birds, marine otters, etc.

The unaltered portion of the shell-midden from where it starts down to where it finishes at 96 cm is 60 cm thick on average at the bank. The base of the shell-midden matrix rests on greenish-grey sandy silt sediments, with inclusions of pebbles and small rock flakes which form part of the base of the cave.

No human bone remains were recorded. According to our information on the previous, undocumented archaeological excavations, we considered that taking more samples from a funerary context that has been dismantled would not produce any new information, since there is no detailed analysis of the 12 individuals who were collected or of the criteria under which they were selected. A series of samples of shells, carbon specks and faunal remains were recovered for dating. A fauna column 10 by 10 cm was excavated in the south-west profile to characterise the composition of the faunal component of the site. No lithic material was recovered. Small fragments of schist and some small rounded pebbles were recorded; however, all can be explained as having fallen from the support which makes up the shelter. There was no record of any other kind of raw material.

Analysis of the fauna column (San Román et al. 2016b) produced a NISP representation of sea lion (n = 2), marine otter (n = 3), coypu (n = 20) and huemul (n = 3) among the mammal bone remains that could be identified at species level. For birds, we found a NISP representation of cormorants (n = 2), petrels (n = 1) and steamer duck (n = 4) among the bird bone remains that could be identified at species level. Remains of 30 fish were found that could be identified taxonomically (Torres 2016). The largest NISP was found for snoek (50%), followed by blenny (30%), jurel (16.7%) and finally Patagonian redfish or ‘chancharro’ (3.3%). The remains were found between 40 and 100 cm. The number of species represented in the levels remained constant, although the majority of remains were found at depths between 40 and 50 cm.

A number of samples were taken from different stratigraphic levels. A pair of samples was taken from human ribs from the remains collected at the surface of the cave identified as BEN 01 and BEN 02 in the 1990s, and reported with their datings in an earlier work (Reyes et al. 2013). Subsequently, based on the stratigraphy observed in the excavation unit, a sample of huemul bone was obtained from the 60–70 cm level (BEN 1–1) and one of coypu in the base level of the shell-midden, 90–100 cm (BEN 1–2); their ages were around ~1400 and 1700 years cal. BP, respectively. It was decided to date a sample of marine otter recovered on the surface of the cave to compare it for Reservoir Effect and to obtain ^{13} N and ^{13}C levels to estimate diet, finding that it was modern (Reyes et al. 2019a). All the datings obtained showed stratigraphic consistency in the successive occupations since the start of the deposit in the cave at least 1700 years cal. BP (Reyes et al. 2019b, Table 4.1).

The excavated area in the site (1 m^2) and the human bones obtained on the surface suggest that at the beginning, in the first occupations recorded around ~1700 years cal. BP, the cave was used as a shelter where we find discarded waste produced mainly by the gathering and consumption of malacological fauna, giving rise to the shell-midden matrix over successive occupations. Based on the faunal record, other activities can be inferred such as fishing, hunting for birds and smaller species like coypu and marine otter, and hunting for larger marine fauna like sea lions. The presence of huemul is interesting, since the habitat of this species of deer is restricted to the mainland, inhabiting the forests of the Andes down to the mainland coast of the archipelago. The nearest point on this coast is 100 km in a straight line from the Isla Benjamín 1 site. We interpret its presence as evidence of the mobility and exchange circuit of the human groups settled in the area. There is no evidence of lithic material and no information on the previous archaeological work which would enable us to specify more precisely other activities carried out at this site.

Subsequently, at the end of the occupation sequence, the cave was used for funerary purposes; a series of individuals (MNI = 12) were deposited there around ~300 to 500 years cal. BP towards the end of the sequence, a period of almost one thousand years of discontinuous use of this space. The extreme calibrated ranges of the dates overlap, so this ossuary may have been formed by a synchronic event or several events over a short lapse of time which cannot be differentiated chronologically, and clearly show reiterated use of the same site to deposit the dead. The isotope values obtained in the dated individuals allowed us to establish the consumption of an exclusively marine diet (see Chapter 6, Evaluation and Discussion of the Evidence) (Reyes et al. 2019a).

4.3.2 Isla Benjamín 2

The Isla Benjamín 2 site is located in the central part of Benjamín Island, on the south coast of an inlet 20 km long and 2 km wide running east–west, which almost cuts the island into two. It is 200 m north-east of Isla Benjamín 1, on the other side of a freshwater stream. The site stands on a marine terrace 2 m high, adjacent to a sand beach and covered by thick evergreen and peat bog forest, shrubby vegetation typical of the area and the layer of plant material which has formed over time. The site has a slight slope of 10° and the archaeological visibility is nil. It was detected by a network of cores drilled in the terrace to discover buried anthropic deposits.

The site consists of three separate areas of shell-midden distributed on the marine terrace, in the forest and away from the shoreline. The total dimensions of these shell-middens are 30 m from east to west and 25 m from north to south. The current area of the site is 750 m^2, with a cultural deposit maximum 0.8 m deep.

We dug a network of test pits to determine the concentrations of shell-midden present on the marine terrace. We later excavated a pit 1 m by 1 m to obtain samples and the stratigraphic record. The site was also surveyed and a plan was made of its position and relation to existing sea level.

The first shell-midden identified in the test cores is 1.5 m above the current sea level and 10 m from the shoreline. It appeared under a thick layer (30 cm) of humus, covered with roots. The shell-midden was recorded between 30 cm and 1.2 m, when we reached the sandy matrix which characterises the marine terrace.

The second shell-midden identified in the test cores is 2 m above the current sea level and 25 m from the shoreline. It appeared under a thick layer (30 cm) of humus, covered with roots. The shell-midden was recorded between 30 and 65 cm, where we reached the sandy matrix which characterises the marine terrace. We decided to open an excavation of 1 m by 1 m in this shell-midden.

Excavation of this shell-midden produced the following stratigraphy: first, the thick layer of humus of 20 to 30 cm depth. Once this layer of humus had been cleared, the top of the shell-midden matrix appeared. From this point we took a new level 0. The 0–10 level is the start of the shell-midden, included in a matrix of coarse to medium sand, with rounded pebbles of $\leq$ 1 cm diameter. Clam valves and mytilid periostracum were recorded. No artefacts or bone remains were found. After 20 cm, the density of shells and periostracum increases. Among the molluscs and crustaceans identified were *Megabalanus sp.*, Mytilidae, Veneridae and *Concholepas concholepas*, as well as birds' bones. At 30 cm we recorded a discrete deposit of carbon sediments; a sample was taken at 32 cm (from our level 0, and 50 cm from the surface). Finally at 40 cm, medium sand and whitish clam shells appeared, giving the impression of the remains of a palaeo-beach. Then between 40 and 50 cm, the remains suggesting a natural origin (palaeo-beach) proved to have been deposited immediately on a pasty, very loose shell-midden, with dark sediments. Below this lens of shell-midden, towards the bottom of the level, we recorded a layer of sand and rounded pebbles of 1 cm diameter. Level 50–60 cm consists of archaeologically sterile sands. The sterile sands continue from 60 to 68 cm, and then a layer of shell-midden appears, very loosely distributed in the sand matrix. Samples were taken of valves and carbon specks recovered by sieving. This lens of shell-midden is clearly of anthropic origin, given the presence of carbons and the dark colour of the associated sediments (Fig. 4.14). A few very splintered bones and remains of fish bones were recorded. Finally, at 75 cm, the lens of the shell-midden disappears leaving a stratum of sterile, coarse to medium sand, with large clasts. A shovel test was done below the level, which confirmed that the sterile sands continued to a depth of approximately 1 m.

The third shell-midden identified is 2 m above the current sea level and 15 m from the shoreline. It is a shell-midden deposit exposed by the fall of a tree (*Nothofagus sp.*). The area exposed by the uplifted roots is a semicircle of 2 m radius. The sediments were cleared to evaluate the presence of artefacts and/or bone remains. The sediment adhering to the roots is dark in colour and similar in appearance to that recorded in the excavation unit.

No archaeological material was recovered either on the surface or in the adjacent intertidal. Samples of shells and carbon were taken at different levels of the excavation, especially the occupation floors adjacent to the intrusions of sand detected in the matrix.

The shell and carbon samples (IBEN 2–1) obtained from the same depth (50 cm) show stratigraphic and chronological consistency for the first occupation settled on the first palaeo-beach encountered; they produce similar datings of around

Fig. 4.14 View of the pit dug in the second shell-midden of the Isla Benjamín 2 site. Red arrows mark the cultural levels dated, separated by an intrusive deposit of 10 cm of fine sands indicating an event of abandonment and reoccupation on the newly formed beach

~2300 years cal. BP (considering the reservoir effect of the shell sample). The carbon sample obtained at 88 cm, IBEN 2–2, produced a date of ~2400 years cal. BP for the base occupation of the site on the second (lower) palaeo-beach. The hiatus of sandy sediments which form a new palaeo-beach arose from an energy event which led to the formation of a new floor before the second occupation (Reyes et al. 2018, Table 4.1).

The Isla Benjamín 2 site consists of a series of camps formed by successive reoccupations distributed and arranged on old palaeo-beaches; the clearest evidence is the discard of waste from the gathering and consumption of malacological fauna. This consumption reflects local gathering of shellfish from different types of shoreline (rocky, sandy, mixed) and sea bed. Additionally, the scarce remains of birds show hunting activities. The start of these occupations goes back to around ~2400 years cal. BP. There was then a hiatus, with deposition of sandy sediments due to a high-energy event to form a new palaeo-beach, before the area was reoccupied (Reyes et al. 2018).

4.3.3 Isla Benjamín 3

The site is located in the central part of Benjamín Island, on the south coast of an inlet 20 km long and 2 km wide running east–west, which almost cuts the island into two. It is situated on the west bank of the mouth of a freshwater stream, in a small bay facing north-east. The site is on a fine sand beach, probably formed by silt from the stream. It is covered by thick evergreen and peat bog forest, the shrubby vegetation typical of the area and an abundant layer of plant matter. The site has a slight slope of 20° and the archaeological visibility is nil. It was detected by a network of cores drilled to discover buried anthropic deposits. It lies almost 2 km east of Isla Benjamín 1 and 2.

The site is an archaeological shell-midden covered by a layer of plant material 30 cm thick, 8 m from east to west and 4 m from north to south. The current area of the site is 32 m^2, with a cultural deposit at least 1 m deep. The base of the deposit is 2 m above current sea level, and 8 m from the present shoreline. The test cores exposed a layer of dark (black) humus 34 cm thick, and then a layer of shell-midden more than 115 cm deep (maximum length of the cores available at the time). No cultural materials were recovered from this site. A sample of shells was taken at a depth of 98 cm for radiocarbon dating. The sample was dated to 2610 ± 25 years BP (2090–2343, 2-sigma cal. BP), the minimum age of this deposit (Reyes et al. 2019b, Table 4.1).

The Isla Benjamín 3 site is a camp formed by discards from gathering and consumption of malacological fauna; successive reoccupations of the area are recorded from ~2300 years cal. BP. According to the record and the archaeological activities carried out, no other activities are inferred for the site.

4.3.4 Isla Benjamín 4

The Isla Benjamín 4 site is located in the central part of Benjamín Island, on the south coast of an inlet 20 km long and 2 km wide running east–west, which almost cuts the island into two. The site is in a small shelter formed in a rock wall at medium height.

The shelter is 8 m above the current sea level and 3 m from the shoreline, in a narrow, north-east facing bay. The area around the site is covered with thick evergreen forest vegetation and peat bogs, which come down to sea level and hide the rock shelter, making access difficult. The archaeological visibility from the shore is nil. It was discovered when we were checking the walls and folds of rock in search of possible refuge sites. The coast is rocky and near vertical, making normal movement impossible (Fig. 4.15).

The shelter lies below a high rock wall, 30 m high; the upper part is visible above the forest. It faces north-west. The site is 15 m long from north-east to south-west, and almost 3 m deep from north-west to south-east, maximum total area 45 m^2. The height varies between 10 and 30 m. Immediately below the drip-line, the bank falls sharply (60°) to the rocky shore. The shelter is long, and offers minimal protection from rain due to its height; there are some protected spots of up to 1 m^2 in some parts. The forest covering the shelter affords some protection. The shelter floor slopes at about 40° and there is a waterfall at the north-east end.

The surface of the shelter was completely overgrown; however, scattered postcranial human bones from one female adult individual and one subadult were found in the central sector and on the bank. Traces of sub-present historical occupation were

Fig. 4.15 View of the shelter in the forest 8 masl, on a steep slope. The superficial vegetation was cleared away to recover the human bone remains from the surface and dig a test pit 1 × 1 m

also observed, attributable to fishermen, for example: a piece of half-burnt wood; very fragmented, charred fauna bones; and areas with ash sediments. These may be related to the disturbance of the human remains on the surface.

In view of the topographical nature of the site and the superficial material observed, we decided to excavate a pit 1 × 1 m near the centre of the shelter, away from the emergent rocks and possible sub-present disturbances (activity of fishermen or cholga gatherers), in the spot that we considered best preserved. The plant layer—mainly creepers—was cleared from the whole shelter, exposing the human bones and leaving them in situ. The same was done on the bank to a distance of three metres, finding more human remains that had fallen or been thrown from the shelter. After cleaning and recording the surface, the site of the excavation unit was selected.

The excavation revealed the following stratigraphy in the site: in artificial level 0–10 cm, we observed pale brown, friable silty sediments with inclusion of organic material like leaves and roots. In the northern corner, at 3 cm, we observed a lens of a sub-present historical hearth. Complete and fragmented human bone remains belonging to an adult and a subadult were found, some charred. Very little malacological fauna was recorded, some fragments of *Mytilus* and clam (*Protothaca thaca*) shells. We also recovered a small metal cylinder (similar to a bullet casing), and remains of small fauna such as coypu and birds. Between 10 and 20 cm, more malacological fauna, in a less fragmented state, began to appear, e.g. choro mussels. Silty sediments still predominated over shell fragments. Further fragments of subadult and adult human bones continued to appear. Between 20 and 30 cm, the matrix turned to very dark sandy silt, with an increased number of shells of picorocos (*Austromegabalanus psittacus*), cholgas (*Aulacomya ater*), choro mussels (*Choromytilus chorus*) Chilean mussels (*Mytilus edulis chilensis*) and snails (*Tegula atra*). There were still roots at this depth. At the 30–40 cm level, the sediments, still dark, became finer and more friable, with inclusion of gravel. The number of shell fragments decreased and at 37 cm the matrix turned to a yellowish brown sandy silt coinciding with the first appearance of the base rock. At 50 cm, the base rock occupied the whole unit; however, dark sediment containing more fragments of human bone was still recovered from the interstices of the rock between 40 and 50 cm. The excavation was closed at this point.

The human bone remains found on the surface and in the excavation were recovered. Samples were also taken of shells and carbon specks from the levels excavated, and of vertebrate fauna found in the test pit. No lithic material was recovered. Small fragments of schist and some small rounded pebbles were recorded; however, all can be explained as having fallen from the support which makes up the shelter.

The high degree of fragmentation of the bone material recovered, and its dispersion on the surface and vertically (present in all the levels of excavation, except 30–40 cm), presented extra difficulties in defining and describing the bio-anthropological component recovered from the shelter. The state of the remains was very variable, from 100% completeness in bones found on the surface to high fragmentation and damage by exposure to fire of the remains excavated. The object of our analysis of the bone remains was to differentiate individuals, giving priority to assigning bone elements to hypothetical individuals rather than analysing them as ossuary elements

(Galimany 2016). The bone remains recovered at the surface and in the excavation belong to a minimum of 8 individuals: one perinate (between third foetal term and 6 months of age); three infants (1 to 2 years, 3 to 6 years and 6 to 10 years of age) and four juvenile and adult individuals (one female between 18 and 30 years, one female between14 and 21 years, one male probably over 14 years and finally one adult of indeterminate age and sex).

The human bone remains were not only highly fragmented, but also presented soot and evidence of charring. We think that this alteration occurred in the historical/sub-present occupation when the funerary area was severely disturbed. The archaeofaunal record shows the presence of birds (cormorant and steamer duck), coypu and other rodents, as well as one specimen of blenny and one of jurel (San Román et al. 2016b).

The samples taken in the field from an adult human rib at the surface (without later reassignation in the laboratory) produced a date of 820 ± 25 years BP (~323–516, 2-sigma cal. BP, Reyes et al. 2019b, Table 4.1).

According to the surface record and the excavated area of the shelter (1m^2), and the human bones recovered both on the surface and from the excavation, we infer that the shelter was used principally as a funerary site or ossuary. The deposition of various individuals (n = 8) of both sexes and all age ranges point to recurrent use over a period around ~400 years cal. BP.

Later use of this restricted shelter for other domestic activities (occasional camp), when the remains of malacological fauna, birds and coypu were discarded, appears to have been a historical rather than a sub-present occupation, when fishermen and/or *cholgueros* used the site as a refuge. This later occupation could explain the high dispersion, fragmentation and burning of human bone remains both at the surface and in the stratigraphy. Given the number of individuals identified in the laboratory versus the quantity of bones recovered and their great fragmentation and disturbance (charring), it may be reasonably suspected that, on the one hand, the site was 'cleaned' for the purposes of the latest occupation, and on the other that the bodies found on the surface, once reduced to skeletons, gradually migrated and fell down the steep bank below the shelter, as found in our record of the site. This is particularly likely to occur to crania, which would roll easily once they reached the bank. Finally, in view of the disturbance of the context, we cannot determine whether all the individuals were deposited on the surface or whether some were buried, especially the infants and subadults. The isotope values obtained in the dated individual allowed us to establish the consumption of an exclusively marine diet (see Chapter 6, Evaluation and Discussion of the Evidence) (Reyes et al. 2019a).

4.3.5 Isla Benjamín 5

The Isla Benjamín 5 site is located in the central part of Benjamín Island, on the south coast of an inlet 20 km long and 2 km wide running east–west, which almost cuts the island into two. The site is in a small shelter formed in a rock wall at medium height, and lies 1.2 km north-west of the Isla Benjamín 4 site.

The shelter is 7 m above current sea level and 2 m from the shoreline, on a narrow channel 600 m wide between Benjamín Island and a small island in the inlet. The area around the site is covered with thick evergreen forest vegetation and peat bogs, which come down to sea level and hide the rock shelter, making access difficult. The archaeological visibility from the shore is nil. It was discovered when we were checking the walls and folds of rock in search of possible refuge sites. The coast is rocky and near vertical, making normal movement impossible.

The shelter lies below a high rock wall, approximately 20 m high; the upper part is visible above the forest. It faces north-west. The site is 6 m long from north-east to south-west, and almost 1.5 m deep from north-west to south-east, maximum total area 9 m^2. The height varies between 10 and 20 m. Immediately below the drip-line, the bank falls abruptly (70°) to the rocky shore. The shelter is long, and affords minimal protection from rain due to its height; there are some protected spots of up to 1 m^2 in some parts. The forest covering the shelter provides some protection. The slope of the site is around 10°, while the access is up a 50° slope. No watercourse was observed in the vicinity. The surface of the shelter was completely overgrown; however, cranial and postcranial human bones could be observed in an area 2.5 m by 1 m.

In view of the topography of the site and the superficial material observed, we decided to collect all the human remains present on the surface. Unlike the Isla Benjamín 4 site, in this shelter there was no evidence of later occupations or sub-present interventions. The shelter is too narrow for habitation and offers little protection from the weather.

First, the whole shelter was cleared of the plant cover, mainly creepers, by which it was covered. The same was done to a distance of 4 m down the bank (where possible); more human remains were found that had fallen from the shelter. Cranial and postcranial human remains of at least 3 individuals (diagnosed in situ) were found among the vegetation, enveloped in roots, either subsuperficially or partly buried by the accumulation of fine silts and organic matter. The crania had rolled down the bank from the shelter and become caught in the vegetation. We found that the bone remains had been moved and dispersed by plant growth, an important factor for evaluating non-anthropic disturbance of human remains at the surface.

The majority of the human bone remains on the surface were surveyed by assemblages that could later be rearranged in the laboratory. We also recovered two pieces of hand-worked bark (Fig. 4.16). To date, no formal cultural element has been recovered or reported archaeologically associated with this type of funerary context. No lithic material was recovered.

The MNI was determined as 4 individuals, with one infant (aged between 8 months and 2 years) and three adults (one adult male aged between 25 and 40, one adult female older than 20 years and one adult of indeterminate age and sex). In the laboratory, it was decided to reassociate the elements, assigning them to hypothetical male and female individuals based on a significant difference in the size of the remains and the possibility of articulating some of the elements. A third adult individual was identified by the presence of a mandible which does not match either of the crania

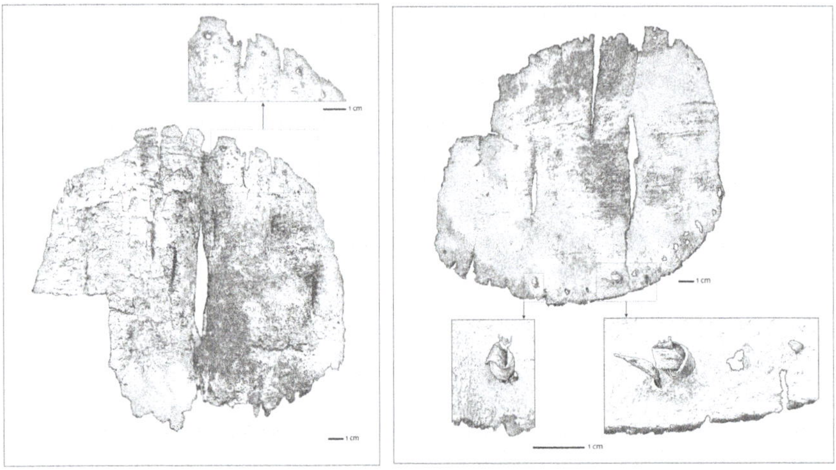

Fig. 4.16 Drawing of the two pieces of bark with evidence of cultural work recovered in the Isla Benjamín 5 site. (left) Bark with perforations round the edge. The straight edge shows where a fragment was cut off for direct dating. (right) Fragment of bark with perforations round the edge and preserved tendons tied in knots

Fig. 4.17 View of Cholguero Isla Izaza 2. Part of the shells are seen to be deposited in the intertidal zone and part of the area is covered by vegetation. The test pit dug in the intertidal can also be seen, revealing that the base of the shell-midden projects below sea level

present and a rib that could not be associated with either of the two rib assemblages reassociated (Galimany 2016).

It was not possible to carry out anthracological analysis of the pieces of bark recovered because no wood adhered to them, which is necessary to identify the species. Both pieces have been cut and rounded to a similar diameter (approximately 21 cm). Their edges are eroded, however, there are perforations or 'eyelets', and in two of them pieces of animal tendon tied in a knot. Their function is unknown; they might be part of an offering, funeral trappings or some kind of instrument. To date, this find is the only cultural element recovered from an ossuary that has clearly been placed in association with the human remains deposited.

The samples of bark (IBEN5-1) and human rib (IBEN5-2) were taken from the surface in the field (and not reassigned in the laboratory). Subsequently another set of dates was obtained for the bone remains reassigned to 4 individuals, all of which were dated: IBEN5-3 (maxillary right third molar of an adult male, aged between 25 and 40 years), IBEN5-4 (right rib of infant, aged between 8 months and 2 years), IBEN 5–5 (mandibular left central incisor of an adult of indeterminate age and sex) and IBEN 5–6 (maxillary left second premolar of an adult female aged over 20 years, the same individual as in the rib sample noted above IBEN5-2).

The Isla Benjamín 5 site is a shelter used exclusively for funerary purposes. Despite the scant shelter offered for human habitation (e.g. comfortable movement, effective protection), the site served for funerary purposes as a narrow shelter away from the water, partly protected from the weather by the rock overhang and the forest cover. The site had not been altered by later anthropic activities. No other types of activity were recorded or could be inferred from the cultural and bio-anthropological materials recovered.

The results of the dates obtained from all the individuals deposited on the surface of the shelter gave a range between ~570 and 230 years cal. BP (Reyes et al. 2019b, Table 4.1). The extreme calibrated ranges of the dates overlap, so this ossuary may have been formed by a synchronic event or several events over a short lapse of time which cannot be differentiated chronologically, and clearly show reiterated use of the same site to deposit the dead.

The bark deposited in this shelter, with mean dating of almost ~400 years cal. BP, may be part of an offering (or funerary trappings if it is part of clothing or a body adornment) synchronic with the event in which the bodies were deposited. The oldest dating of all the skeletons may be altered by a reservoir effect produced by the consumption of an exclusively marine diet, the consequences of which for the chronology we have documented elsewhere (Reyes et al. 2019a). Future finds of funerary packets with direct material and/or ecofactual association could help us to identify these items more precisely.

The isotope values obtained in the dated individuals, i.e. the whole bio-anthropological assemblage, allowed us to establish the consumption of an exclusively marine diet (see Chapter 6, Evaluation and Discussion of the Evidence).

4.3.6 Cholguero Isla Level 1

This site is located on the south-east coast of Level Island, on the north shore of the channel which separates it from Izaza Island. The area is forested with evergreen forest and peat bogs, and the shoreline is formed by a pebble beach. It is backed by a hill immediately behind the site, enclosing the cholguero between the coast and the hill. The site is 2 m above current sea level and 0 m from the shoreline, exposed to the tides and the formation of sandbanks by storms. It stands on the shoreline facing south-east. The site was found due to the exposure of its south profile, giving it medium archaeological visibility. There was no archaeological material either on the surface or in the adjacent intertidal zone. The site presents a slope of 40°.

The site is a cholguero covered by a layer of plant material 10 cm thick, beneath which is a shell deposit, principally of mytilids, 1 m thick. Below this shell-midden appears the palaeo-beach. According to estimates based on drilling cores, the area is 20 m from east to west and 5 m from north to south, covering approximately 100 m^2. The shell deposit is 1 m deep. No occupation or archaeological material was found below the cholguero, nor on the surface or in the nearby intertidal.

The cores indicated an uninterrupted deposit of shell-midden matrix from 100 cm consisting mainly of cholgas (*Aulacomya ater*), choro mussels (*Choromytilus chorus*) and Chilean mussels (*Mytilus edulis chilensis*), although clams (*Protothaca thaca, Venus antiqua*), locos (*Concholepas concholepas*), keyhole limpets (*Fisurella nigra*) and sea urchins (*Loxechinus albus*) were also observed. Carbon samples from the base of the cholguero at 100 cm were recovered from the core for radiocarbon dating. The date assigned to the sample was 120 ± 25 BP (1–253, 2-sigma cal. BP, Reyes et al. 2019b, Table 4.2).

Cholguero Isla Level 1 is a campsite developed by reiterated seasonal visits, related to intensive, selective gathering of mytilids for the canning industry. This activity probably implied other associated practices, like hunting and fishing for the fishermen's and mussel gatherers' subsistence. The start of occupation may be dated to the beginning or middle of the twentieth century. The radiocarbon dating obtained from a speck of carbon from the base of the deposit confirms our estimated chronology.

4.3.7 Cholguero Isla Izaza 1

The site is situated on the north-east coast of Izaza Island, on the southern shore of the channel which separates it from Level Island, in a small, rocky, north-east facing bay. The area is forested with evergreen forest and peat bogs, and the shoreline is formed by a pebble and rock beach. The site is adjacent to the shore and a part of it lies 60 cm below current sea level. It is 0 m from the current shoreline and is exposed to tidal action and to the formation of banks caused by storms. The site was found due to the exposure of its south profile, giving it medium archaeological visibility. There was no archaeological material either on the surface or in the adjacent intertidal zone. The site presents a slope of 10°.

It is a cholguero port covered by a layer of plant material 10 cm thick, beneath which is a shell deposit, principally of mytilids, 38 cm thick. Below this shell-midden appears the palaeo-beach. According to estimates based on drilling cores, the area is 20 m from east to west and 10 m from north to south, covering approximately 200 m^2. The shell deposit is discontinuous and no more than 40 cm deep.

No occupation or archaeological material was found below the cholguero, nor on the surface or in the nearby intertidal. The test cores indicated an uninterrupted deposit of shell-midden matrix to 100 cm, consisting mainly of cholgas (*Aulacomya ater*), choro mussels (*Choromytilus chorus*) and Chilean mussels (*Mytilus edulis chilensis*). Carbon and shell samples were recovered from the base of the cholguero for radiocarbon dating. The samples obtained at the base of the deposit, at 38 cm (beneath the surface level which is 60 cm below current sea level), were dated to 490 $\pm$ 25 years BP (out of range, 2-sigma) and 110 $\pm$ 25 years BP (out of range, 2-sigma cal. BP) (Reyes et al. 2018, Table 4.2).

Cholguero Isla Izaza 1 is a campsite developed by seasonal visits, possibly reiterated, related to intensive, selective gathering of mytilids for the canning industry. This activity probably implied other associated practices, like hunting and fishing for the fishermen's and mussel gatherers' subsistence. The start of occupation may be dated to the beginning or middle of the twentieth century. The radiocarbon datings obtained from a speck of carbon and a shell from the base of the deposit confirm our estimated chronology. We also observed geomorphological changes subsequent to the formation of the cholguero site, with part of the anthropic sediments below the current sea level (Reyes et al. 2018).

4.3.8 Cholguero Isla Izaza 2

This site is located on the north-west coast of Izaza Island, on the shore of the channel which separates it from Level Island. The area is forested with evergreen forest and peat bogs, and the shoreline is formed by a pebble and rock beach. The site is adjacent to the shore and a part of it lies 30 cm below current sea level. It is 0 m from the current shoreline and exposed to tidal action and to the formation of banks caused by storms. The site was found due to the exposure of its north profile, giving it medium archaeological visibility (Fig. 4.17). There is an old hearth in a ring of rocks, a grassy surface and the remains of a wooden structure. The site presents a slope of 30°.

It is a cholguero covered by a layer of plant material 10 cm thick, beneath which is a shell deposit, principally of mytilids, 65 cm thick. Below this shell-midden appears the palaeo-beach. According to estimates based on drilling cores, the area is 30 m from east to west and 8 m from north to south, covering approximately 240 m^2. No occupation or archaeological material was found below the cholguero, nor on the surface or in the nearby intertidal.

The test cores indicated an uninterrupted deposit of shell-midden matrix to 100 cm, consisting mainly of cholgas (*Aulacomya ater*), choro mussels (*Choromytilus chorus*) and Chilean mussels (*Mytilus edulis chilensis*). The samples of shell and carbon obtained at the base of the deposit, at 65 cm (30 cm below present sea level), were dated to 570 ± 25 years BP (1–255, 2-sigma cal. BP) and 70 ± 25 years BP (out of range, 2-sigma cal. BP, Reyes et al. 2018, Table 4.2).

Cholguero Isla Izaza 2 is a campsite developed by seasonal visits, possibly reiterated, related to intensive, selective gathering of mytilids for the canning industry. This activity probably implied other associated practices, like hunting and fishing for the fishermen's and mussel gatherers' subsistence. The start of occupation may be dated to the beginning or middle of the twentieth century. The radiocarbon datings obtained from a speck of carbon and a shell from the base of the deposit confirm our estimated chronology. We also observed geomorphological sinking of the area subsequent to the formation of the cholguero site (Reyes et al. 2018).

4.4 South-Central Area of the Chonos Archipelago (44°15′–46°15′S)

Below, we describe the archaeological sites recorded and studied in the south-central area of the Chonos Archipelago (Fig. 4.18).

4.4.1 Isla Elena 1

The site is situated on the south-west coast of Elena Island, towards the north-west mouth of Aysén Fjord where it joins the Moraleda Channel. It is a small cave formed by rocky outcrops on a steep shore with no intertidal zone. The area around the site is covered with thick evergreen forest vegetation and peat bogs, which come down to sea level and hide the cave, making access difficult. The archaeological visibility is nil (Fig. 4.19). It is 1 m above current sea level and 0 m from the shoreline. It was found thanks to information provided by fishermen.

The Isla Elena 1 site is a long, narrow cave lying north-west to south-east, estimated to be 11 m deep (as far as it was possible to enter the crack). Its dimensions are variable. In general terms, it is high (approx. 4 m) and very narrow (approx. 50 to 150 cm); the height diminishes towards the end of the crack, while it becomes broader. The estimated total area is approximately 60 m^2. Access is very complicated as there are no beaches or flat areas to land. The only possibility is by tying the boat to the trees which hang down to the sea. At high tide, the water advances some 3 m past the entrance. The cave is very damp and water has filtered in.

Human remains were recorded scattered on the surface of the cave. There are also remains adhering strongly to the walls, because the water that has filtered in

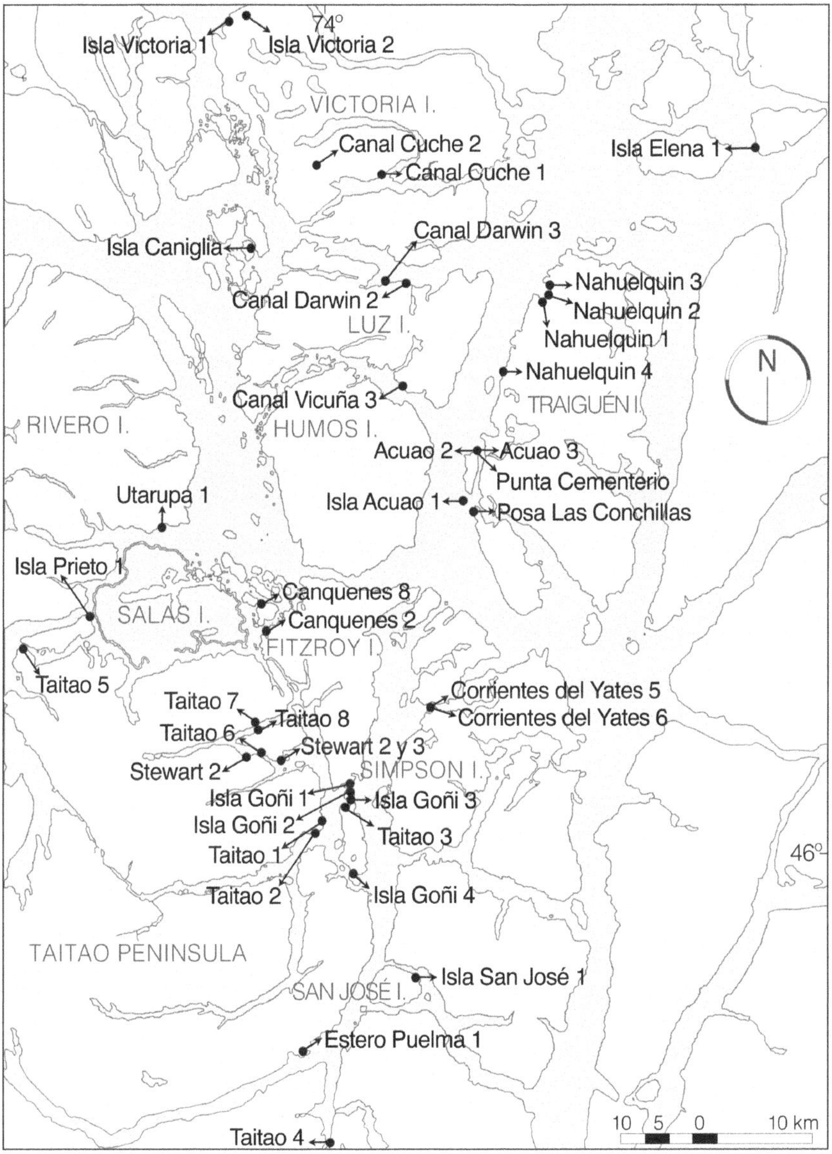

Fig. 4.18 Detail of the central-south portion of the Chonos Archipelago and the distribution of the sites recorded and studied in the area

Fig. 4.19 Coast where the Isla Elena 1 site is located, where human remains were deposited. It is almost inaccessible and the visibility is nil. View from outside the narrow cave

has deposited carbonates which have petrified over time, forming a film which has covered the bones and fixed them in contact with the stone. Inside the cave, four sectors were found with human bone remains; only three were surveyed as the fourth was inaccessible, since the entry crack was very narrow. The bone remains were separated into packets for analysis in the laboratory and to deduce the MNI, and to avoid over-dimensioning the sample in view of the evident deterioration and disturbance of the bone remains by fishermen. There is an old story that the family who discovered the cave found four bodies 'sitting up'.

In the cave, we surveyed the principal measurements and related them to the location and dispersion of the bone remains identified on the surface (packets). A high degree of disturbance, movement and alteration (trampling and fragmentation) was immediately apparent. We also cleared the sediments from the sectors where most of the bones were concentrated (packets 2 and 3) in order to sieve them and recover small fragments of bone present in the matrix.

Analysis of the bone remains found in the Isla Elena 1 site allowed us to determine a total MNI of 4: one adult, probably female, between 35 and 44 years of age; one adult, probably female, of indeterminate age; a young infant aged between 1.5 and 3 years; and one subadult between 14 and 16 years of age. All were identified by the most represented bone, the clavicle.

A total of 147 human bone remains were recovered, of which 108 were identified or assigned to a specific element. Packet 1 consists of only 2 remains, packet 2 of 16 remains, while packet 3 presents the largest number with a total of 129 specimens. The possibility of sexing the individuals was very low. The absence of crania and the poor preservation of specific features of the hip bones found meant that only a few indicators could be observed. Age was easier to determine, but only within fairly broad ranges in the majority of cases.

The Isla Elena 1 site is a narrow cave used exclusively for funerary purposes. It is a rock shelter protected from direct rain (but not from humidity and current tidal range) and covered by forest. The site has been altered by later anthropic activities. No artefactual or ecofactual material was recorded to indicate whether the site was occupied apart from funerary rituals. There is no evidence of rock art on the walls of the cave, and no remains of funerary offerings or trappings.

Samples were taken from each of the individuals identified, i.e. all the individuals recovered. The results of the dates obtained from all the individuals deposited on the surface of the shelter gave a mean range between ~1200 and 1500 years cal. BP (Reyes et al. 2013, Table 4.1). The extreme calibrated ranges of the dates overlap one another, so this ossuary may have been formed by a synchronic event or several events over a short lapse of time which cannot be differentiated chronologically, and clearly show reiterated use of the same site to deposit the dead. The isotope values obtained in the dated individuals, i.e. the whole bio-anthropological assemblage, allowed us to establish the consumption of an exclusively marine diet (see Chapter 6, Evaluation and Discussion of the Evidence) (Reyes et al. 2019a).

4.4.2 Nahuelquín 1

The Nahuelquín 1 archaeological site is on the north-west coast of Traiguén Island. The 'Nahuelquín Delgado' Huilliche indigenous fishing community has settled in this area, to take advantage of its geomorphology which offers excellent protection from the west winds. The site itself is on the south shore of a small bay facing south-west. The coast forms a rocky peninsula on which the archaeological shell-midden stands. On the east side of the site, there are two sandy-bottomed beaches connected by a strip of earth allowing access to the site from both the north and south sides of the peninsula. There is a permanent watercourse close to the north-facing beach. The base of the shell-midden deposit is 2 masl, and its edge is 5 m from the high tideline. The shell-midden slopes at 20° and is set in thick evergreen and peat bog forest. Its archaeological visibility is high. The shell-midden matrix can be seen at the surface since the activities of the indigenous community have exposed surfaces where the matrix appears. Part of the west profile can also be seen where it is exposed and has become eroded.

The site is an extensive shell-midden deposit approximately 30 m from north to south and varying between 8 m (northern end) and approximately 15 m (southern end) from east to west. It covers approximately 345 m^2 and follows the shape of

Fig. 4.20 (left) Thick cross-section lanceolate stone projectile head, made of grey obsidian from the Chaitén Volcano source, located almost 300 km further north. (right) View of the projectile head in situ, recovered in prospection of the intertidal zone at low tide.

the rocky peninsula on which it is situated. The core samples taken at the site gave a maximum thickness of 3 m, diminishing towards the periphery and the ends. It should be noted that the site extends beyond the dimensions of the shell-midden: a lithic instrument (thick cross-section lanceolate stone projectile head) was recorded on the beach to the south of the shell-midden (Reyes et al. 2011).

In general terms, the site is an archaeological occupation with the presence of a thick cross-section lanceolate stone projectile head, made of grey obsidian from the Chaitén Volcano source located 300 km further north (Fig. 4.20), and human remains appearing in the exposed profile of the west side of the shell-midden. Two radiocarbon datings were carried out on samples from the site: one on samples of shell obtained from the base of the shell-midden at a depth of 3 m, giving 1820 ± 25 years BP (1461 – 1216, 2-sigma years cal. BP, NAH 1–2); and one on a sample of the human remains found in the exposed profile (75–90 cm), with 430 ± 25 years BP (505 – 330, 2-sigma years cal. BP, NAH 1–1; Reyes et al. 2011, 2019b, Table 4.1).

The new work undertaken at the Nahuelquín 1 site focused on clearing the exposed profile and excavating individual N°1, dated previously (Reyes et al. 2011) to the post-contact period and still visible in the west profile of the shell-midden. In this field trip, we discovered new bones and carried out a second excavation, recovering two more individuals (N°2 and N°3), buried in the east end of the same shell-midden.

Clearing and recording of 300 cm of the exposed profile to a height of 145 cm, where individual N°1 was buried, revealed 5 stratigraphic units: the first was 40 cm of a thick layer of plant matter and fine silt composed of organic earth, with medium and thick roots, dark brown in colour, with a few small clasts, which covers the top of the archaeological shell-midden. Next came an interface matrix between the organic layer and the start of the shell-midden, 10 cm thick (down to 50 cm); it consisted of grey-brown organic sandy silt, with inclusion of shell fragments. This interface matrix is followed by another, much deeper matrix between 50 and 300 cm depth, composed of the shells discarded after consumption of malacological fauna.

This matrix consists mainly of picorocos (*Austromegabalanus psittacus*), cholgas (*Aulacomya ater*), choro mussels (*Choromytilus chorus*), Chilean mussels (*Mytilus edulis chilensis*), oysters (*Ostrea chilensis*), clams (*Protothaca thaca, Venus antiqua*), locos (*Concholepas concholepas*), keyhole limpets (*Fissurella nigra*), snails (*Tegula atra*), chitons (*Chiton sp.*) and sea urchins (*Loxechinus albus*). Fishbone remains (vertebrae and other bones) were also recorded.

The stratigraphy is not homogeneous: it presents very hard, compact layers consisting of lenses of picorocos (less fragmented than the *Mytilus*), as well as places where the shells are mingled with organic sandy silt sediments corresponding to the formation of soils during hiati between occupations of the site. Towards the base of the shell-midden deposit, a fourth stratigraphic unit can be identified, another interface matrix 5 to 10 cm thick in which sandy silt sediments are mixed with gravel and fragmented shell-midden, greyish in colour and slightly compacted; this is the contact floor where formation of the shell-midden began. Samples were taken from this layer to date the base of the deposit (Reyes et al. 2011). The last stratigraphic unit is sandy silt mixed with organic matter, which is the floor on which the shell-midden was deposited. It is approximately 5 cm thick and lies on the solid rock substrate of the small peninsula where the site stands. No clearly discernible hearth lenses were distinguished in the profile, nor were any cultural materials observed during sieving of the sediments.

4.4.2.1 Excavation of bones of individual N°1

The remains of individual N°1 observed in the exposed profile consisted of a weathered human cranial dome exposed to the high ambient humidity, white (weathering) and green (moss-stains) in colour. Due to erosion and fragmentation because of its exposure, the cranium came away from the maxillofacial portion, which remained in the profile. Other anatomical elements were also recorded, like a fragment of scapula, and two vertebrae: one cervical and one dorsal. The excavation of unit 1 (individual N° 1) was carried out in the cleared profile; the initial excavation was 80 cm from north to south and 60 cm from east to west.

Individual N° 1 was deposited between 75 and 90 cm from the present surface. We excavated to 110 cm to leave the skeleton fixed in its context while it dried, due to the high humidity of the matrix and the saturation of the bones, in order to remove the remains subsequently. These were highly fragmented and deteriorated by the humidity of the deposit and the roots in the matrix. The excavation was extended to 1.20 m in the north–south axis, as part of the skeleton remained in the profile.

No trappings of any kind were observed. We thought that an entire loco shell (*Concholepas concholepas*) found near the individual's thorax might have been placed as an offering. The shells in the deposit were highly fragmented, while this loco shell, placed with the opening downwards, was whole. As we remark below, the other two burials excavated in the site presented what we believe to be similar offerings.

To judge from the anatomical elements present and the details observed in the excavation, this was a single, primary, individual burial. The body was placed in flexed left lateral decubitus, with the head to the west and the pelvic girdle to the east. Observation of the morphology of the cranium, mandible and pelvic girdle suggests a female individual aged 20–25 years at death. The cranium presents features compatible with mongoloid ancestry, e.g. elliptical dental arch, oval auditory canal and complex major sutures, as well as shovel incisors.

Preliminary mitochondrial DNA results for this individual indicate that she belonged to the infrequent D4h3a sub-haplogroup, detected in modern Kawésqar individuals (Moraga et al. 2010) and inhabitants of Laitec Island, in the Chiloé Archipelago (Moraga et al. 2000), suggesting some relationship with island populations from Chiloé or the Kawésqar people (Reyes et al. 2011).

4.4.2.2 Bone excavation of individuals N°2 and N°3

An excavation 2 m east–west by 1.5 m north–south was made on the southern side of this extensive shell-midden, where today there is a vegetable garden, 20 m south of the first set of bones (individual N°1) and 30 m from the present high tideline. Part of the human remains removed when the vegetable garden was created fell down the 30° slope to the base of the shell-midden. Unlike the burial of individual N°1, the bodies deposited in this sector are practically in the subsurface, at depths of between 4 and 14 cm. In this area of the shell-midden, the plant layer has been moved systematically due to the cultivation of crops.

While we were excavating and removing individual N°2, we noticed the presence of another adjacent burial, which we called individual N°3. Based on the opposing positions of the two bodies, and the depth and stratigraphic integrity of the burials, we though this was a synchronic multiple burial, or two individual burials within a short space of time, since the second body was placed close to the first without affecting it, suggesting previous knowledge; it is impossible to say which came first (unless significant differences in radiocarbon dating are observed). The mere coincidence of two adjacent contexts is more difficult to test in this extensive shell-midden. We therefore suspect a multiple burial.

Individual N° 2 is lying in hyperflexed right lateral decubitus, with the pelvis to the west and the cranium to the east. The thorax and pelvis are lying more to ventral than to lateral. The head is looking north, the arms are flexed with the left hand between the legs, while the right is parallel with the cranium. Individual N° 3 is also in hyperflexed right lateral decubitus, with both arms flexed and in the same position as individual N° 2, but with the general orientation of the head and pelvis reversed so that the bodies are in a sense 'opposed', in what we suppose to be joint deposition in a single burial event (Fig. 4.21).

Predominantly male characteristics can be observed in the fragments of the cranium of individual N°2: the greater sciatic foramen is closed in the left hip, and the pubis is triangular. These observations suggest that the individual was male. From the cranial and postcranial age indicators, we estimate the age at death as

Fig. 4.21 Plan showing the flexed lateral decubitus position of the two bodies—Individual 2 (left) and 3 (right), placed opposed to one another

20 ± 2 years. The cranium presents features compatible with mongoloid ancestry, e.g. elliptical dental arch, oval auditory canal and complex major sutures, as well as shovel incisors.

Most of the bony parts of the cranium and postcranium (scapular girdle, backbone, thorax, pelvic girdle and upper and lower limbs) are present in individual N°3. The only complete bones are the humerus, clavicle, fibula and femur, all on the right side. The cranium presents predominantly male characteristics, with large mastoid processes, prominent zygomatic arches extending behind the external auditory canal; the sockets are square with rounded edges and the superciliary arches are marked. Pronounced muscle insertions are observed in the occipital process. The mandibular body is robust, with a square chin and thick ascending rami, with straight edges and everted gonions. The greater sciatic foramen is closed in both hips and the pubis is triangular. These observations suggest that the individual was male. Age at death of this individual is also estimated as 20 years. The cranium presents features compatible with mongoloid ancestry, e.g. elliptical dental arch, oval auditory canal and complex major sutures, as well as shovel incisors.

A sample of rib was taken from individual N°2 to date the context of this multiple burial, giving a result of close to ~600 years cal. BP (683 – 537, 2-sigma years cal. BP). With the datings obtained previously (Reyes et al. 2011) from the base of the deposit and the first burial recorded, we can say that the area around the site was occupied from nearly 1400 years cal. BP. The different occupations occurring in this shell-midden indicate a sequence of at least one thousand years (Table 4.1).

No funerary trappings were recorded; however, when the remains of individual N° 3 were lifted up, two large, complete loco shells (*Concholepas concholepas*) were found, with the opening downwards, at the level of the cranium and the foot bones.

Both shells were filled with shell-midden matrix (dark silt and ground shell) and with a high concentration of sea urchin fragments.

The Nahuelquín 1 site was a camp where various activities are recorded from around ~1400 years cal. BP. The most evident is the discarding of waste from the gathering and consumption of molluscs, reflecting local resources in which different types of shore and sea-floor are represented in the general gathering strategy. The scarce presence of lithic material (thick-section lanceolate stone projectile head) obtained in the intertidal (Reyes et al.) also indicates other activities in the camp and elsewhere, such as hunting fauna for food. The obsidian recorded, from the Chaitén Volcano source, also associates part of the groups who settled in the area with a wide network of circulation and/or exchange of raw materials. The recurrence of these activities allowed the formation of a dense shell-midden, 3 m deep, over nearly one thousand years of reoccupations of the site.

Over the thousand years of occupation, at least three bodies were buried within quite a short period—less than ~200 years—with common patterns such as the arrangement of the body, offerings (presence of inverted locos beside the body?) and deposition in the edges of the shell-midden, on the periphery of the camp. The diet consumed, on the other hand, shows a remarkable change. Individual N°1 was buried in the post-contact period. Although no other type of evidence than the late chronology was detected in the excavation of either of the burial areas in the shell-midden, or in the stratigraphy, the isotope values of this post-contact burial reflect dietary changes in which the diet was no longer exclusively marine (Reyes et al. 2011, 2013). The isotope values obtained in the individual dated to the pre-contact period (NAH1-2) allowed us to establish the consumption of an exclusively marine diet (see Chapter 6, Evaluation and Discussion of the Evidence) (Reyes et al. 2019a).

4.4.3 Canal Vicuña 3

The site lies on the north coast of the Vicuña Channel, formed by Luz Island in the north and Humos Island in the south. On the south coast of Luz Island, there is a small peninsula 200 m north to south and 100 m east to west), connected to the rest of the island by an earth isthmus 50 m long, with an intertidal zone of sandy sediments and small pebbles on either side. The site is located at the western end of this land bridge. The area is heavily forested with evergreen forest and peat bog, however, the vegetation has been cleared by fishermen, and the surface is grassed over. The archaeological visibility is medium; the profiles of the north side of the site can be seen under large arrayan trees (*Luma apiculata*), exposed by tidal action. The adjacent shoreline is a pebble and sand beach; however, the rest of the isthmus has sheer, rocky sides.

The site stands on the beach formed by the land bridge, 1 m above the current sea level and 1 m from the shoreline. The two beaches face north-west and south-east. The slope of the site is 20° to 30°, however, the eroded profile of the shell-midden is cut at almost 90°. It is a large archaeological shell-midden covered by a layer of plant matter 10 cm thick on average. According to estimates based on drilling cores, the area is 20 m from north-west to south-east and 40 m from north-east to south-west, covering approximately 800 m^2. The shell deposit is 2 m deep. No watercourses were recorded nearby.

Clearing of the south profile revealed an uninterrupted deposit of shell-midden matrix from 30 to 200 cm. The base of the deposit lies on the palaeo-beach composed of pale brown sandy silt sediments with inclusion of clasts and medium-sized pebbles, forming the base of the land bridge. The pale-coloured shell-midden matrix, 1.5 m deep, consists mainly of picorocos (*Austromegabalanus psittacus*), cholgas (*Aulacomya ater*), choro mussels (*Choromytilus chorus*), Chilean mussels (*Mytilus edulis chilensis*), oysters (*Ostrea chilensis*), clams (*Protothaca thaca, Venus antiqua*), locos (*Concholepas concholepas*), keyhole limpets (*Fissurella nigra*), snails (*Tegula atra*), chitons (*Chiton sp.*) and sea urchins (*Loxechinus albus*). Below this is 50 cm of a dark matrix, composed mainly of fine silty sediments and *Mytilus* periostracum, which rests on the palaeo-beach. A bird bone punch was recovered from sand in the intertidal to the south of the site, under water. Curry (1997) recovered a net weight from the adjacent intertidal (Reyes et al. 2015).

The Canal Vicuña 3 site is an extensive camp formed by discards from gathering and consumption of malacological fauna; successive reoccupations of the area are recorded. The bird bone punch may be related to domestic activities, working in leather or bark, as well as hunting birds. The net weight is clearly related with fishing in the area. According to the record and the archaeological work carried out, no other activities are inferred for the site.

4.4.4 *Punta Cementerio*

The site is located on the north coast of Acuao Island, separated from the southwest coast of Traiguén Island by a narrow channel. It is on a mixed shore where an intertidal with sandy sediments and pebbles alternates with steep rocks. The area is heavily forested with evergreen and peat bog forest. The site is 0 m above the current sea level and 0 m from the shoreline. It is on the beach adjacent to the shoreline, facing north. The archaeological visibility is nil, the site is invisible to the passing eye. From a very close-up, the tidal erosion of one of profiles enabled us to confirm the presence of an anthropic shell-midden. There was no archaeological material either on the surface or in the adjacent intertidal zone. The surface presents a slope of 10° to 20°; however, the edge of the profile is cut off abruptly at exactly 90° by tidal erosion.

The site is an extensive archaeological shell-midden covered by a layer of plant matter 10 to 20 cm thick on average. On this shell-midden, we found 6 crosses and a historical *animita* (small shrine to the dead) in memory of those who died in a shipwreck. These crosses give the area its name Punta Cementerio (Cemetery Point). According to estimates based on drilling cores, the area is 15 m from east to west and 20 m from north to south, covering approximately 300 m^2. The shell deposit is 3 m deep. Several watercourses were observed around the site.

Clearing of the exposed north-east profile revealed an uninterrupted deposit of shell-midden matrix from 10–20 cm down to 3 m. The base of the deposit lies on the palaeo-beach, composed of grey silt sediments with inclusion of clasts and medium-sized pebbles. The shell-midden matrix consists mainly of picorocos (*Austromegabalanus psittacus*), cholgas (*Aulacomya ater*), choro mussels (*Choromytilus chorus*) Chilean mussels (*Mytilus edulis chilensis*), oysters (*Ostrea chilensis*), clams (*Protothaca thaca, Venus antiqua*), locos (*Concholepas concholepas*), keyhole limpets (*Fissurella nigra*), snails (*Tegula atra*), chitons (*Chiton sp.*) and sea urchins (*Loxechinus albus*).

No cultural or ecofactual materials were recovered from this site. However, Curry (1997) recovered six stone axes from the adjacent intertidal (Reyes et al. 2015).

The Punta Cementerio site is an extensive camp formed by discards from gathering and consumption of malacological fauna; successive reoccupations of the area are recorded. The presence of stone axes suggests that they were used to obtain forest resources, such as cutting trunks or as wedges for removing their bark, for inferred purposes like the maintenance and construction of canoes, and/or other activities involving the use of wood. According to the record and the archaeological work carried out, no other activities are inferred for the site.

4.4.5 Acuao 2 Isolated Find and Acuao 3

Both the isolated find and the site are located on the north coast of Acuao Island, which is separated from the south-west coast of Traiguén Island by a narrow channel. They lie just to the east of the Punta Cementerio site. They are on a mixed shore where an intertidal with sandy sediments and pebbles alternates with steep rocks. The area is heavily forested with evergreen and peat bog forest. The site lies 0 m above the present sea level and 0 m from the adjacent shoreline, facing north; it is invisible to the passing eye. From a very close-up, the tidal erosion of one of profiles enabled us to confirm its presence. There was no archaeological material either on the surface or in the adjacent intertidal zone. The site presents a slope of 10°.

The Acuao 3 site is an archaeological shell-midden covered by a layer of plant matter 10 cm thick. According to estimates based on drilling cores, the area is 4 m from east to west and 20 m from north to south, covering approximately 80 m^2. The shell deposit is 50 to 150 cm deep. Immediately to the north of the shell-midden runs a small stream which flows all year round. The site is a camp formed by discards from gathering and consumption of malacological fauna; successive reoccupations of the area are recorded.

The isolated find, which we called Acuao 2, appears to represent an extension of the activities derived from nearby camps, such as Acuao 3 and Punta Cementerio. It is a core with patina, possibly of andesite, made from the pebbles available locally in the intertidal. It was found in the intertidal formed of pebbles and shingle east of the Punta Cementerio site.

4.4.5.1 Canal Utarupa 1

The site is located on the north-east coast of the Utarupa Channel, formed by Salas Island to the south and Rivero Island to the north. It stands on the south-east tip of the latter on a small, sandy intertidal beach exposed to the south-east. The area is heavily forested with evergreen forest and peat bog, however, the vegetation around the site has been cleared by fishermen, and the surface is grassed over. The archaeological visibility is nil and it was discovered from cores drilled at various points. The shoreline consists of a sand beach. The site is 1 m above the current sea level and 2 m from the shoreline. It stands right on the shoreline facing south-east. There was no archaeological material either on the surface or in the adjacent intertidal zone. The site presents a slope of 30°.

The test cores showed 20 cm of organic cover, and beneath it a shell-midden matrix composed mainly of fragments of *Mytilus* shells and dark silt 20 cm thick, with inclusion of specks of carbon. This matrix of shells lies on a dark brown clay-silt matrix, with inclusion of gravel, 20 cm thick. Finally, below this matrix, the gravel which formed part of the palaeo-beach appeared at 60 cm depth. According to estimates based on cores, the area is 10 m from east to west and 10 m from north to south, covering approximately 100 m^2. The shell deposit is 20 cm deep. No watercourses were recorded nearby.

The Canal Utarupa 1 site is a camp formed by discards from gathering and consumption of malacological fauna, mainly *Mytilus*; successive reoccupations of the area are recorded. According to the record and the archaeological work carried out, no other activities are inferred for the site.

4.4.6 *Canquenes 2*

The site is located in the group of islands called Canquenes, between Salas Island and Fitzroy Island. Just off the west coast of the latter is an islet separated by a narrow channel which dries at low tide. The site is on the west coast of this islet. The area is heavily forested with evergreen forest and peat bog, however, the vegetation around the site has been cleared by fishermen, and the surface is grassed over. Nevertheless, the archaeological visibility is nil; it was only by drilling cores and clearing profiles adjacent to the sea that we discovered eroded profiles in which the shell-midden matrix could be seen. The adjacent shoreline consists of a pebble and sand beach, rocky in places. The site is 1 m above the current sea level and 2 m from the shoreline. It is on the beach adjacent to the shoreline, and is exposed to the north and the south-east. The slope of the surface is 10° to 20°, while the exposed profile falls at 70°. It is a large archaeological shell-midden covered by a layer of plant matter 10 cm thick on average. On the surface, there is an old cholguero camp with a wooden structure (fishermen's shelter). Grapevines and apple trees have been planted, and on the surface are the remains of an accumulation of Chilean mussels and cholgas.

According to estimates based on drilling cores, the area is 50 m from east to west and 20 m from north to south, covering approximately 1000 m^2. The shell deposit is 2 m deep. No watercourses were observed nearby.

Clearing of the exposed profile on the east side revealed an uninterrupted deposit of shell-midden matrix from 10 cm to 6 m; the base of the shell-midden deposit terminates in a thin, dark, very silty layer 3 cm thick. This lies on the palaeo-beach composed of sand-silt sediments with inclusion of clasts and medium-sized pebbles. The shell-midden matrix consists mainly of picorocos (*Austromegabalanus psittacus*), cholgas (*Aulacomya ater*), choro mussels (*Choromytilus chorus*), Chilean mussels (*Mytilus edulis chilensis*), oysters (*Ostrea chilensis*), clams (*Protothaca thaca, Venus antiqua*), locos (*Concholepas concholepas*), keyhole limpets (*Fissurella nigra*), snails (*Tegula atra*), chitons (*Chiton sp.*) and sea urchins (*Loxechinus albus*). Fish and sea lion bones were also found, as well as specks of carbon. An andesite core made from a local pebble was recovered from the sandy intertidal.

The Canquenes 2 site is an extensive camp formed by discards from gathering and consumption of malacological fauna; successive reoccupations of the area are recorded. Lithic knapping of pebbles obtained locally from the intertidal can be inferred. Hunting of sea lions and fishing can be inferred from observation of the remains of fauna present in the profile.

4.4.7 Canquenes 8

The site is located in the group of islands called Canquenes, between Salas Island and Fitzroy Island. It was discovered and recorded on the east shore of a small islet in a south-facing bay. The area is heavily forested with evergreen and peat bog forest. The archaeological visibility is nil; it was detected by exploring under bunches of roots at low tide, where we observed an undermined shell-midden matrix. The adjacent shoreline is rocky. However, the intertidal around all the nearby islands is mixed, with pebbles and/or sandy sediments.

The site is 0 m above the current sea level and 1 m from the shoreline. Part of the base has been eroded by tidal action. It faces east and lies between the third marine terrace and the adjacent beach on the shoreline. There was no archaeological material either on the surface or in the adjacent intertidal zone. The slope of the exposed side is 30° (Fig. 4.22).

The site is an archaeological shell-midden covered by a layer of plant matter 10 cm thick. It is difficult to establish its size because of the thick vegetation which hinders movement and observation. According to our estimates, the area is 15 m from east to west and 10 m from north to south, covering approximately 150 m^2. The shell deposit is 2 m deep. No watercourses were recorded nearby. Clearing of the exposed profile revealed an uninterrupted deposit of shell-midden matrix down to 2 m. The

Fig. 4.22 Canquenes 8 shell-midden on an islet surrounded by rocky intertidal. Tidal action has undermined the base of the shell-midden. It is possible to crawl under it at low tide

base of the deposit lies on the intertidal, composed of medium-sized pebbles and a substrate of large rocks.

The Canquenes 8 site is a camp formed by discards from gathering and consumption of malacological fauna; successive reoccupations of the area are recorded. According to the record and the archaeological work carried out, no other activities are inferred for the site.

4.4.8 Isla Prieto 1

The site lies on the north-east coast of Prieto Island on a small, wooded peninsula at the junction of two channels which separate Prieto Island from Rivero and Salas Islands. The south-east side of this little peninsula forms a narrow bay with an intertidal of sandy silt and small pebbles. The site is on this intertidal. The archaeological visibility is nil. It is overhung by thick evergreen and peat bog forest, and abundant shrubby vegetation. When the roots and plant layer were cleared away, they revealed part of the northern profile of the site, exposed by tidal action. The surface slopes markedly (60°). No watercourses are observed nearby.

The site is an archaeological shell-midden covered by a layer of plant material 30 cm thick, 8 m from east to west and 4 m from north to south. Its current area is 32 m^2 with a maximum cultural deposit of 0.5 m. The base of the deposit is 0 m above current sea level and 1 m from the shoreline, and was exposed by tidal erosion.

The eroded profile shows a thick layer of dark brown plant matter (30 cm), with abundant roots. Between 30 and 80 cm, there is a thick layer of mollusc valve

Fig. 4.23 Location of the Isla Prieto 1 site, completely covered by forest. Tidal action has undermined the base of the shell-midden. When the profile was being cleared, abundant periostracum of mytilid valves was found in dark organic matrices which contained no shells

periostracum; where the profile has collapsed under tidal action, pockets of air are left behind due to the consistency and compaction of the matrix. Below 80 cm, there was a thick layer of gravel, pebbles and sandy sediments, making up the intertidal on which the first occupants of the site settled. The periostracum sample obtained at the base of the deposit when clearing the profile to a depth of 90 cm gave a date of 1710 ± 30 years BP (1316–1078, 2-sigma cal. BP, Table 4.1).

The shell-midden record shows that it was formed exclusively by periostracum of mytilid valves (e.g. choro mussels, cholgas). No remains of shells of malacological fauna or any other species were observed. Differentiated conservation, with only periostracum being found, was documented by Curry (1997) in some shell-middens of the archipelago (Fig. 4.23).

The Isla Prieto 1 site is a camp formed by discards from gathering and consumption of malacological fauna, especially Chilean mussels and cholgas; successive reoccupations of the area are recorded from ~1200 years cal. BP. According to the record and the archaeological work carried out, no other activities are inferred for the site.

4.4.9 *Taitao 5*

The site is on the north-west coast of Taitao Peninsula at the south-west mouth of a channel running east–west that separates Prieto Island from the peninsula. It is on the intertidal formed by a small beach of large pebbles in a little rocky bay, facing north. The archaeological visibility is nil. It is overhung by thick evergreen and peat bog forest, and abundant shrubby vegetation. When the roots and plant layer were cleared away, they revealed part of the northern profile of the site, exposed by tidal action. It slopes at 40° and there is a small watercourse.

The site is an archaeological shell-midden covered by a layer of plant material 20 cm thick, 30 m from east to west and 10 m from north to south. Its current area is 300 m^2 with a maximum cultural deposit of 2 m. The base of the deposit is 0 m above current sea level and 0 m from the shoreline, and was exposed by marine erosion.

The shell-midden consists of the remains of malacological fauna, either whole or fragmented, in different levels and consisting mainly of picorocos (*Austromegabalanus psittacus*), cholgas (*Aulacomya ater*), choro mussels (*Choromytilus chorus*), Chilean mussels (*Mytilus edulis chilensis*), oysters (*Ostrea chilensis*), clams (*Protothaca thaca, Venus antiqua*), locos (*Concholepas concholepas*), keyhole limpets (*Fissurella nigra*), snails (*Tegula atra*) and chitons (*Chiton sp.*). The shell-midden continues without interruption from 10 cm mean depth to the base at 2 m, where it lies on a pebble beach typical of the area.

Close to this profile, we recovered an anvil made of a flattened local stone with slight bore marks on both sides. It was found in the intertidal adjacent to the exposed profile, and had probably fallen from this matrix. The shell sample obtained at the base of the deposit when clearing the profile to a depth of 200 cm gave a date of 2850 ± 30 years BP (2695–2363, 2-sigma cal. BP, Reyes et al. 2019b, Table 4.1).

The Taitao 5 site is a camp formed by discards from gathering and consumption of malacological fauna; successive reoccupations of the area are recorded from ~2500 years cal. BP. We also detected stone knapping activities, such as indirect percussion, inferred from the anvil recovered. According to the record and the archaeological work carried out, no other activities are inferred for the site.

4.4.10 Corrientes Del Yates 5

The site is on the west coast of Simpson Island, at the south-west entrance of a narrow, fast-flowing channel which runs east–west between Simpson and McPherson Islands. It is on a small, wooded peninsula which has practically no foreshore. It is no more than a rocky promontory covered in vegetation which comes down to the sea. There is no intertidal and the shore exposed at low tide is rock. The archaeological visibility is nil; however, it is one of the sites mentioned in her prospection by Patricia Curry (1997), so we were able to find it in the area. It is overhung by thick evergreen and peat bog forest, shrubby vegetation and an abundant layer of plant matter. The slope is 40° and no watercourses were observed in the vicinity. Opposite this site, on McPherson Island, on the other side of the fast-flowing channel, is the Corrientes del Yates 6 site.

The site is an archaeological shell-midden covered by a layer of plant material 10 cm thick, 10 m from east to west and 20 m from north to south. The current area is 200 m^2 with a maximum cultural deposit of 1.3 m. The base of the deposit is almost at sea level and adjacent to the shoreline; tidal action has exposed part of the north profile (still covered by vegetation). Closer examination of the exposed profile revealed human bone remains which had fallen from the profile, with some anatomical portions still articulated. Several individuals were buried here. There were also remains of marine fauna (sea lion, birds' bones) and hearth lenses.

A 2 m wide stretch of the east–west profile was cleared. The base of the shell-midden was established on very fine sandy silt, formed from the sediment carried by the currents in the channel and the protection given by the rocky point where the sediment could be deposited. The shell-midden is almost entirely covered by a thick layer of reddish brown organic earth, maximum 50 cm deep, composed of leaves and impenetrable roots. Below this layer, the whitish/greyish shell-midden appears. It is 1.3 m deep, very compact and continues without interruption down to the base where it rests on the sandy silt sediments, which in turn lie on the base rock. The shell-midden consists of the remains of malacological fauna, either whole or fragmented, in different levels and consisting mainly of picorocos (*Austromegabalanus psittacus*), cholgas (*Aulacomya ater*), choro mussels (*Choromytilus chorus*), Chilean mussels (*Mytilus edulis chilensis*), oysters (*Ostrea chilensis*), clams (*Protothaca thaca, Venus antiqua*), locos (*Concholepas concholepas*), keyhole limpets (*Fissurella nigra*), snails (*Tegula atra*) and chitons (*Chiton sp.*). This matrix includes dark sandy silt sediments interspersed with the remains of malacological fauna. Different degrees of compaction were noted according to the malacological components present. Human bone remains were found 70 cm below the top of the shell-midden (1.2 m below the present surface, including the 50 cm organic layer). They are all at the same level (between 70 and 80 cm) and in a space no more than 1 m in diameter. As we continued clearing the profile, we observed a small hearth lens, blackened and with abundant carbon, 18 cm across by 5 cm deep. It lay 1.3 m below the top of the shell-midden (1.8 m below the current surface) and 17 cm above the base of the shell-midden.

Human bone remains belonging to 4 individuals were recovered. We determined that at least one individual had been buried in situ in hyperflected lateral decubitus (individual identified in the laboratory as N°1); in the other individuals, we assumed a similar position based on the anatomical portions recovered and the small space in which they were lying.

We also recovered an anvil made of a local stone, probably obtained from the immediate rocky intertidal. Samples were taken from the hearth and the base of the shell-midden.

4.4.10.1 Individual N°1

The following skeletal parts were recovered: a fragment of left mandibular corpus with dental pieces in their alveoli (lateral incisor, first and second premolar and first, second and third molar); a fragment of scapula including part of the body and the acromion; the medial portion, including the acromial border, of the right clavicle; the hyoid bone; left humerus missing its distal third; a carpal phalange and three medial phalanges of a hand; a cervical vertebra and five dorsal vertebrae; six fragments of right rib; the fragmented body of the sacrum; and right femur missing its epiphysis and proximal metaphysis. The clearest morphological indicators of the gender of the individual, both in the pelvic girdle and in the cranium and mandible, are missing; however, the marked muscle insertions and the robustness of the diaphyses of the long bones lead us to the assumption that this individual was probably male.

The age at death of the individual can be estimated from observation of the state of fusion of the epiphyses of the long bones. Moreover, the first vertebra of the sacrum is not fused. The morphology of the articular facets of the vertebral bodies are observed to be ridged and in an early phase of transformation, with no signs of ageing (e.g. vertebral lipping or development of osteophytes). Finally, all the permanent teeth are erupted, but still preserve part of the cusps and the dentin of the crowns, showing only slight wear. Based on the above, this individual is thought to be a young adult, with age at death between 20 and 25 years.

4.4.10.2 Individual N°2

Represented only by proximal metaphysis of the right humerus. However, to judge by the fusion of the epiphyses of the long bones, it belonged to a subadult aged approximately 10–15 years, of indeterminate sex.

4.4.10.3 Individual N°3

A perinate (between third foetal term and 6 months of age) of indeterminate sex. Fragments of cranium (frontal, parietal and malar) and mandible (fragment of the right corpus) were recovered, as well as two deciduous teeth (right molar and canine) in formation, with only the crown formed.

4.4.10.4 Individual N°4

Represented by a fragment of the left mandibular corpus, with ascendant ramus, containing both premolars and the second molar in their alveoli; the two acromial extremes of both clavicles; four cervical vertebrae, the sternal edge of a rib; and a metacarpal and three phalanges of one hand. The clearest morphological indicators of gender, both in the pelvic girdle and in the cranium and mandible, are missing. The ascendant ramus of the mandible is observed to be broad and robust, as are the acromial extremes, leading us to the conclusion that this individual was probably male.

The age at death of individual N°4 can be estimated from observation of the state of the morphology of the articular facets of the vertebrae. Formation of osteophytes and lipping are observed on the ventral face of the vertebral bodies and the transverse processes. Finally, all the permanent teeth are fully erupted and observation of oral health indicates tooth loss, infections and generalised wear. Based on the above, this individual is thought to be a mature adult, with age at death over 30 years.

The sample of human bone obtained at 120 cm, taken from individual N°1, gave a date of 1420 ± 20 BP (1088 – 880, 2-sigma cal. BP), while the carbon sample from the hearth at 180 cm, close to the base of the shell-midden matrix, gave a date of 1920 ± 30 BP (1890–1734, 2-sigma cal. BP) (Reyes et al. 2019a and b, Table 4.1).

The Corrientes del Yates 5 site was a camp where various activities are recorded from around ~1800 years cal. BP. The most evident is the discarding of waste from the gathering and consumption of molluscs, reflecting local malacological fauna obtained from the different types of shore and sea-floor found in the area. At least four burials were carried out on the edge of the shell-midden, on the periphery of the camp, around ~1000 years cal. BP. This was almost a thousand years after the first occupation was started on the sediments deposited by the contiguous channel. The burials appear to belong to a single event, or perhaps events occurring within a short lapse of time, in which the individuals were deposited in the same area (edge of the shell-midden, practically together) in what is estimated to be hyperflexed lateral decubitus. Other subsistence activities inferred from the faunal record are hunting of birds and sea lions. We also detected stone knapping activities such as indirect percussion, based on the anvil recovered, and the use of hearths for heating and/or cooking. The isotope values obtained in the dated individual allowed us to establish the consumption of an exclusively marine diet (see Chapter 6, Evaluation and Discussion of the Evidence) (Reyes et al. 2016b).

4.4.11 Corrientes Del Yates 6

The site is on the south coast of Simpson Island, at the south-west entrance of a narrow, fast-flowing channel which runs east–west between Simpson and McPherson Islands. It stands on a small, wooded bay 20 m wide which has practically no shore; there is a small beach of large pebbles in the intertidal. The archaeological visibility is nil; however, it is one of the sites mentioned in her prospection by Patricia Curry (1997), so we were able to find it. It is overhung by thick evergreen and peat bog forest, shrubby vegetation and an abundant layer of plant matter. When the roots and plant layer were cleared away, they revealed part of the west profile of the site. The slope is 40° and no watercourses were observed in the vicinity. Opposite this site, on Simpson Island, on the other side of the fast-flowing channel, is the Corrientes del Yates 5 site.

The site is an archaeological shell-midden covered by a layer of plant material 40 cm thick, 10 m from east to west and 10 m from north to south. Its current area is 100 m^2 with a maximum cultural deposit of 1.5 m. The base of the deposit is 1 m above current sea level and 2 m from the shoreline.

The shell-midden consists of the remains of malacological fauna, either whole or fragmented, in different levels and consisting mainly of picorocos (*Austromegabalanus psittacus*), cholgas (*Aulacomya ater*), choro mussels (*Choromytilus chorus*),

Chilean mussels (*Mytilus edulis chilensis*), oysters (*Ostrea chilensis*), clams (*Protothaca thaca, Venus antiqua*), locos (*Concholepas concholepas*), keyhole limpets (*Fissurella nigra*), snails (*Tegula atra*) and chitons (*Chiton sp.*).

In the exposed profile, close to the base of the shell-midden, we recovered a pendant made of a polished fragment of rib (*Mammalia*), fined down at both ends and with one end pierced for hanging.

Corrientes del Yates 6 is a camp formed by discards from gathering and consumption of malacological fauna; successive reoccupations of the area are recorded, with domestic activities such as the production of bone adornments. The adjacent site, on the other side of the channel, presents first occupation around ~1800 years cal. BP. According to the record and the archaeological work carried out, no other activities are inferred for the site.

4.4.12 Isla Goñi 1

Our knowledge of the variety of stone materials collected and deposited in the Patagonian Institute (Curry 1997) enabled us to analyse them in advance (Morello 2014; Reyes et al. 2015); we then reidentified the site, searched it and carried out new interventions.

The Isla Goñi 1 site is on the north-west coast of Isla Goñi, where a land bridge at low tide connects it with another smaller island. This was once part of Goñi Island but has been separated by erosion. There is a historical careenage close to the site, where all the large stones have been cleared from the intertidal to avoid damage to boats during careening. The site, discovered by Curry (1997), stands on an old sand and pebble beach facing west. Today, the adjacent shore consists of an area of sand and medium-large pebbles no more than 10 m wide at low tide. The area is heavily forested with evergreen and peat bog forest. The archaeological visibility of the site is nil, and it was detected by the find of a series of stone instruments (large, thick-section, lanceolate stone projectile points, net weights) and knapping waste in the intertidal (Curry 1997; Reyes et al. 2015). The site is flat and no watercourses are observed nearby.

The site is 0 m above the current sea level and 0 m from the shoreline. It is completely exposed to tidal erosion; lithic material is scattered across the sandy intertidal and in a long exposed earth profile, where knapping waste can be seen (Fig. 4.24).

Clearing of the exposed profiles and the use of test cores determined the presence of dark organic sediment in which cultural materials and carbon specks were found. We defined a long north–south profile of 70 m, while the east–west width is just 1–2 m. The archaeological occupation detected, which suffers permanent, continuous erosion, has an area of 140 m^2 and a depth of 20 cm. The site has been almost completely eliminated by tidal action, which we understand to be related with big changes in sea levels leading to erosion of the island's coastline. The sediments

Fig. 4.24 View of the degraded Isla Goñi 1 site. The dark horizontal band above the intertidal is the site profile exposed by the tides. The associated intertidal where the site has been 'washed' presents various lithic pieces, especially preforms of axes and lanceolate points in rhyolithic obsidian

exposed to tidal action drop onto the intertidal in columns, and we could sieve 'slices' of the site right on the beach.

We carried out superficial archaeological collection in the adjacent intertidal. The exposed profiles were searched for cultural material. We cleared an area of the profile 1.3 m by 25 cm to characterise the stratigraphic deposits and dug a test pit 50 × 100 cm, from which we recovered lithic material and carbon specks. We also defined the stratigraphic units.

The record in the exposed profiles is as follows: first, abundant plant remains and a layer of burnt peat over 1 m thick. When this irregular plant component is cleared away, the stratigraphic sediments are reached where a surface level can be defined. From this surface, a black layer is observed between 0 and 10 cm; it consists of very organic, almost black sandy silt, and includes a large quantity of knapping waste, the majority being grey obsidian from Chaitén Volcano. The knapping waste could only be recovered because we were sieving on the beach with seawater, allowing us to break up very compact lumps of sediment. Between 10 and 20 cm the same dark, semi-compact organic matrix was recorded, including knapping waste. Below this, between 20 and 30 cm, there is a change in the sediments to a coarser, inorganic sand matrix, dark brown to chestnut in colour, from which material was recovered (a fragment of stone projectile head). In the 30–40 cm level, the inorganic coarse sand turns to a reddish brown colour and no cultural material is recovered. Finally, the 40–50 cm level is identical to the previous level. A sample core was taken adjacent to this unit and a 20 cm section was extracted containing the cultural layer in order to obtain carbon speck samples.

Fig. 4.25 Preform of a stone axe recovered from the intertidal. Isla Goñi 1 site

When test cores were drilled a few metres inland from the eroded profile, in the immediate forest, we found no evidence of the dark layer which defines this site, showing that all that remains is part of the eroded profile. Lithic material was recovered from the cleared profile (n = 16), the excavation unit (n = 331) and the adjacent intertidal (n = 44). Carbon samples were also recovered to date the base of the occupation.

Of the lithic material collected in the intertidal (n = 44), we recovered two preforms of stone axes, one in a greenish, talc-type raw material, like tuff (measuring 12 cm by 5 cm) (Fig. 4.25), and the other in grey basalt (10 cm by 6 cm). These are raw materials of moderate to good quality obtained locally, probably from the intertidal; they have been shaped and thinned, and the edge and sides polished. We also recovered a hammerstone with a battered end, made from a beach pebble (9.5 cm); a core of prepared platform basalt; and two wide-edged side-scrapers, one on tuff and the other on rhyolithic obsidian from Chaitén Volcano. Other items were a preform of a lanceolate stone projectile head 7 cm by 3 cm with a thick cross-section (1 cm) and the remnant of the heel of the flake, on rhyolithic obsidian; a medial fragment of a narrow biface with serrated edges and a medium cross-section (0.8 mm); two proximal fragments of thick bifaces; and a mesodistal fragment of a preform for a thick cross-section stone projectile head on rhyolithic obsidian.

Together with this assemblage of instruments and preforms (stone axes and projectile heads), 6 pieces of rhyolithic obsidian were recovered (some with their cortex), three pieces of other raw materials (sandstone, tuff?), 10 flakes of rhyolithic obsidian, a flake of andesite, three fragments of initial phases (broad heel and cortex), a retouch flake of rhyolithic obsidian, a facial thinning flake (simple marginal retouching) and 8 flakes of different raw materials in early phases of matrix formation (thick, with cortex), two on rhyolithic obsidian and the other six on different raw materials.

The analysis of the Curry Collection carried out in advance shows that in Curry's archaeological prospection (1997), she collected a similar quantity of lithic material from the intertidal (n = 40); more than half were remains associated with the manufacturing chain of lanceolate projectile points on obsidian from Chaitén (Morello 2014; Reyes et al. 2015). There were 3 thick cross-section lanceolate stone projectile heads, two thick bifacial preforms, knapping waste (n = 27), four cores, a side-scraper, a notched net weight and two doubtful items. All the stone projectile heads but two were made on rhyolithic obsidian from the Chaitén Volcano source.

Turning to the lithic materials recovered in stratigraphy, from the clearing of the eroded profile (0–30 cm), we recovered 11 fragments of rhyolithic obsidian (one with cortex), a thick cross-section biface on rhyolithic obsidian, a flake derived from a core, possibly on tuff, a marginal knapping flake on rhyolithic obsidian, a waste flake from bifacial knapping on rhyolithic obsidian and a piece of knapping waste, possibly sandstone.

The 1 m by 0.5 m test pit was divided into artificial levels. From the 0–10 cm level, we recovered two waste flakes from bifacial knapping, one marginal knapping flake, four pieces of retouching waste and 14 fragments, all on rhyolithic obsidian with the exception of one on quartz. The 10–20 cm level yielded 96 stone fragments of which 93 are rhyolithic obsidian: six flakes derived from cores on rhyolithic obsidian and three flakes from bifacial knapping also on rhyolithic obsidian, as well as 26 knapping waste pieces of rhyolithic obsidian, 11 retouch flakes and two marginal flakes indicating bifacial knapping on rhyolithic obsidian. Finally, the 20–30 cm level produced 111 fragments of rhyolithic obsidian, three pieces of obsidian, one flake derived from a sandstone core, seven waste pieces from bifacial knapping on rhyolithic obsidian, 35 pieces of retouch waste on obsidian, three retouch flakes on rhyolithic obsidian, one flake from an obsidian core and one fragment of a stone projectile head preform on rhyolithic obsidian.

We note the almost exclusive use of rhyolithic obsidian from the Chaitén Volcano source as raw material in the manufacture of instruments (stone projectile heads, bifaces and a side-scraper), represented both in the waste and in the bifacial preforms recovered. The evidence of the assemblages obtained from the intertidal and the stratigraphy do not differ in the majority use of obsidian, only in the size of the waste recovered. The pieces recovered from the stratigraphy were smaller than 1 cm, while in the intertidal they were always larger than 3 cm; the axe preforms, the hammerstone, one side-scraper and one core were worked on locally obtained resources. The waste of other materials also accounts for a minority of the total obtained.

The carbon sample obtained at a depth of 20 cm gave a date of 4070 ± 30 years BP (4583–4417, 2-sigma cal. BP, Table 4.1).

The rediscovery and search of the Isla Goñi 1 site, plus our previous study of the materials of Curry Collection, enable us to report a camp where multiple activities were carried out around ~4500 years cal. BP. The association of the instruments recovered in the intertidal, the composition and predominance of the raw materials used, and the tidal erosion of the site lead us to the conclusion that the lithic materials recovered date from the same deposition events, neither older nor more modern than those recovered in the record of the profiles and in the unit excavated.

The archaeological work carried out reveals that abundant lithic waste, cores and preforms of lanceolate stone projectile heads, in different initial stages of manufacture, were deposited in the site, but that there is an absence of finished instruments (thinning and edge-formatting stages, and to a lesser degree retouching). No doubt priority was given to making thick cross-section lanceolate projectile heads of the rhyolithic obsidian from Chaitén Volcano. This raw material was brought to the site in the form of lumps with cortex, which was only possible by canoe. Its use is clearly for making instruments for hunting.

Although there is absolutely no faunal record, the activities inferred from the material recovered also suggest fishing activities (net weights), while the presence of axe preforms, produced at the site, suggests that they were used to obtain forest resources by cutting tree trunks or as wedges to peel the bark, for inferred functions such as canoe building/maintenance and other activities involving the use of wood.

The large quantity of flakes and microflakes obtained in the excavation unit (almost exclusively of rhyolithic obsidian), unprecedented in the other sites explored in the archipelago, we explain by the fact that the areas of the site searched show evidence of the manufacture and maintenance of stone projectile heads and other instruments (e.g. side-scrapers).

Stern and Curry (1995) reported the use of rhyolithic obsidian from the Chaitén Volcano source, nearly 400 km away to the north-east. Our record supports the almost exclusive use of this raw material for the manufacture and maintenance of lanceolate stone projectile heads, associating it with the wide network of circulation and/or exchange of raw materials of groups who settled in the area.

This site presents similar behaviour to others from the Mid-Holocene in the Guaitecas Archipelago (GUA-010 Terraza and Gran Guaiteca 2 Terraza), where lithic material is present in blackened organic soils and can only be discerned by tests and excavations (Reyes et al. 2016b, 2019b).

4.4.13 Isla Goñi 2

The site is located on the north coast of Goñi Island, 1 km from Isla Goñi 1. It lies in a small bay formed by small islands which are connected at low tide. The area is heavily forested with evergreen and peat bog forest. The site is 1 m above the present sea level, on the intertidal formed of sandy sediments, facing north-east. The archaeological visibility is nil. A series of finds were made on the narrow beach lying between the forest and the rocky promontories at either end. Inland from the beach, in the forest, cores produced no evidence of dark cultural sediments (as in Isla Goñi 1). The materials are scattered along the beach for 180 m from north-west to south-east, and were recovered at low tide. The intertidal is almost 3 m wide, limited by the sea in front and the forest immediately behind where the salt water does not reach.

The material recovered from the sandy intertidal consists of a preform of a thick cross-section (1.5 cm) lanceolate projectile head made of light brown tuff, 8 cm long and 3.5 cm wide, a bifacial axe preform on basalt, two pieces of rhyolithic obsidian and two flakes derived from cores also of the same rhyolithic obsidian from Chaitén Volcano, according to our macroscopic observation.

The Isla Goñi 2 site represents an extension of activities derived from the camp at Isla Goñi 1 (1 km further north). Flakes and two preforms (one of a projectile head for hunting and the other an axe for working wood) discarded during occupation were found in the intertidal.

4.4.14 Cholguero Isla Goñi 3

The site lies on the north-west coast of Isla Goñi, on the eastern shore of a small, north-facing bay. The cholguero port is in the partially cleared forest, and on the intertidal composed of small pebbles and large rocks. The area is forested with evergreen forest and peat bogs, and lies close to the shoreline formed by a beach of sand and pebbles interspersed with rocky outcrops.

The site is 1 m above the current sea level and 2 m from the shoreline, facing north. The forest in the area has been felled and cleared, so that now there are grassy clearings covered with scrub. The archaeological visibility is medium. We recorded the remains of an old wooden structure with a hearth in a ring of stones, all covered and grassed over. The site presents a slope of 40°.

The site is an abandoned historical cholguero, partly covered by a layer of plant matter 10 cm thick in some areas. In others the shell-midden is completely exposed, showing the intensity of extraction of just a few species of malacological fauna (Fig. 4.26). No pre-Hispanic occupation or archaeological material was found below the cholguero, nor on the surface or in the nearby intertidal. The test cores indicated a deposit of shell-midden matrix 200 cm thick, consisting mainly of cholgas (*Aulacomya ater*), choro mussels (*Choromytilus chorus*) and Chilean mussels (*Mytilus edulis chilensis*). Below this shell matrix is the base rock which forms the foundation of the cholguero. According to estimates based on drilling cores, the area is 10 m from east to west and 10 m from north to south, covering approximately 100 m^2.

Cholguero Isla Goñi 3 is a campsite developed by seasonal visits, possibly reiterated, related to intensive, selective gathering of mytilids for the canning industry. This activity probably implied other associated practices, like hunting and fishing for the fishermen's and mussel gatherers' subsistence. The start of occupation may be dated to the beginning or middle of the twentieth century.

4.4.15 Cholguero Isla Goñi 4

The site lies on the south-west coast of Isla Goñi, on the eastern shore of a small, north-facing bay. The cholguero is in the partially cleared forest, and on the intertidal composed of small pebbles and large rocks. The area is forested with evergreen forest

Fig. 4.26 View of Cholguero Isla Goñi 3, showing the great accumulation of cholga and mussel shells projecting onto the intertidal. They have been thrown down from the camp above, where the wooden structure was constructed after the forest had been cleared

and peat bogs, and lies close to the shoreline formed by a beach of sand and pebbles interspersed with rocky outcrops.

The site is 1 m above current sea level and 2 m from the shoreline, facing north. The forest in the area has been felled and cleared, so that now there are grassy clearings covered with scrub. The site has a slope of 40° and the archaeological visibility is medium.

The site is an abandoned historical cholguero, partly covered by a layer of plant matter 10 cm thick in some areas. No pre-Hispanic occupation or archaeological material was found below the site, nor on the surface or in the nearby intertidal. The test cores indicated a deposit of shell-midden matrix 100 cm thick, consisting mainly of cholgas (*Aulacomya ater*), choro mussels (*Choromytilus chorus*) and Chilean mussels (*Mytilus edulis chilensis*). Below this shell matrix is the base rock which forms the foundation of the cholguero. According to estimates based on drilling cores, the area is 10 m from east to west and 10 m from north to south, covering approximately 100 m^2.

Cholguero Isla Goñi 4 is a campsite developed by seasonal visits, possibly reiterated, related to intensive, selective gathering of mytilids for the canning industry. This activity probably implied other associated practices, like hunting and fishing for the fishermen's and mussel gatherers' subsistence. The start of occupation may be dated to the beginning or middle of the twentieth century.

4.4.16 Cholguero Isla San José 1

The site lies on the west coast of San José Island, off the Taitao Peninsula, in a small, north-facing bay. The cholguero is in the forest, and on the intertidal composed of small pebbles and large rocks. The area is forested with evergreen forest and peat bogs, and lies close to the shoreline formed by a beach of sand and pebbles interspersed with large rocks. The site is 2 m above the current sea level and 2 m from the shoreline. It has a slope of 40° and the archaeological visibility is nil.

The site is an abandoned historical cholguero, totally covered by a plant layer 10 to 20 cm thick. No occupation or archaeological material was found below the cholguero, nor on the surface or in the nearby intertidal. The test cores indicated a deposit of shell-midden matrix 200 cm thick, consisting mainly of cholgas (*Aulacomya ater*), choro mussels (*Choromytilus chorus*) and Chilean mussels (*Mytilus edulis chilensis*). Below this shell matrix is the base rock which forms the foundation of the site. According to our estimates, the area is 20 m from east to west and 20 m from north to south, covering approximately 400 m^2.

Cholguero Isla San José 1 is a campsite developed by seasonal visits, possibly reiterated, related to intensive, selective gathering of mytilids for the canning industry. This activity probably implied other associated practices, like active hunting and fishing and passive fishing using stone fishing pens for the fishermen's and mussel gatherers' subsistence. The start of occupation may be dated to the beginning or middle of the twentieth century.

4.4.17 Cholguero Estero Puelma 1

The site is located on the Taitao Peninsula, on the north shore of the mouth of the Puelma inlet, in a south-facing bay. The cholguero is in the forest and on the intertidal composed of a sand substrate with small pebbles and large rocks. The area is forested with evergreen forest and peat bogs, close to the shoreline formed of flat stones and pebbles. The site is 1 m above the current sea level and 2 m from the shoreline; it has a slope of 20° and the archaeological visibility is medium.

The site is an abandoned historical cholguero, partly covered by a layer of plant matter 10 cm thick in some areas. Below the shell matrix is a grey layer of glacial silt. There are the remains of a hearth, ringed by rocks and very overgrown, and a fragment of a whisky bottle was found in the intertidal. There is also an old apple tree (Fig. 4.27).

According to estimates based on cores taken in the shell-midden, this cholguero is around 30 m from east to west and 10 m from north to south, covering an area of 300 m^2 with a deposit of shell-midden matrix 20 cm deep, composed mainly of cholgas (*Aulacomya ater*), and to a lesser degree choro mussels (*Choromytilus chorus*) and Chilean mussels (*Mytilus edulis chilensis*). No pre-Hispanic archaeological material was found below the cholguero, nor on the surface or in the nearby intertidal.

Fig. 4.27 View of Cholguero Estero Puelma 1. The historical activities left areas of thinned forest, now overgrown with scrub, and a large apple tree beside the intertidal

Cholguero Estero Puelma 1 is a campsite developed by seasonal visits, possibly reiterated, related with intensive, selective gathering of mytilids for the canning industry. This activity probably implied other associated practices, like hunting and fishing for the fishermen's and mussel gatherers' subsistence. Fruit trees were also cultivated. The start of occupation may be dated to the beginning or middle of the twentieth century.

4.4.18 Cholguero Península Taitao 1

The site lies on the east coast of the central section of the Taitao Peninsula, on the south shore of a small, north-facing bay. The cholguero is in the partially cleared forest and on the intertidal composed of small pebbles and large rocks. The area is forested with evergreen forest and peat bogs, and lies close to the shoreline formed by a beach of sand and pebbles interspersed with rocky outcrops.

The site is 0 m above the current sea level and 10 m from the shoreline. The forest has been felled, and fallen trunks have been positioned to tie up boats. We found fishing pens and the remains of an old wooden structure with a hearth ringed by stones, all covered and grassed over. The site has a slope of 10° and the archaeological visibility is medium. It is an abandoned historical cholguero, partly covered by a layer of plant matter 10 cm thick in some areas. The test cores indicated a deposit of shell-midden matrix 50 cm thick, consisting mainly of cholgas (*Aulacomya ater*), choro mussels (*Choromytilus chorus*) and Chilean mussels (*Mytilus edulis chilensis*). Below this shell matrix is the base rock which forms the foundation of the site.

Fig. 4.28 View of Cholguero Peninsula Taitao 1 and fishing pen in a stream, showing that the sea enters at high tide

The semicircular fishing pen is approximately 10 m from east to west and 1 to 1.5 m wide, formed of large and medium-sized stones and rocks in the intertidal. Another smaller pen continues below it, towards the low tideline, 6 m long and 50 cm wide on average, at 1 masl. Both traps are set in the bed of a small tidal stream mouth; they are clearly disused and covered with moss (Fig. 4.28).

The area of the cholguero is about 80 m from east to west and 40 m from north to south, covering approximately 3,200 m^2. No occupation or archaeological material was found below the cholguero, nor on the surface or in the nearby intertidal.

Cholguero Península Taitao 1 is a campsite developed by seasonal visits, possibly reiterated, related to intensive, selective gathering of mytilids for the canning industry. This activity probably implied other associated practices, like active hunting and fishing and passive fishing using stone fishing pens for the fishermen's and mussel gatherers' subsistence. The start of occupation may be dated to the beginning or middle of the twentieth century.

4.4.19 Cholguero Península Taitao 2

The site lies on the east coast of the central section of the Taitao Peninsula, 300 m south of Cholguero Península Taitao 1, at the head of a narrow, east-facing bay. The cholguero is in the forest and on the intertidal composed of small pebbles and large rocks. The area is forested with evergreen and peat bog forest, and lies close to the shoreline formed by a beach of sand and pebbles interspersed with rocky outcrops. The site is 2 m above current sea level and 0 m from the shoreline. It has a slope of 40° and the archaeological visibility is medium.

The site is an abandoned historical cholguero, partly covered by a layer of plant matter 10 cm thick in some areas. Below the shell matrix which defines the site is the base rock which forms the foundation. Approach by boat is restricted. On the south shore of this bay, where the pebble beach begins, a large shell-midden of cholga shells can be seen on the rocks, 2 m above sea level (Fig. 4.29). There are also the remains of an old wooden structure and signs of an old hearth ringed by stones. Following the pebble beach and bed of a small tidal stream, there is a fishing pen approximately 18 m from east to west, rather elliptical in shape.

According to estimates based on drilling cores in the shell-midden, and including the fishing pen, the area is 40 m from east to west and 20 m from north to south, covering approximately 800 m^2. The test cores indicated a deposit of shell-midden

Fig. 4.29 View of Cholguero Peninsula Taitao 2, located immediately above the intertidal

matrix 200 cm thick, consisting mainly of cholgas (*Aulacomya ater*), as well as choro mussels (*Choromytilus chorus*) and Chilean mussels (*Mytilus edulis chilensis*). No occupation or archaeological material was found below the cholguero, nor on the surface or in the nearby intertidal.

Cholguero Península Taitao 2 is a campsite developed by seasonal visits, possibly reiterated, related to intensive, selective gathering of mytilids for the canning industry. This activity probably implied other associated practices, like active hunting and fishing and passive fishing using stone fishing pens for the fishermen's and mussel gatherers' subsistence. The start of occupation may be dated to the beginning or middle of the twentieth century.

4.4.20 Cholguero Península Taitao 3

The site lies on the east coast of the central section of the Taitao Peninsula, almost 2 km south of Cholguero Península Taitao 1 and 2, on the overgrown south-facing coast at the mouth of the Estero Vidal inlet. The cholguero is in the forest and on the intertidal composed of small pebbles and large rocks. The area is forested with evergreen forest and peat bogs, and lies close to the shoreline formed by a beach of sand and pebbles interspersed with rocky outcrops. The site is 0 m above the current sea level and 1 m from the shoreline. It has a slope of 40° and the archaeological visibility is medium. The site is an abandoned historical cholguero, partly covered by a layer of plant matter 10 cm thick in some areas. The test cores indicated a deposit of shell-midden matrix 200 cm thick, consisting mainly of cholgas (*Aulacomya ater*), choro mussels (*Choromytilus chorus*) and Chilean mussels (*Mytilus edulis chilensis*). Below this shell matrix is the base rock which forms the foundation of the cholguero.

On the adjacent pebble beach, there are wooden structures which formed the camp and an area cleared of stones and rocks for hauling boats up without damaging them—a sort of path free of rocks can be seen on the beach (Fig. 4.30).

According to estimates based on drilling cores in the shell-midden, and including the area for hauling up boats, the area is 20 m from east to west and 20 m from north to south, covering approximately 400 m^2. No occupation or archaeological material was found below the cholguero, nor on the surface or in the nearby intertidal.

Cholguero Península Taitao 3 is a campsite developed by seasonal visits, possibly reiterated, related with intensive, selective gathering of mytilids for the canning industry. This activity probably implied other associated practices, like hunting and fishing for the fishermen's and mussel gatherers' subsistence. The start of occupation may be dated to the beginning or middle of the twentieth century.

Fig. 4.30 View of Cholguero Peninsula Taitao 3 and its location close to the shoreline. The cleared path for hauling up boats can be seen in the intertidal, with part of the wooden structure of the camp

4.4.21 Península Taitao 4

The site lies towards the end of the southern coast of the Taitao Peninsula, on the narrow strip of land that separates the end of the south coast of Estero Thompson from the outlet of Presidente Ríos Lake located in the centre of the peninsula. This site is not a cholguero, but a path cut and cleared through the forest, with steps dug out and tree trunks embedded in the soil for portaging boats overland (Fig. 4.31) between two bodies of water. The area is forested with evergreen forest and peat bogs, close to the forested shoreline formed of glacial silt and peat. No beach is observed in this sector.

The site is 8 m above current sea level and between 0 and 40 m from the shoreline. It has a slope of between 20° and 40°, and the archaeological visibility is low. It consists of a path which is now abandoned, or very seldom used. After a climb of some 30 m, it reaches a level forested area, which has been felled and partly cleared, with overgrown hearths covered with organic earth and moss. This cleared area served as a temporary campsite. According to estimates based on cores, there are no shells, only burnt wood. The area of the site is about 30 m from east to west and 30 m from north to south, covering approximately 900 m^2. No pre-Hispanic archaeological materials were recorded.

Península Taitao 4 is a campsite and path used for portaging boats (Borrero 1997, Prieto et al. 2000). It was created for seasonal visits, possibly reiterated, related to the movement of marine otter-hunters, fishermen and/or shellfish gatherers from Estero Thompson to the interior of the Taitao Peninsula by crossing Presidente Ríos Lake by boat. This is one of the historical portages mentioned to avoid going round the peninsula to the west, by the open sea. The other, better-known portage is over

Fig. 4.31 View of Península Taitao 4 site. On the coast of Estero Thompson, there is a portage route made with trunks laid crosswise for portaging boats. (*'envaralado'*) View of the remains of an old stone-ringed hearth, overgrown with grass and moss

the Ofqui Isthmus. Its earliest use may be dated to the beginning or middle of the eighteenth century (Urbina 2010, 2013).

4.4.22 Península Taitao 6

The site is located on a north-facing coast of the Taitao Peninsula, opposite Stewart Island, in a north-facing bay. The cholguero is in the forest and on the intertidal composed of a sand substrate with small pebbles and large rocks. The area is forested with evergreen forest and peat bogs, and the shoreline is formed by a pebble beach. The site is 3 m above the current sea level and 3 m from the shoreline. It presents a slope of 30°. The site is an abandoned historical cholguero, partly covered by a layer of plant matter 10 cm thick in some areas. Test cores indicated a deposit of shell-midden matrix 200 cm thick, consisting mainly of cholgas (*Aulacomya ater*), as well as choro mussels (*Choromytilus chorus*) and Chilean mussels (*Mytilus edulis chilensis*). Below this shell matrix is the base rock which forms the foundation of the site. Stakes have been installed to tie up boats. The archaeological visibility is high.

According to estimates based on drilling cores in the shell-midden, the area is 10 m from east to west and 20 m from north to south, covering approximately 200 m^2. No occupation or archaeological material was found below the cholguero, nor on the surface or in the nearby intertidal.

Cholguero Península Taitao 6 is a campsite developed by seasonal visits, possibly reiterated, related to intensive, selective gathering of mytilids for the canning industry. This activity implied other associated practices, like hunting and fishing for the fishermen's and mussel gatherers' subsistence. The start of occupation may be dated to the beginning or middle of the twentieth century.

4.4.23 Península Taitao 7

The site is located in the southern part of the Taitao Peninsula coast, opposite Stewart Island, in a north-facing bay. The cholguero is in the forest and on the intertidal composed of a sand substrate with small pebbles and large rocks. The area is forested with evergreen forest and peat bogs, and the shoreline is formed by a pebble beach. The site is 2 m above current sea level and 2 m from the shoreline; it has a slope of 20° and the archaeological visibility is medium. The site is an abandoned historical cholguero, partly covered by a layer of plant matter 10 cm thick in some areas. The test cores indicated a deposit of shell-midden matrix 100 cm thick, consisting mainly of cholgas (*Aulacomya ater*), as well as choro mussels (*Choromytilus chorus*) and Chilean mussels (*Mytilus edulis chilensis*). Below this shell matrix is a layer of sand and pebbles which forms part of the palaeo-beach.

According to estimates based on drilling cores in the shell-midden, the area is 10 m from east to west and 10 m from north to south, covering approximately 100 m^2. No occupation or archaeological material was found below the cholguero, nor on the surface or in the nearby intertidal.

Cholguero Península Taitao 7 is a campsite developed by seasonal visits, possibly reiterated, related to intensive, selective gathering of mytilids for the canning industry. This activity implied other associated practices, like hunting and fishing for the fishermen's and mussel gatherers' subsistence. The start of occupation may be dated to the beginning or middle of the twentieth century.

4.4.24 Península Taitao 8

The cholguero is located in the southern part of the Taitao Peninsula coast, opposite Stewart Island, in a south-facing bay. It is in the forest and on the intertidal composed of a sand substrate with small pebbles and large rocks. The area is forested with evergreen forest and peat bogs, and the shoreline is formed by a pebble beach. The site is 2 m above the current sea level and 2 m from the shoreline. It has a slope of 20° and the archaeological visibility is medium.

The site is an abandoned historical cholguero, totally covered by a plant layer 10 cm thick. The shell-midden, around 50 cm thick, is formed almost exclusively of mytilids. Below this shell matrix is a layer of sand and pebbles which forms part of the palaeo-beach. According to estimates based on drilling cores in the shell-midden, the area is 10 m from east to west and 10 m from north to south, covering approximately 100 m^2. No occupation or archaeological material was found below the cholguero, nor on the surface or in the nearby intertidal.

Cholguero Península Taitao 8 is a campsite developed by seasonal visits, possibly reiterated, related to intensive, selective gathering of mytilids for the canning industry. This activity implied other associated practices, like hunting and fishing for the fishermen's and mussel gatherers' subsistence. The start of occupation may be dated to the beginning or middle of the twentieth century.

4.4.25 Cholguero Stewart 1

The cholguero is on the south coast of Stewart Island, in a south-east-facing bay, in the forest and on the intertidal composed of a sand substrate with small pebbles and large rocks. The area is forested with evergreen forest and peat bogs, and the shoreline is formed by a pebble beach.

The site is 1 m above the current sea level and 2 m from the shoreline; it has a slope of 20° and the archaeological visibility is medium. It is an abandoned historical cholguero, partly covered by a layer of plant matter 10 cm thick in some areas. In others the shell-midden is totally exposed, showing that it is formed almost exclusively of mytilids. It is around 50 cm thick. Below the shell matrix is a grey layer of glacial silt. The remains of a wooden structure are also visible, and on the beach an area of the intertidal has been cleared for hauling up boats, making a sort of path between the rocks (Fig. 4.32).

According to estimates based on drilling cores in the shell-midden, the area is 10 m from east to west and 10 m from north to south, covering approximately 100 m^2. No occupation or archaeological material was found below the cholguero, nor on the surface or in the nearby intertidal.

Cholguero Stewart 1 is a campsite developed by seasonal visits, possibly reiterated, related to intensive, selective gathering of mytilids for the canning industry. This activity implied other associated practices, like hunting and fishing for the fishermen's and mussel gatherers' subsistence. The start of occupation may be dated to the beginning or middle of the twentieth century.

4.4.26 Cholguero Stewart 2

The site is on the south coast of Stewart Island, in a west-facing bay. The cholguero is in the forest, and on the intertidal composed of a sand substrate with small pebbles

Fig. 4.32 View of Cholguero Stewart 1. Part of the shell-midden is covered by vegetation and part exposed and subject to tidal action. Note the clear area in the intertidal for hauling up boats

and large rocks. The area is forested with evergreen forest and peat bogs, and the shoreline is formed by a pebble beach.

The site is 1 m above the current sea level and 2 m from the shoreline; it has a slope of 20° and the archaeological visibility is low. The site is an abandoned historical cholguero, totally covered by a plant layer 10 cm thick. The test cores indicated a deposit of shell-midden matrix 30 cm thick, consisting mainly of cholgas (*Aulacomya ater*), as well as choro mussels (*Choromytilus chorus*) and Chilean mussels (*Mytilus edulis chilensis*). Below this shell matrix is the base rock which forms the foundation of the site.

According to estimates based on drilling cores in the shell-midden, the area is 10 m from east to west and 5 m from north to south, covering approximately 50 m^2. No occupation or archaeological material was found below the cholguero, nor on the surface or in the nearby intertidal.

Cholguero Stewart 2 is a campsite developed by seasonal visits, possibly reiterated, related to intensive, selective gathering of mytilids for the canning industry. This activity implied other associated practices, like hunting and fishing for the fishermen's and mussel gatherers' subsistence. The start of occupation may be dated to the beginning or middle of the twentieth century.

4.4.27 Cholguero Stewart 3

The site is on the south coast of Stewart Island, in an east-facing bay. The cholguero is in the forest and on the intertidal composed of a sand substrate with small pebbles

Fig. 4.33 View of Cholguero Stewart 3, partly covered by vegetation. The anthropic sediments (shell-midden matrix) fall towards the intertidal

and large rocks. The area is forested with evergreen forest and peat bogs, and the shoreline is formed by a pebble beach. The site is 3 m above the current sea level and 3 m from the shoreline; it has a slope of 30° and the archaeological visibility is high.

The site is an abandoned historical cholguero, partly covered by a layer of plant matter 10 cm thick in some areas. The test cores indicated a deposit of shell-midden matrix 500 cm thick, consisting mainly of cholgas (*Aulacomya ater*), as well as choro mussels (*Choromytilus chorus*) and Chilean mussels (*Mytilus edulis chilensis*). Below this shell matrix is the base rock which forms the foundation of the site. There are also the remains of a bedstead, an overgrown hearth and old stakes for tying up boats (Fig. 4.33).

According to estimates based on drilling cores in the shell-midden, the area is 20 m from east to west and 50 m from north to south, covering approximately 1000 m^2. No occupation or archaeological material was found below the cholguero, nor on the surface or in the nearby intertidal.

Cholguero Stewart 3 is a campsite developed by seasonal visits, possibly reiterated, related to intensive, selective gathering of mytilids for the canning industry. This activity implied other associated practices, like hunting and fishing for the fishermen's and mussel gatherers' subsistence. The start of occupation may be dated to the beginning or middle of the twentieth century.

4.5 Offshore Islands and Pacific Border area of the Chonos Archipelago (43°31′–44°55′S)

Below, we describe the prospection of the westerly area of the archipelago exposed to the Pacific (Figs. 1.2 and 4.12).

In this area of prospection, no cultural material or archaeological sites were recorded. Nevertheless, we will describe briefly the areas covered in order to include them in the general archaeological discussion, and subsequently assess the implications of this result.

We prospected the north-west and north-east coasts of Guafo Island, covering 32 km of coastline. These coasts are fairly well protected from the prevailing south-westerly winds, and are closer to Chiloé and the Guaitecas Archipelago, 40 km away. Our choice was therefore conditioned by two variables to determine possible prehistoric access to the area: relative protection from the wind for sea travel and relative proximity to possible starting points. We also considered—as in the rest of the archipelago—the important factor of tectonic subsidence which has modelled the whole region and resulted in sinking of the central area of the archipelago and emergence of the Pacific border zone down to the Taitao Peninsula (Plafker and Savage 1970). In this island in particular, the Valdivia earthquake of 1960 caused the land to rise by more than 5 m, causing a tsunami recorded as 30 m high (SHOA 2000). Thus apart from prospecting the immediate shoreline, which was under water 57 years ago, we went further inland in bays and sectors with accessible slopes up to 30 masl. In most of the island the shore is steep, with slopes of more than 40° and cliffs in some places, making access difficult (Fig. 4.34).

In the absence of results, we were able to interview several fishermen who were familiar with the area to obtain information on finds which they might know about (ossuaries, cholguero ports and archaeological shell-middens are generally known if they are exposed and easily seen). The results of the interviews were also negative. It may be added that we used this method whenever we had the chance of talking to seamen/fishermen in any part of the archipelago.

We subsequently visited parts of the east coast of Ipún Island and the west coast of Stokes Island, finding that both had risen by approximately 2 m (Fig. 4.35). The results were again negative, as they were in the western islets off Johnson Island and the south coast of Forsyth Island. The total length of shoreline prospected in these islands was 40 km.

It should be noted, however, that despite this absence of a record of archaeological sites in our prospection, information exists of skeletal remains recovered from caves and/or rock shelters used as ossuaries on Ipún, Johnson and Sin Nombre Islands. The bone remains were recovered in the context of judicial investigations after reports of assumed accidents made by local fishermen. The remains were collected by the Chilean Police and Navy, who kept no contextual records. It was, therefore, impossible to locate the sites in the field to complete our record. We know, however, that

Fig. 4.34 View of the north-east coast of Guafo Island. Cliffs can be seen which emerged in the 1960 earthquake, and the emerged sea bottom with exposed sedimentary rock

the area of islands exposed to the Pacific presents at least some archaeological occupations determined by the presence of funerary assemblages in the form of ossuaries. We will discuss these assemblages in the following section.

4.5.1 Unpublished and Partly Reported Sites

Many published sites present unpublished information, as these publications pursued different objects (e.g. diet and subsistence, description of materials and geomorphological changes). We will summarise briefly the descriptions of the archaeological sites recorded in order to organise the information in the context of the data presented above.

Fig. 4.35 Protected bay on the east coast of Ipún Island. In this bay we can observe sedimentary rocks of the sea bed and a shingle beach

4.5.2 Northern Area of the Archipelago. Repollal 02, Repollal Caverna and Puquitín 01 Sites

These sites are in the northern area of the Chonos Archipelago, the Guaitecas Archipelago, specifically on the Puquitín Channel which runs between Gran Guaiteca and Ascensión Islands (Fig. 4.10). The sites were searched and their basic features reconstructed, based mainly on the information provided by the anthropologist Eugenio Aspillaga (personal communication 2012) about earlier work in the area (Ocampo and Aspillaga 1984, 1991). Only the Repollal 02 site (shell-midden with burials) could be identified in the field, and subsequently reassigned to the work done and records kept by us in the site which we called Gran Guaiteca 2 Conchal.

The human remains recovered in the earlier work (Ocampo and Aspillaga 1984, 1991), and partially described by bio-anthropological studies (Aspillaga et al. 1990; Constantinescu and Aspillaga 1991), form part of the Chonos Osteological Collection held in the University of Chile. We know, from a database that has now been created, that the Repollal Caverna site was an ossuary containing around 17 individuals, while Puquitín 01 was a cave with shell-midden matrix, containing human remains of 19 individuals deposited on the surface (Aspillaga personal communication 2012). The contexts had been moved and suffered anthropic disturbance. In both cases, only the osseous materials on the surface were recovered. In this way, we obtained samples of bone fragments to situate these two contexts chronologically and evaluate the isotope results of the human samples selected (Reyes et al. 2013, 2015). The isotope values obtained from the individuals dated enabled us to establish consumption of an exclusively marine diet (see Chapter 6, Evaluation and Discussion of the Evidence). An adult individual of indeterminate age from each site was dated; the results were ~400 years cal. BP (Repollal Caverna) and ~600 years cal. BP (Puquitín 01) (Reyes et al. 2013). Whether these ossuaries were formed synchronically, with no statistical difference, or whether this was a funerary practice maintained over a longer time scale can only be determined with a set of direct, representative dates for each site.

4.5.3 Mainland Coast Area of the Archipelago—Seno Gala 1 Site

At the start of our investigation, we recorded an archaeological site on the mainland coast of the Chonos Archipelago, the details of which we can now publish at leisure. This is the Seno Gala 1 site (Reyes et al. 2007, Table 4.1), located in the Melimoyu Peninsula (Fig. 4.18) very close to the funerary sites mentioned in this study (Osario Melimoyu, Isla Yalac 1 and Estero Sur). It is the only archaeological site recorded to date as a camp on the mainland coast, with archaeological materials both in stratigraphy and the adjacent intertidal, in addition to the funerary record of bones. We will briefly summarise its principal features in order to include it in the discussion.

The Seno Gala 1 site is located on the east shore of Gala Sound, 2.5 km from the head of the sound. It stands on a small peninsula at the entrance to a bay, opening to the west, which connects via an inland channel with a small lake ~3 km inland. It is an evergreen and peat bog forest, and the archaeological visibility is nil. The two components of the site, lithic remains in the intertidal and shell-midden above the foreshore, are, respectively, -1 m below and 2 m above current sea level, and 0 and 10 m from the shoreline. The intertidal is 40 m from north to south and 60 m from east to west.

The site is divided into two sectors: the first is a shell-midden located on the peninsula, on which a house has been built. This promontory has a steep shoreline to the north and east; to the south it adjoins a forested hill while to the west it looks onto a small pebble beach. Cores indicated the presence of a shell-midden, maximum 120 cm thick in the centre (i.e. 1.8 masl) and diminishing to the west where the stratum slopes down to the pebble beach. The second sector is the intertidal, consisting of a pebble beach (between -1 m and 0.5 m), adjacent to the peninsula and very well protected by the granite on which it stands, forming a small bay. A series of stone instruments and lithic waste were found under water on this beach (Fig. 4.36). The situation is similar to that described at the start of this chapter in the area of the Cucuquén Stream, on Gran Guaiteca Island, where some shell-middens and occupations on higher terraces can be related to the lithic material recorded in

Fig. 4.36 Seno Gala 1 site. View from the shell-midden towards the intertidal where lithic material was recovered (flakes, cores, bifaces and medium cross-section lanceolate points)

the intertidal adjacent to the sites. We therefore dug a 1m^2 pit in the shell-midden and collected superficial material in the intertidal.

The lithic sample analysed consisted of 69 pieces recovered either on the surface (n = 53) or in stratigraphy (n = 16). The evidence recovered shows a significant variety of raw materials selected (n = 9); analysis by major petrographic groups shows that the most frequently used types were siliceous rocks and andesites. We note, however, the presence of obsidian in the assemblage samples (n = 7, of a total of 69 pieces); its chemical composition shows that it is rhyolithic obsidian from Chaitén Volcano, 220 km further north. Among the lithic categories represented, we found priority given to bifacial designs, particularly fragments of thick cross-section lanceolate projectile heads in different stages of manufacture. A smaller number of edged tools of various designs was observed, including denticulate edges and notches possibly used for working wood (Reyes et al. 2007).

In the excavation, we recorded 335 bone remains of marine and terrestrial mammals, birds and fish; fish remains predominated. In general terms, no clear pattern of abundance was observed, with homogeneous presence in all the levels of the excavation. Jurel (*Trachurus murphyi*, 80% of total NISP) were identified by their sagittal crests. The differences in the sizes of these anatomical features indicate the capture of a wide spectrum of specimens, which might reflect a strategy of non-selective capture. The most common mammals were pinnipeds (13.3% of total

NISP), although no diagnostic anatomical elements could be detected for identification at species level; we identified remains of crania, ribs and phalanges, especially in the lowest level of the test pit (50–60 cm). Terrestrial mammals were also found in the lower levels, e.g. a juvenile pudú (*Pudu pudu*, 5% of total NISP). One bone instrument was recorded, a punch or awl, made from a cormorant tarsometatarsus (*Phalacrocorax sp.*; 1.6% of total NISP). Turning to the malacological material, we identified the presence of locos (*Concholepas concholepas*), cholgas (*Aulacomya ater*), choro mussels (*Choromytilus chorus*), keyhole limpets (*Fissurella sp.*) and other gastropods (Reyes et al. 2007). Samples were taken of carbon and pudú bone (*Pudu Pudu*) from the base of the shell-midden deposit, giving a range around 1200 years cal. BP for the formation of this deposit.

The Seno Gala 1 site was a camp where many activities were carried out and there were successive reoccupations from almost 1200 years cal. BP. The most evident is the discard of waste from the gathering of molluscs, reflecting local resources obtained in the general gathering strategy. The archaeofaunal record also enabled us to establish the use of a wide range of resources including terrestrial (pudú) and marine mammals (pinnipeds), fish (jurel) and to a lesser degree birds. Given the ubiquity of most of the prey species consumed, only the fact that the pudú was a juvenile, and the characteristics of the habitat for jurel suggest—for the moment—seasonal summer use of the site. Jurel comes in to the coast in large shoals, especially in summer, and can be fished with nets or hooks (Meléndez et al. 1993).

The lithic material recovered from the intertidal reflects a limited set of activities carried out at the site, predominantly the production of large, thick cross-section lanceolate stone projectile points. The small proportion of other evidence may imply that the set produced was kept and not discarded in the site, or else that sampling factors specific to the area where the pieces were exposed mean that they do not necessarily reflect the whole camp. In view of the evidence, we incline to the latter alternative. The processing of lithic resources, their significant diversity, the representation of final phases of operating chains—suggesting functional relations with other camps—and the lithic categories observed suggest a strong influence of transport, and consequently high mobility. This is supported by the use of rhyolithic obsidian from Chaitén Volcano, 120 km north of the site.

4.5.4 Central Area of the Archipelago. Isla Harris, Seno Canalad and Cueva Benjamín 1

In the area defined in this study as the centre of the Chonos Archipelago (Fig. 4.21), Ocampo and Aspillaga (1984, 1991) studied sites which were funerary contexts like caves and/or rock shelters with human remains on the surface. Only the Cueva Benjamín site could be identified in the field and subsequently reassigned to the work which we carried out and recorded as Isla Benjamín 1, the sediments of which also contain an anthropic shell-midden matrix. We know that the Isla Harris archaeological site was an ossuary containing around 34 individuals, while the Seno Canalad context reportedly presented the human remains of at least 11 individuals, similar to Isla Benjamín 1 (n = 12). The funerary contexts had apparently been moved and suffered anthropic disturbance. In the cases of Isla Harris and Seno Canalad, only the bone material on the surface was recovered. In Isla Benjamín 1, an archaeological excavation of at least 2 m^2 was carried out, as we confirmed in the field; however, no information on this exists. We obtained samples of bone fragments from these three contexts to situate them chronologically and evaluate the isotope results of the human samples selected (Reyes et al. 2013, 2015, 2019a). The isotope values obtained from the individuals dated enabled us to establish consumption of an exclusively marine diet. One adult individual each, of indeterminate age, from Isla Harris and Seno Canalad, and two adults of indeterminate age from Isla Benjamín 1 were dated with results around ~400 years cal. BP. Whether these ossuaries were formed synchronically, with no statistical difference, or whether this was a funerary practice maintained over a longer time scale can only be determined with a set of direct, representative dates for each site.

4.5.5 Central-South Area of the Archipelago

In the central-south prospection area of the Chonos Archipelago (Fig. 4.27), in the first systematic prospections, we made detailed records of a series of archaeological sites, some of them being cholguero ports (Reyes et al. 2011, 2013, 2015). The area covered included the shorelines of the west coast of Traiguén Island, the north coast of Quemada Island and the south coast of Luz Island, as well as the north-west coast of Cuche Channel in Victoria Island, part of the Smith Group and the Quetros Islets in the Darwin Channel, and Elena Island, at the entrance to Aysén Fjord, a total of more than 250 km of shoreline. Some of the sites form part of this unpublished work, since further activities were carried out after the initial reports (e.g. excavations, records), as in the Nahuelquín 1 and Isla Elena 1 sites.

There is also a set of sites that were mentioned and described briefly in accordance with the specific objectives of a previous publication (Reyes et al. 2015). Here, we summarise the information collected in order to homologate it with the data obtained from the site records of this work, so as to include all the information available in the general archaeological discussion (Table 4.1).

4.5.6 Nahuelquín 2

This site is an old fishing pen located on the north-east coast of Traiguén Island, 500 m north of Nahuelquín 1. It is in a small, north-facing bay where the tide goes out a long way due to the gentle slope of the intertidal, around 20°. There is a line of large

Fig. 4.37 View of the Nahuelquín 2 fishing pen. The land has risen, grass and rushes are growing round the rocks and they are covered with lichen

and medium rocks obtained from the immediate vicinity (rocky and stony shoreline), running from south-west to north-east in an elliptical curve for nearly 25 m. Part of this anthropic accumulation of rocks is in the water, but about two-thirds are more than 2 m above the existing sea level, with rushes growing round the rocks and lichens on their surfaces (Fig. 4.37). This bears witness to the change produced by vertical tectonic movements and/or sea level changes, recorded previously for other sites in the region (Reyes et al. 2011, 2018). The archaeological visibility of the site is high.

The function of the site was for passive fishing, using the geomorphology of the coast. The gentle slope of the intertidal and the arrangement of rocks in a line create a pen where fish can enter at high water and remain trapped when the tide goes down. The presence of stone axes suggests that these were used to obtain forest resources, such as cutting trunks or as wedges for stripping their bark, for inferred purposes like the maintenance and construction of canoes, and/or other activities involving the use of wood. The chronology of the site is unknown, however, Benjamín Nahuelquín—a resident of the area—says that when he arrived more than 50 years ago, the pen already looked as it does now.

4.5.7 Nahuelquín 3

This site is an old fishing pen located on the north-east coast of Traiguén Island, 1.2 km south of Nahuelquín 1. The coast in this area forms a very enclosed bay, almost closed, facing north-west. Like the pen described above, it is formed by a line of big rocks in an elliptical curve extending for 70 m roughly north-east to south-west. Here also, changes can be observed in the sea level, as shown by the proliferation of lichens on the rocks of the pen and the grass growing round them. The pen is 2 m above the current sea level and between 1 and 70 m from the shoreline. The archaeological visibility of the site is high (Reyes et al. 2011, 2018).

The site was used for passive fishing, using the geomorphology of the coast. The gentle slope of the intertidal and the arrangement of rocks in a line create a pen where fish can enter at high water and remain trapped when the tide goes down. The chronology of the site is unknown, however, Benjamín Nahuelquín—a resident of the area—says that when he arrived more than 50 years ago, the pen already looked as it does now.

4.5.8 Nahuelquín 4

The site is located on the north-east coast of Traiguén Island, 8 km south of the area of Nahuelquin 1, 2 and 3. It is a shell-midden covered by vegetation, discovered by cores drilled in the area, on a small, south-facing bay. The shell-midden is oriented along the beach where it is deposited, 30 m wide from east to west and 10 m from north to south, total 300 m^2. The site presents a slope of 30°. The base of the shell-midden deposit is 2 m above current sea level and it is 2 m from the high tideline. The archaeological visibility is nil.

The anthropic shell-midden was detected under a 5 cm layer of plant matter; it is 30 to 50 cm deep and consists mainly of highly fragmented *Mitylus*. Below this matrix of shells is the palaeo-beach, composed of sandy sediments including pebbles. No cultural materials were detected. This site is a camp formed by discards from gathering and consumption of malacological fauna; successive reoccupations of the area are recorded. According to the record and the archaeological work carried out, no other activities are inferred for the site (Reyes et al. 2011).

The archaeologist Pat Curry (1997) carried out an archaeological prospection in the south and west parts of Traiguén Island in which she identified a series of shell-middens, lithic materials in the intertidal and human bone remains in shell-middens exposed and degraded by tidal action. Her identification and brief description of the archaeological shell-middens made it possible to draw up a panorama of archaeological distribution which was of the utmost importance for our investigations. Based on this information, we visited two archaeological shell-middens recorded previously (Curry 1997; Stern and Curry 1995; Simpson 1875). One known to the fishermen as Las Pampas in the south-west of Traiguén Island (called Posa Las Conchillas 1 by

Curry), and Las Conchillas in the south of Acuao Island (called SP by Curry 1997 and recorded again by us as Isla Acuao 1) (Reyes et al. 2015).

4.5.9 Posa Las Conchillas 1

This site is a large shell-midden on the south coast of Traiguén Island (Stern and Curry 1995). Off this coast is a large group of small islets which protect it from the currents, making it a protected site despite its exposure to the west winds. Several shell-midden mounds are visible around, also recorded by Curry (1997).

The shell-midden is completely covered by vegetation, in an environment of evergreen and peat bog forest. The top has been cleared by fishermen who camped there. The whole front has been eroded by high tides. A large quantity of shell remains and flakes can be seen under water. The archaeological visibility is medium. The change in sea level has caused permanent erosion along the whole west face. The shell-midden is 6.6 m above the high tideline. It measures approximately 40 m from north to south and 15–20 m from east to west, a total of 800 m^2. The slope of the shell-midden face is almost 80°. The dimensions are approximate due to the difficulty presented by the dense plant cover growing on the shell-midden, which can only be opened with machetes.

Test cores drilled below the high tideline showed the presence of shell-midden down to a depth of 120 cm; this, added to the remains of the mound, would give an original height of 780 cm. Core samples were taken at 60 cm, 740 cm and 780 cm (Fig. 4.38), and sediment samples at 160 cm and 360 cm; they were all sent for dating to estimate the chronology and deposit rate of the shell-midden (Reyes et al. 2013, 2015). No human remains were found, although Curry (1997) identified some fragments in eroded profiles. Lithic material was also recovered from the intertidal; a total of 12 lithic pieces deposited in the Patagonian Institute (Punta Arenas): six axes, four anvils, one side-scraper and a pestle/hammerstone (Morello 2014; Reyes et al. 2015), in addition to knapping waste—some of which came from the rhyolithic obsidian source of Chaitén Volcano (Stern and Curry 1995).

The Posa Las Conchillas 1 site is an extensive camp where various activities were carried out. It was formed by discards from gathering and consumption of malacological fauna, in successive reoccupations over at least two thousand years. The five shell samples and four of carbon enabled us to establish the occupation sequence of the site, starting from around ~3400 years cal. BP and continuing to ~1000 years cal. BP (Reyes et al. 2013). The stone axes recovered in the intertidal would probably have been used to obtain forest resources, e.g. for cutting trunks or as wedges for removing their bark, for inferred purposes like the maintenance and construction of canoes, and/or other possible activities involving the use of wood. The knapping waste recorded (some in rhyolithic obsidian) is explained by the manufacture and maintenance of instruments (e.g. hammerstone, side-scrapers) and indirect percussion (anvil). The obsidian recorded, from the Chaitén Volcano source (Stern and Curry 1995) located more than 330 km further north, associates

Fig. 4.38 Measurement of the shell-midden gave a total height of 7.8 m, 6.6 m above the high tideline and 1.2 m below it. The photograph shows the west front of the Posa Las Conchillas 1 site

the groups who settled in the area with a wide network of circulation and/or exchange of raw materials.

4.5.10 Isla Acuao 1

At the southern end of Acuao Island, there is a small point facing south. On the long adjacent beach stands a shell-midden exposed by tidal action, 40 m east to west and 20 m north to south. The deposit is 3 m deep and the slope is 40°. The shell-midden is completely covered by vegetation, except for the eroded front, in an environment of evergreen and peat bog forest. The top has been cleared by fishermen who have camped there, and there is an abandoned cholguero camp with stakes, a hearth ringed with stones and scattered pieces of metal. The archaeological visibility is medium.

The intertidal, exposed at low tide, shows a 5 to 10 m stretch of flooded shell-midden, the base of which is 1 m below the present sea level. On the surface of the intertidal, in the flooded shell-midden, Curry (1997) recovered four stone axes. In our prospection, we recorded a fragment of an instrument in rhyolithic obsidian, the distal part of a bifacial projectile point—the fracture may be due to other factors than knapping. The piece has been thinned with a soft hammerstone and retouched with a compressor. Retouching can be seen that does not pass the central axis of the piece; on one edge only it is parallel and regular. The piece may have been discarded before it was finished. It is 10 mm thick. Seven flakes were also recovered, one on rhyolithic obsidian, three on andesite, one on tuff and two on coarse-grained rock. Among the flakes recovered, there is no evidence of bifacial production; five are derived from cores and two are knapping waste. Fragments of sea lion and bird bone were also recorded in the intertidal, as well as a fragment of human parietal and one of scapula which reveal burial practices in the shell-midden. Human and archaeological remains are scattered on the beach due to erosion. Such events have been recorded previously by Simpson (1875) and Curry (1997). Four stone axes were also collected from the intertidal (Morello 2014; Reyes et al. 2015).

Isla Acuao 1 site is an extensive camp in which many activities were carried out throughout its sequence. It was formed by discards from gathering and consumption of malacological fauna, in successive reoccupations over at least two thousand years. The burial in the shell-midden, by then already substantial, dates to around ~1200 years cal. BP by direct dating of the fragment of cranium, giving a minimum period of occupation of the site (Reyes et al. 2013). The isotope values obtained from the individual dated enabled us to establish consumption of an exclusively marine diet (see Chapter 6, Evaluation and Discussion of the Evidence). The presence of stone axes, meanwhile, suggests that these were used to obtain forest resources, e.g. for cutting trunks or as wedges for removing their bark, for inferred purposes like the maintenance and construction of canoes, and/or other activities involving the use of wood. The fragment of stone projectile point and the knapping waste and core derivatives suggest manufacture and maintenance of formatted instruments, and other hunting-related activities. The presence of rhyolithic obsidian (fragment of stone projectile point) from the Chaitén Volcano source (Stern and Curry 1995) located more than 330 km further north, associates the groups who settled in the area with a wide network of circulation and/or exchange of raw materials.

4.5.11 Canal Cuche-1

On the south coast of Victoria Island, in the Cuche Channel, we recorded a shell-midden in a south-west-facing bay. This shell-midden is set on a beach of sandy sediments, with a small channel to the east which ends in a sound 3 km to the north-east. The shell-midden is completely covered by vegetation, in an environment of evergreen and peat bog forest. The top has been cleared by fishermen who camped there. The shell-midden measures 30 m from east to west and 15 m from north to south, with a depth of 3.6 m. It is 5 m from the high tideline. The base of the shell-midden deposit is 2 m above the present sea level. The site has a slope of 40° and the archaeological visibility is nil. To the east of this shell-midden, there is a small, permanent freshwater stream. No cultural material was recorded. Samples of carbon and shell were taken for ^{14}C dating from the base of the shell-midden, at a depth of 3.6 m (Reyes et al. 2013, Table 4.1).

The Canal Cuche-1 site is an extensive camp formed by discards from the gathering and consumption of varied malacological fauna; successive reoccupations of the area are recorded from ~2000 years cal. BP according to the sample obtained from the base of the shell-midden.

4.5.12 Canal Darwin-2

This is a big shell-midden on the north shore of Luz island, in the Darwin Channel. The site is in a small, sheltered, east-facing bay, 40 m wide and 20 m long. The shell-midden is 1 m above current sea level, immediately above the high tideline. It is approximately 5 m deep, with a steep slope of 70°. Despite its size, it cannot be seen because of the dense plant cover; the archaeological visibility is nil. The area is heavily forested with evergreen forest and peat bogs; nevertheless, high tides have eroded the exposed front of the shell-midden, revealing the mollusc valves. No cultural material was recorded on the surface. The profile was cleared to take a sample from the base of the shell-midden for ^{14}C dating, giving a chronology of ~3200 years cal. BP.

This site is a camp formed by discards from gathering and consumption of malacological fauna; successive reoccupations of the area are recorded. According to the record and the archaeological work carried out, no other activities are inferred for the site (Reyes et al. 2013).

4.5.13 *Isla Victoria-2*

The Isla Victoria-2 site is on the north coast of Victoria Island. It is a small rock shelter close to sea level and 15 m from the current shoreline (Osorio 2011). The context was recorded and surveyed in the context of a legal investigation. The bones were incorporated into the CIEP Osteological Collection once it had been confirmed that they belonged to an archaeological context. The site is covered by evergreen and peat bog forest, with nil archaeological visibility from the coast. The shelter, according to our visual inspection, is quite small, offering very little protection against the rain; nonetheless, it was used to deposit the remains of at least three individuals, according to our analysis (Reyes et al. 2015). It was not used as a human dwelling place, due to the restrictions on movement and the high humidity (although it might have served in an emergency caused by bad weather), so it must be considered to have been selected in line with the custom of using caves and rock shelters, sometimes quite small, as funerary sites in which to deposit bodies (e.g. Isla Ipún 1, Isla Benjamín 5).

The context presents clear anthropic disturbance. It contained at least three individuals deposited on the surface; no cultural material or remains of shells were recorded. Samples for radiocarbon dating were obtained from two of the individuals identified; the male adult, aged between 30 and 40 years, was dated to ~1400 years cal. BP, while the female, of indeterminate age, was dated to ~2000 years cal. BP, the oldest body dated so far in the Chonos Archipelago (Reyes et al. 2015). The exclusive use of the shelter for funerary purposes extended throughout the sequence of use, i.e. it was used repeatedly for individual depositions, as occurred in Isla Ipún 1.

4.5.14 *Cholguero Canal Cuche-2*

This is a small shell-midden of historical origin, or cholguero, located on the south coast of Victoria Island. The site is on a small pebble beach which faces both south-east and west. The forest has been thinned by fishermen; there is sub-present rubbish and the remains of an old camp with wooden stakes, containing a central hearth lined with stones. The site presents a slope of 20°. The archaeological visibility of the site is high.

The shell-midden measures 10 m from north to south and 12 m from east to west; it is 3 m above the present sea level and 5 m from the shoreline. Test cores showed a 30 cm layer of plant material and beneath it 70 cm of shell-midden. Below this shell matrix is a layer of inorganic earth mixed with fine sand 10 cm thick, and then the palaeo-beach. The test cores indicated a discrete deposit of shell-midden matrix 20 cm thick, consisting mainly of cholgas (*Aulacomya ater*), choro mussels (*Choromytilus chorus*) and Chilean mussels (*Mytilus edulis chilensis*). No occupation or archaeological material was recorded below the cholguero, nor at the surface or in the intertidal.

Cholguero Canal Cuche-2 is a campsite developed by seasonal visits, possibly reiterated, related to intensive, selective gathering of mytilids for the canning industry. This activity implied other associated practices, such as hunting and fishing for the fishermen's and mussel gatherers' subsistence. The start of occupation may be dated to the beginning or middle of the twentieth century (Reyes et al. 2013).

4.5.15 Cholguero Caniglia-1

On the south coast of Caniglia Island, at the western end of the Darwin Channel, we recorded a small shell-midden formed by occupation by fishermen and/or cholgueros in historical times. It is in a small, very well-protected, rocky bay, facing east. Like other camps with historical use, the trees have been felled to clear the site and had no time to grow back; however, it is covered with low scrub and grass. The archaeological visibility is low. The shell-midden measures approximately 10 m from north to south and 7 m from east to west. A small, permanent watercourse flows to the north of the site.

Test cores confirmed a deposit of 50 to 60 cm of shells, consisting mainly of cholgas (*Aulacomya ater*), choro mussels (*Choromytilus chorus*) and Chilean mussels (*Mytilus edulis chilensis*) under a layer of humus 10 cm deep. Below this is the base rock of the bay. No sub-present material was recorded on the surface. A sample was taken from the core at 40 cm for radiocarbon dating, which confirmed a modern date (Reyes et al. 2013). No occupation or archaeological material was found below the cholguero, nor on the surface or in the nearby intertidal.

Cholguero Caniglia-1 is a campsite developed by seasonal visits, possibly reiterated, related to intensive, selective gathering of mytilids for the canning industry. This activity implied other associated practices, such as hunting and fishing for the fishermen's and mussel gatherers' subsistence. The start of occupation may be dated to the beginning or middle of the twentieth century (Table 4.2).

4.5.16 Cholguero Canal Darwin-1

The site is on the south shore of the Darwin Channel on Luz Island, in a small, west-facing but very protected bay, with a small brook nearby that runs year round. The area of the shell-midden measures approximately 30 m from north to south and 15 m from east to west. In the northern part there is a large mound, the base of which is 1 m above sea level and 5 m from the shoreline. It has been cleared for use by fishermen (typical felling of trees) and sub-present rubbish is observed on the surface and in the subsurface. The archaeological visibility is medium. Test cores showed a fragment of glass at 40 cm. At 80 cm the shell-midden, consisting mainly of cholgas (*Aulacomya ater*), choro mussels (*Choromytilus chorus*) and Chilean mussels (*Mytilus edulis chilensis*), rests on a matrix of shingle.

Despite our findings in the northern part of this small bay, we investigated a thick shell-midden almost immediately on the shoreline which has been degraded by tidal action and overgrown with large trees. In the exposed profile, we observed periostracum of *Mytilus* valves as well as shells to a depth of 2 m. In addition to the shell-midden, we also found lithic material with patina, a flake and three net weights, in the intertidal (which consists of pebbles).

Shell samples were taken from the exposed base of this shell-midden for ^{14}C dating, giving a calibrated age close to 200 years (Reyes et al. 2013, 2019b Table 4.2).

Cholguero Canal Darwin is a campsite developed by seasonal visits, possibly reiterated, related to intensive, selective gathering of mytilids for the canning industry. This activity implied other associated practices, such as hunting and fishing for the fishermen's and mussel gatherers' subsistence. The start of occupation may be dated to the nineteenth or twentieth century. The identification of three net weights suggests fishing with nets. The weights were made quite quickly, with simple, hard blows of direct percussion to the ends. They are probably archaeological, but in the absence of any other kind of evidence we cannot discard the possibility that they are historical and were made for use by fishermen. The origin of the flake (on tuff, obtained by hard, direct percussion) may be interpreted in the same way as the weights, in the absence of any clear evidence to the contrary.

4.5.17 Cholguero Canal Darwin-3

This is a small shell-midden of historical origin, or cholguero, located on the south coast of Luz Island in the Darwin Channel. The site is located on an east-facing beach of sandy sediments in a small sound, offering good shelter from the west winds. Like other camps with historical use, the trees have been felled to clear the site and had no time to grow back; however, it is covered with low scrub and grass. The vegetation has been cleared, as is typical of these occupations, and smashed shells can be observed on the surface, as well as plastic, remains of metal and stakes which once formed part of a basic shelter. The archaeological visibility is medium. The base of the deposit shell-midden lies 3 m above the present sea level, and 5 m from the high tideline. No occupation or archaeological material was recorded below the cholguero, nor at the surface or in the intertidal.

Cholguero Darwin-3 is a campsite developed by seasonal visits, possibly reiterated, related to intensive, selective gathering of mytilids for the canning industry. This activity implied other associated practices, such as hunting and fishing for the fishermen's and mussel gatherers' subsistence. The start of occupation may be dated to the beginning or middle of the twentieth century.

4.5.18 *Offshore Islands and Pacific Border Area of the Chonos Archipelago. Isla Ipún and Isla Sin Nombre Sites*

As stated above, although no archaeological site was found during our prospection of the offshore islands and those exposed to the Pacific (Figs. 4.2 and 4.21), incomplete information exists of the presence of funerary sites in Ipún, Johnson and Sin Nombre Islands. This information derives from another important source of data, judicial investigations. These occur fairly regularly whenever a fisherman or salmon industry worker travelling along a coastline discovers human bone remains. These are reported to the police or the navy, who collect the bones. Once a sub-present origin has been discarded, they are handed over to different institutions in the region. In this context, we were able to examine the osteological remains collected from the surface at Isla Ipún (handed over to the CIEP Osteological Collection in Coyhaique) and Isla Sin Nombre (handed over to the Chonos Osteological Collection of the University of Chile). We carried out basic bio-anthropological analysis of the two individuals identified from the Isla Ipún 1 site, and determined minimum chronologies for the occupation of both sites. We also evaluated the isotope results of the human samples tested (Reyes et al. 2013, 2015, 2019a and b, Tables 4.1 and 4.2). The isotope values obtained from the individuals dated enabled us to establish consumption of an exclusively marine diet (see Chapter 6, Evaluation and Discussion of the Evidence). Isla Ipún presents an extended sequence of reuse of a shelter for funerary purposes, spread over nearly ~600 years. The two individuals presented datings of ~600 and ~1200 years cal. BP, respectively. The sample obtained from an adult of indeterminate age from the ossuary at Sin Nombre Island produced a dating of ~400 years cal. BP; however, a total of 13 individuals were recovered from this site. Whether this ossuary was formed synchronically, with no statistical difference, or whether this was a funerary practice maintained over a longer time scale can only be determined with a set of direct, representative dates.

4.5.19 *Chiloé Archipelago*

Another site incorporated into our work is the Puqueldón archaeological site (Sáez 2008), in Lemuy Island in the Chiloé Archipelago. Although it lies outside the study area, it presents a late chronological context, associated with human bones, which allows comparison of changes in the diet and subsistence of populations in the northern archipelago post-contact with the Spanish. The site is a Hispanic indigenous cemetery, with at least 20 individuals buried in the matrix of an archaeological shell-midden. The human bone remains were deposited in the Osteological Collection of the University of Chile. We dated two adult individuals (~200 years cal. BP; Table 4.1, Reyes et al. 2013); this allowed us to compare the isotope results of

Table 4.1 Summary of total ^{14}C datings obtained from archaeological sites in the Chonos Archipelago presented in this study and earlier works

Site	Laboratory code	Sample code	Depth of sample (cm)	^{14}C yr BP	$\delta^{13}C$ ‰	Material dated	Calibrated range (cal BP) 2-sigma
GUA-010 Terraza	BETA-34404[d]	GUA-010	60[base]	5020 ± 90	–	Carbon	5918–5583
GUA-010 Terraza	BETA-355644	GUA-10T1	60[base]	5370 ± 30	−24.6	Carbon	6210–5994
GUA-010 Conchal	BETA-355645	GUA-10C1	80[base]	2170 ± 30	1.5	Shell	1840–1554
Gran Guaiteca 1	BETA-355643	GGT1	30	1120 ± 30	2.2	Shell	723–531
Gran Guaiteca 2 Terraza	D-AMS 6073	GG2T-1	50[base]	3958 ± 30	−22.8	Carbon	4428–4243
Gran Guaiteca 2 Conchal	UGAMS 8294[a]	REP-01	20	830 ± 30	−13.9	Human bone	624–491
Gran Guaiteca 2 Conchal	D-AMS 6069	GG2C-1	100[base]	1963 ± 25	−6.1	Shell	1585–1329
Gran Guaiteca 3	D-AMS 006,070	GG3	160[base]	1929 ± 28	−0.6	Shell	1540–1300
Gran Guaiteca 4	D-AMS 006,071	GG4	230[base]	2259 ± 26	−5.3	Shell	1946–1664
Alero Low	UGAMS 26,283	ALOW 1–1	0–10	820 ± 25	−11.39	Human bone	531–362
Alero Low	UGAMS 26,284	ALOW 1–2	30–40	990 ± 25	−11.16	Human bone	650–510
Alero Low	D-AMS 017,799	ALOW 1–5	150–160	1671 ± 28	–	Carbon	1583–1430
Alero Low	UGAMS 26,285	ALOW 1–3	160–170	1940 ± 25	−22.57	Coypu (*Myocastor coypus*)	1906–1747
Alero Low	UGAMS 26,286	ALOW 1–4	180–190 [base]	2050 ± 25	−19.35	Coypu (*Myocastor coypus*)	2012–1901
Repollal Caverna	UGAMS 8295[a]	REP-02	Surface	730 ± 25	−11.7	Human bone	482–300

(continued)

Table 4.1 (continued)

Site	Laboratory code	Sample code	Depth of sample (cm)	^{14}C yr BP	$\delta^{13}C$ ‰	Material dated	Calibrated range (cal BP) 2-sigma
Isla Solitaria	D-AMS 006,072	ISOL-1	35[base]	607 ± 22	−23.2	Carbon	629−529
Isla Marta 1	UGAMS 26,290	IMART1-2	120−130 cm	1520 ± 25	−22.29	Pudú (*Pudu pudu*)	1406−1310
Isla Marta 1	D-AMS 006,068	IMART-1	175[base]	2074 ± 30	3.3	Shell	1733−1428
Repollal Bajo 2	BETA-355646	RB2-1	70	1960 ± 30	1.7	Shell	1586−1322
Puquitín 01	UGAMS 8293[a]	PUQUIT-01	–	1050 ± 30	−10.1	Human bone	666−508
Isla Sin Nombre	UGAMS 8290[a]	ISN-01	Surface	870 ± 30	−10.4	Human bone	546−335
Isla Ipún 1	UGAMS 10,450[a]	IPUN 1	Surface	870 ± 25	−12.8	Human bone	625−490
Isla Ipún 1	UGAMS 10,451[a]	IPUN 2	Surface	1590 ± 25	−9.3	Human bone	1223−962
Estero Sur	UGAMS 10,818	ES1	Surface	1980 ± 25	−11.5	Human bone	1677−1412
Estero Sur	UGAMS 10,819	ES2	Surface	2030 ± 25	−11.9	Human bone	1736−1507
Estero Sur	UGAMS 10,820	ES7	Surface	2030 ± 25	−11.9	Human bone	1736−1507
Estero Sur	UGAMS 10,821	ES9	Surface	2030 ± 25	−11.8	Human bone	1734−1501
Estero Sur	UGAMS 10,822	ES11	Surface	1960 ± 25	−12.4	Human bone	1683−1428
Osario Melimoyu	UGAMS 26,279	MEL 1−1	Surface	1010 ± 25	−11.65	Human bone	667−528
Osario Melimoyu	UGAMS 26,280	MEL 1−3	Surface	920 ± 25	−12.17	Human bone	633−498
Isla Yalac 1	UGAMS 26,281	YAL1-1	Surface	710 ± 25	−12.23	Human bone	479−301
Isla Yalac 1	UGAMS 26,282	YAL1-6	Surface	780 ± 25	−11.92	Human bone	514−330
Seno Gala 1	BETA 230,515[c]	SG1-01	55 [base]	1340 ± 40	−23.6	Pudú (*Pudu pudu*)	1299−1093
Seno Gala 1	BETA 230,493[c]	SG1-02	56 [base]	1430 ± 40	−25.5	Charcoal	1366−1187

(continued)

Table 4.1 (continued)

Site	Laboratory code	Sample code	Depth of sample (cm)	^{14}C yr BP	$\delta^{13}C$ ‰	Material dated	Calibrated range (cal BP) 2-sigma
Isla Harris	UGAMS 8289[a]	HAR-01	Surface	650 ± 25	−12.5	Human bone	450–277
Seno Canalad	UGAMS 8288[a]	CAN-01	Surface	740 ± 30	−12	Human bone	493–308
Isla Benjamín 1	UGAMS 8286[a]	BEN-01	Surface	700 ± 30	−10.6	Human bone	463–239
Isla Benjamín 1	UGAMS 8287[a]	BEN-02	Surface	770 ± 30	−10.5	Human bone	482–288
Isla Benjamín 1	UGAMS 26,289	BEN 1–3	Area	Modern	−9.87	Marine otter (*Lontra felina*)	Out of range
Isla Benjamín 1	UGAMS 26,287	BEN 1–1	60–70	1590 ± 25	−22.67	Huemul (*Hippocamelus bisulcus*)	1518–1371
Isla Benjamín 1	UGAMS 26,288	BEN 1–2	90–100 [base]	1820 ± 25	−16.74	Coypu (*Myocastor coypus*)	1748–1607
Isla Benjamín 2	UGAMS 21,287	IBEN2-1	50	2580 ± 25	−0.3	Shell	2322–2055
Isla Benjamín 2	UGAMS 21287ch	IBEN2-1	50	2200 ± 25	−26.8	Carbon	2306–2082
Isla Benjamín 2	UGAMS 21,288	IBEN2-2	88 [base]	2400 ± 25	−25.8	Carbon	2486–2327
Isla Benjamín 3	UGAMS 21,289	IBEN3	98	2610 ± 25	1.0	Shell	2343–2090
Isla Benjamín 4	UGAMS 21,292	IBEN4	Surface	820 ± 25	−10.9	Human bone	516–323
Isla Benjamín 5	UGAMS 21,290	IBEN5-1	Surface	340 ± 25	−29.9	Bark	450–302
Isla Benjamín 5	UGAMS 21,291	IBEN5-2	Surface	890 ± 25	−11.1	Human bone	573–433
Isla Benjamín 5	UGAMS 21,451	IBEN5-3	Surface	770 ± 25	−11.0	Human bone	489–302
Isla Benjamín 5	UGAMS 21,452	IBEN5-4	Area	850 ± 25	−10.9	Human bone	542–361
Isla Benjamín 5	UGAMS 21,453	IBEN5-5	Surface	790 ± 25	−11.4	Human bone	508–231

(continued)

Table 4.1 (continued)

Site	Laboratory code	Sample code	Depth of sample (cm)	^{14}C yr BP	$\delta^{13}C$ ‰	Material dated	Calibrated range (cal BP) 2-sigma
Isla Benjamín 5	UGAMS 21,454	IBEN5-6	Surface	830 ± 25	−11.1	Human bone	459–290
Isla Elena 1	UGAMS 8119[a]	ELE 1–1	Surface	1880 ± 25	−10.6	Human bone	1517–1298
Isla Elena 1	UGAMS 8120[a]	ELE 1–2	Surface	1880 ± 25	−10.6	Human bone	1517–1298
Isla Elena 1	UGAMS 8121[a]	ELE 1–3	Area	1750 ± 30	−11.2	Human bone	1393–1188
Isla Elena 1	UGAMS 8122[a]	ELE 1–4	Surface	1820 ± 30	−10.4	Human bone	1475–1244
Nahuelquín 1	UGAMS 04,949[b]	NAH 1–1	75–90	430 ± 25	−21	Human bone	505–330
Nahuelquín 1	UGAMS 04,950[b]	NAH 1–2	300 [base]	1820 ± 25	−1.2	Shell	1461–1216
Nahuelquín 1	UGAMS 13,127	NAH1-2	10–20 cm	1030 ± 25	−11.8	Human bone	683–537
Posa Las Conchillas	UGAMS 7751[a]	PLC 1–1	60	1580 ± 25	1.3	Shell	1215–955
Posa Las Conchillas	UGAMS 7752[a]	PLC 1–2	160	1670 ± 25	1.4	Shell	1280–1053
Posa Las Conchillas	UGAMS 7752ch[a]	PLC 1–2	160	1320 ± 25	−26	Carbon	1277–1172
Posa Las Conchillas	UGAMS 7753[a]	PLC 1–3	360	1760 ± 25	0	Shell	1371–1157
Posa Las Conchillas	UGAMS 7753ch[a]	PLC 1–3	360	1450 ± 25	−25.6	Carbon	1352–1281
Posa Las Conchillas	UGAMS 7754[a]	PLC 1–4	740	1810 ± 25	1.4	Shell	1417–1186
Posa Las Conchillas	UGAMS 7754ch[a]	PLC 1–4	740	3180 ± 25	−24.8	Carbon	3446–3251

(continued)

Table 4.1 (continued)

Site	Laboratory code	Sample code	Depth of sample (cm)	^{14}C yr BP	$\delta^{13}C$ ‰	Material dated	Calibrated range (cal BP) 2-sigma
Posa Las Conchillas	UGAMS 7755[a]	PLC 1–5	780 [base]	1800 ± 25	0.4	Shell	1403–1182
Posa Las Conchillas	UGAMS 7755ch[a]	PLC 1–5	780 [base]	3110 ± 25	−25.5	Carbon	3364–3206
Canal Cuche 1	UGAMS 7749[a]	CUCHE-1	360 [base]	2420 ± 25	1.1	Shell	2133–1861
Cuche Channel 1	UGAMS 7749ch[a]	CUCHE-1	360 [base]	1960 ± 25	−25.4	Carbon	1927–1818
Isla Victoria 2	UGAMS 10,452[a]	VIC 2–1	Surface	1820 ± 25	−10.6	Human bone	1473–1257
Isla Victoria 2	UGAMS 10,453[a]	VIC 2–2	Surface	2330 ± 25	−11.2	Human bone	2073–1820
Darwin Channel 2	UGAMS 7750[a]	DAR-2	500 [base]	3360 ± 25	−0.2	Shell	3315–2989
Isla Acuao 1	UGAMS 8117[a]	ACUAO 1	Surface	1630 ± 25	−10.8	Human bone	1280–1167
Corrientes del Yates 5	UGAMS 13,128	CY5-1	120	1420 ± 20	−11.4	Human bone	1088–880
Corrientes del Yates 5	BETA 350,305	CY5F	180 [base]	1920 ± 30	−24.1	Carbon	1890–1734
Isla Prieto 1	BETA 350,307	IP1	90 [base]	1710 ± 30	−13.2	Shell periostracum	1316-–1078
Isla Goñi 1	BETA 350,306	IG1U2	10–20 [base]	4070 ± 30	−23.7	Carbon	4583–4417
Península Taitao 5	BETA 350,308	TAI5	200 [base]	2850 ± 30	0.4	Shell	2695–2363
Puqueldón	UGAMS 8291[a]	PUQUEL-01	–	210 ± 25	−20.1	Human bone	297/ -1
Puqueldón	UGAMS 8292[a]	PUQUEL-02	–	210 ± 30	−19.7	Human bone	299-/-1

Source BETA: Beta Analytic Inc.; UGAMS: Center for Applied Isotope Studies, University of Georgia; D-AMS: Direct AMS Radiocarbon Dating Services; [a]Reyes et al. (2013), [b]Reyes et al. (2011), [c]Reyes et al. (2007), [d]Porter (1993). Human bone with no asterisk, Reyes et al. (2019a); Carbon and shells with no asterisk Reyes et al. (2019b)

Table 4.2 Summary of total ^{14}C datings obtained from historical sites in the Chonos Archipelago presented in this study and earlier works

Site	Laboratory code	Sample code	Depth of sample (cm)	^{14}C yr BP	$\delta^{13}C$ ‰	Material dated	Calibrated range (cal BP) 2-sigma
Isla Level 1	UGAMS 21,284[b]	ILEV1	100 [base]	120 ± 25	−25.0	Carbon	253−1
Isla Izaza 1	UGAMS 21,286[b]	IIZA1	38 [base]	490 ± 25	−0.7	Shell	Out of range
Isla Izaza 1	UGAMS 21286ch[b]	IIZA1	38 [base]	110 ± 25	−25.0	Carbon	Out of range
Isla Izaza 2	UGAMS 21,285[b]	IIZA2	65 [base]	570 ± 25	0.5	Shell	255−1
Isla Izaza 2	UGAMS 21285ch[b]	IIZA2	65 [base]	70 ± 25	−25.5	Carbon	Out of range
Isla Caniglia	UGAMS 8118[a]	CAN-1	40 [base]	Modern	0.9	Shell	Out of range
Canal Darwin 1	UGAMS 8116[a]	DAR-1	80	520 ± 25	0.7	Shell	229−1

Source BETA: Beta Analytic Inc.; UGAMS: Center for Applied Isotope Studies, University of Georgia; D-AMS: Direct AMS Radiocarbon Dating Services; [a]Reyes et al. (2015), [b]Reyes et al. (2018)

post-contact groups from the northern archipelago with the records of archaeological human samples generated previously in the Chonos Archipelago (Reyes et al. 2013, 2019a). The isotope values obtained from the individuals dated enabled us to establish consumption of a terrestrial diet (see Chapter 6, Evaluation and Discussion of the Evidence).

References

Aspillaga E, Castro M, Ocampo C (1990) Paleopathology and life-style: the Chono and Fuegian examples (Abstract). Am J Phy Anthropol, Supplement

Buikstra J, Ubelaker D (1994) Standards for Data Collection from Human Skeletal Remains: Proceedings of a Seminar at the Field Museum of Natural History, Fayetteville: Arkansas Archeological Report Research Series

Cárcamo J (2015) El consumo de moluscos en el conchal GUA-10, Archipiélago de las Guaitecas, Patagonia septentrional: *Un Estudio Arqueomalacológico*. Universidad de Tarapacá, Memoria de Título

Constantinescu F, Aspillaga E (1991) Paleopatalogia y anatomia funcional de la columna de una muestra de indigenas Chonos. Santiago, Chile. Actas del XI Congreso Nacional de Arqueologia Chilena, 237–241

Curry P (1997) Manuscrito. Archaeological exploration of the central Chonos Archipelago. Informe Arqueológico Consejo Monumentos Nacionales, Santiago

Galimany J (2016) Manuscrito. Informe bioantropológico. Informe Manuscrito, Proyecto FONDECYT 1130151. Santiago

Galimany J (2017) Manuscrito. Informe bioantropológico del sitio Alero Low1, Melimoyu 1 y Yalac 1. Proyecto FONDECYT 1130151. Santiago

Jiménez J (2010) Southern pudu, Pudu puda (Molina 1782). Neotropical Cervidology: Biology and Medicine of Latin American Deer. Funep y IUCN, Jaboticabal y Gland, 140–150

Meléndez R, Falabella F, Vargas L (1993) Manuscrito. Osteometría e identificación de restos ictio-arqueológicos de Chile Central. Informe final Proyecto FONDECYT 0089–91. Santiago

Mena F (2015) Manuscrito. Informe pericial restos humanos Melimoyu. CIEP, Coyhaique

Moraga M, Rocco P, Miquel J, Nervi F, Llop E, Chakraborty R, Rothammer F, Carvallo P (2000) Mitochondrial DNA polymorphisms in chilean aboriginal populations: implications for the peopling of the southern cone of the continent. Am J Phys Anthropol 113:19–29

Morello F (2014) Manuscrito. Colección Patricia Curry: Avances sobre la tecnología lítica de los canoeros del Archipiélago de los Chonos central. Informe Proyecto Fondecyt 1130151. Santiago

Ocampo C, Aspillaga E (1984) Breves notas sobre una prospección arqueológica en los archipiélagos de las Guaitecas y de los Chonos. Revista Chilena de Antropología 4:155

Ocampo C, Aspillaga E (1991) Problemas del registro arqueológico de los sitios del archipiélago de los Chono y las Guaitecas. Actas del XII Congreso Nacional de Arqueología Chilena 1:17–18

Plafker G, Savage JC (1970) Mechanism of the Chilean Earthquakes of May 21 and 22, 1960. Geol Soc Am Bull 81(4):1001–1030

Porter C (1993) GUA-010, un sitio costero erosionado en una zona sísmica activa. En: Actas del XII Congreso Nacional de Arqueología Chilena. Boletín del Museo Regional de la Araucanía 4:81–88

Reyes O, Méndez C, San Román M, Cárdenas P, Velásquez H, Trejo V, Morello F, Stern C (2007) Seno Gala 1; Nuevos Resultados En La Arqueología De Los Canales Septentrionales (44° S, XI Región de Aisén, Chile). Magallania. 35(2):105–119

Reyes O, San Román M, Moraga M (2011) Archipiélago de los Chonos: Nuevos registros arqueológicos y bioantropológicos en los canales septentrionales. Isla Traiguén, XI Región de Aisén. Magallania 39(2):295–303

Reyes O, Moraga M, Aspillaga E (2013) El registro bioantropológico y las evidencias de ocupación en el Archipiélago de los Chonos (XI Región de Aisén, Chile). Avances en la arqueología de los canales septentrionales del extremo sur. In: Tendencias Teórico-Metodológicas y Casos de Estudio en la Arqueología de la Patagonia, pp 227–232. Argentina

Reyes O, Moraga M, Méndez M, Cherkinsky A (2015) Maritime Hunter-Gatherers in the Chonos Archipelago (43°50'–46°50' S), Western Patagonian Channels. J Island Coastal Archaeol 10(2):207–231

Reyes O, San Román M, Morello F (2016b) Search for maritime hunter-gatherer archaeological record in the shifting shorelines of the South Pacific Coast (Chonos and Guaitecas Archipelago, Chile). In: Marine Ventures: Archaeological Perspectives on Human-Sea Relations, pp 141–155. Equinox Publishing, United Kingdom. https://doi.org/10.1558/equinox.24563

Reyes O, Tessonne A, San Román M, Méndez C (2019a) Dieta e isótopos estables de cazadores recolectores marinos en los canales occidentales de Patagonia Chile. Latin Am Anti 30(3):550–568

Reyes O, Méndez C, San Román M (2019b) Cronología de la ocupación humana en los canales septentrionales de la Patagonia occidental Chile. Intersec Antropol 20(2):195–210

Reyes O, Trejo V (2010) Manuscrito. Informe antropológico físico de restos óseos recuperados en la localidad de Repollal, isla Ascensión, Archipiélago de las Guaitecas, XI Región de Aisén. Proyecto FONDECYT 1130151. Santiago

San Román M, Reyes O, Méndez C, Cárcamo J, Torres J (2016b) Western Patagonia subsistence strategies: zooarchaeological studies of marine hunter-fisher-gatherers of the Chonos archipelago, Chile. Libro Resúmenes 81 The Annual Meeting of the Society for American Archaeology. Orlando

Schaefer M, Black S, Scheuer L (2009) Juvenile Osteology: a laboratory field manual. Academic Press, USA

Simpon E (1875) Esploraciones hechas por la Corbeta Chacabuco en 1870 .... Anuario Hidrográfico de la Marina, Tomo I

Stern C, Porter C (1991) Obsidiana en yacimientos arqueológicos de Chiloé y las islas Guaitecas. Anales del Instituto de la Patagonia. Serie Ciencias Humanas 20:205–209

Stern C, Curry P (1995) Obsidiana del sitio Posa Las Conchillas, Isla Traiguén (45°30′S), Archipiélago de los Chonos. *Anales delInstituto de la Patagonia.* Serie Ciencias Humanas. 23:119–124

Torres J (2016) Manuscrito. Informe análisis ictioarqueológico de seis sitios del archipiélago de las Guaitecas y los Chonos, XI Región de Aisén. Proyecto FONDECYT 1130151. Santiago

Urbina X (2010) La navegación por los canales australes en la Patagonia occidental insular en los siglos coloniales: la ruta del istmo de Ofqui. Magallania, Punta Arenas 38–2:41–67

Urbina X (2013) Expediciones a las costas de la Patagonia Occidental en el período colonial. Magallania, Punta Arenas 41–2:51–84

Chapter 5
Evaluation and Discussion of the Evidence

Abstract This Chapter discusses the archaeological information generated by the surveys and the records searched. We discuss: types of site recorded in the Chonos Archipelago, visibility and preservation of the archaeological record, coastal remodelling as an agent of change, spatial distribution of archaeological sites, characteristics of lithic technology, circulation of obsidian pieces in western Patagonia, the diversity of the faunal record, human and faunal isotopes and the interpretation of the diet consumed, chronology of the archaeological record and finally the use of space and function of sites.

Keywords Western patagonia · Maritime hunter-gatherers · Coastal survey · Sea-level change · Holocene chronology · Archaeological sites · Archipelagic settlements · Human marine reservoir effect · Subsistence · Lithic technology

The prospections that we planned in the Chonos Archipelago, separated into different areas so as to cover a broad spectrum, provided results that differed with the number of kilometres of coastline prospected and the intensity of cover over these years. When the distance travelled between points of interest and the repeat journeys along the same stretch of coast are excluded, the 1,100 km of coastline actually travelled are reduced to an effective archaeological prospection of 567 km.

From the distribution and the types of sites recorded along this whole shoreline, we can build up a regional panorama with manifest differences in both the distribution of human occupations and the archaeological features by which they are identified. To date, we have compiled information on 52 pre-contact archaeological sites, as well as 25 post-contact archaeological sites (*cholgueros*) and a historical path which was a portage route.

Despite the intensity and cover of our prospection, there may still be many undiscovered sites. We established several factors that have a direct impact on our ability to find sites. The first is related to the modelling of today's coastal topography by glacial ice. The shorelines tend to be abrupt, with cliffs or steep gradients, restricting access to some areas. The second factor is the climate, which frequently prevented sailing or access to some study sectors. The third is that both the coast and the interior

O. Reyes, *The Settlement of the Chonos Archipelago, Western Patagonia, Chile*, The Latin American Studies Book Series,
https://doi.org/10.1007/978-3-030-54326-6_5

of the islands present thick plant cover which comes right down to the high tide line; this hinders not only access but also clear visibility of archaeological sites, even when using archaeological techniques (cores, test pits). The fourth factor affecting finds is the degradation and loss of archaeological sites due to changes in the coastline, principally due to uplifting and subsidence caused by the intense seismic activity of the area, as well as world and regional changes in sea levels already reported in previous studies (Porter 1993; Reyes et al. 2015, 2016b, 2018). This factor is evident when we observe the shell-middens of cultural origin which have subsided below sea level and been damaged by tidal erosion, with lithic material hidden under water and highly fragmented bio-anthropological remains scattered in intertidal areas; these effects imply high levels of energy acting on the exposed profiles (Curry 1997; Simpson 1875). Another testimony of these changes are the fishing pens which are now high and dry (Reyes et al. 2011). Paradoxically, in this context of nil visibility, the shell-middens are sometimes made more visible by their eroded profiles, which show up amongst the vegetation.

5.1 Types of Sites Recorded in the Chonos Archipelago

We have recorded a series of archaeological sites that represent the full material and ecofactual range to be found in the Chonos Archipelago.

The first type are *fishing pens* or "intertidal management areas" (Caldwell et al. 2012). These are stone structures arranged along the intertidal in lines adapted to the coastal geomorphology (Figs. 4.28 and 4.37). The rocks are obtained from the intertidal and arranged in straight, semi-circular or circular lines, of differing dimensions, across the line of tidal rise and fall (e.g. stream mouths, restricted beaches) as passive fishing devices. Fish that enter these constructions when the tide is high remain caught in the pen when it goes out. Pens are found throughout the western archipelagos of Patagonia and both historical and contemporary use is recorded, although no direct dating is possible (Álvarez et al. 2008, 2019; Munita et al. 2005, 2016; Reyes et al. 2011; Vásquez and Zangrando 2017). In this study, we have found that their archaeological visibility is relative, since it can vary from high to nil depending on the state of the tide. Today we know that they are also good indicators of geomorphological changes that have caused variations in sea levels (e.g. Reyes et al. 2011, 2018). In this investigation, we have included four of these sites.

The second type of site are *pre-contact archaeological shell-middens*, extensively described in world and regional archaeological literature (e.g. Álvarez et al. 2010; Curry 1991; Claassen 1998; Dìaz and Garretón 1972/73; Fabier Dubois and Borella 2007; Fabier Dubois et al. 2015; Gaete et al. 2004; Hammond 2015; Legoupil 2005; Munita et al. 2016; Ocampo and Aspillaga 1984; Orquera and Piana 1991, 1992, 2000; Piana and Orquera 2010; Roksandic et al. 2014; Stein 1992; Waselkov 1987; Zubimendi 2012). They are sites formed by large deposits or mounds of mollusc valves, in this area between 1 and 8 m deep and extending for up to tens of metres. They are located close to the shore, either in the open or in some cases in caves and rock shelters where they are smaller (Figs. 4.5, 4.6, 4.7, 4.13, 4.14 and 4.38). They are formed principally by human consumption of varied malacological fauna gathered close to the camp, and subsequent discard of the valves. In this area, they consist mainly of mytilids (*Choromytilus chorus, Aulacomya ater*), keyhole limpets (*Fissurella sp.*), locos (*Concholepas concholepas*), crustaceans (*Austromegabalanus psittacus*) and echinoderms. In general, the waste deposits are stratified due to the gathering and discard of molluscs together with bone remains of hunted mammals, either marine (sea lion), estuarine (coypu, marine otter) or occasionally terrestrial (fox, pudú, huemul), as well as sea birds (cormorants, ducks) and fish. We also observed areas with other activities like lithic knapping, burials and hearths, resulting from a succession of archaeological occupations in which different camp activities were carried on in the open or under shelter.

Given the high humidity and the acid podzol soils of the archipelago (Holdgate 1961; Reyes et al. 2016b), the valves are indistinguishable in many of these shell-midden deposits and form a degraded mass of calcium carbonate. In others only the periostracum of *Mytilus* valves remains, creating friable black layers in the soil (Curry 1997; in this study Isla Prieto 1, Fig. 4.23). Because of their formation and content, they tend to turn the acid pH of the soil (typically ~pH 2.9 to 4.5) more alkaline. The effect of this is differential conservation of the cultural and ecofactual materials deposited, especially bone remains, and the vegetation growing on the site is also affected. Apart from anthropised (cleared) areas, these sites mostly present nil archaeological visibility due to the forest cover which has grown up since they were abandoned. In this investigation, we have included 33 of these sites, 22 of which present one or more radiocarbon datings.

A third type of archaeological site is defined by the distribution of *lithic material in the intertidal*. This is the area between the high and low tide lines, where the sediments are exposed when the tide goes down and covered with water when it rises again. In some sites we found cultural materials in this zone, either singly or in larger assemblages, including axes, lanceolate projectile points, bifaces, net weights, side-scrapers and knapping waste, almost always with a patina produced by tidal friction against the sandy and rocky sediments of the context (Curry 1997; Porter 1993; Reyes et al. 2007; Figs. 4.20, 4.24 and 4.36). Human bone fragments, which had fallen from degraded shell-middens, were also encountered sporadically (Reyes et al. 2015; Simpson 1875). These sites are commonly associated with occupations adjacent to the shoreline, and therefore are formally considered to be extensions of activities in the archaeological pre-contact shell-midden camps and/or in the open-air sites with *lithic remains in organic soils* without shell-midden matrices. However, this contiguity does not necessarily mean that the materials scattered in the intertidal are contemporary with recorded and dated surface events; evidently, they derive from activities carried out in the environs, which may serve to fix a minimum chronology. In some specific cases, however, the contextual and stratigraphic evidence clearly shows that the lithic materials present in the intertidal correspond to and proceed from the same dated event in the adjacent site (e.g. Isla Goñi 1, Corrientes del Yates 5, Península de Taitao 5). In this study, we have found that their archaeological visibility is absolutely nil. Due to the permanent movement of sediments in the intertidal, these materials may be covered by muddy silt or exposed, or else concealed among pebble beaches. The only way of finding them is to search the intertidal at low tide, when the sediments are exposed, and this does not always coincide with the time when prospection is carried out. In this investigation, we have included 11 of these sites.

A fourth type of archaeological context are the sites with *lithic remains in organic soils*. These are cultural deposits buried below a single stratigraphic pattern, or under a layer of plant matter, 20 cm thick on average, after which appears an organic layer no more than 20 cm thick consisting of dark or grey silt sediments, with acid pH (~2.9–4.5) and with no associated shell-midden matrix to make it "visible". The cultural layer contains hearth lenses, carbon specks, knapping waste and/or bifacial instruments (Figs. 4.2 and 4.24). The function that we interpret for these sites is usually an area for specific activities, and in one case a look-out point. They are identified exclusively by test pits. In this study, we have found that their archaeological visibility is absolutely nil. The lithic materials recorded, the only testimony of the cultural occupation, present signs of degradation due to the highly acid soil (Hiscock 1985). In this investigation, we found four of these sites, all of which were dated.

A fifth type of site are the *ossuaries of humans bones deposited and/or buried in caves and rock shelters* (Figs. 4.5, 4.10, 4.11, 4.13, 4.15 and 4.19). These places, protected and circumscribed by their rock walls, were used as funerary areas for depositing bodies on the surface protected from the weather; they may contain 10 or more individuals (Aspillaga et al. 2006; Ocampo and Aspillaga 1984; Reyes et al. 2015). Bodies may also be deposited singly, but reiterated use over time results in a greater concentration of individuals (e.g. Isla Ipún 1, Isla Victoria 2). Bone remains are mainly recorded inside the shelter, on the surface, scattered or regrouped in no anatomical order, under the action of different transforming agents. They are generally severely deteriorated due to atmospheric (e.g. weathering and humidity), animal (dispersion and gnawing marks on bones), plant (action of roots scattering the bone remains as they grow) and especially anthropic forces, e.g. fragmentation, removal and loss of portions of the anatomy, especially crania. This anthropic action has been reported since the nineteenth century (Simpson 1875) and continues to be caused today by fishermen and salmon industry workers travelling around the coasts. Due to the destruction of these contexts, the lack of records and the absence of associated cultural materials, it has not so far been possible to establish mortuary patterns (e.g. trappings, offerings, original position and arrangement of the bodies). This funerary practice of deposition was mentioned for this region by direct witnesses in the eighteenth century (Byron 1901 [1746]), and later referred to by explorers and scientists (Simpson 1875, Cooper 1946). Some authors originally associated these contexts with the last moments of the archaeological sequence (Aspillaga et al. 2006; Ocampo and Aspillaga 1984, 1991), however our investigations have shown evidence of initial chronologies around ~2000 years cal. BP (Reyes et al. 2015). These caves with funerary evidence sometimes present a shell-midden matrix, going back to domestic occupations at the start of the formation of the site, with hearths and the consumption and discard of molluscs and fauna obtained by hunting and fishing. Subsequently, the use and function of the sites changed, and they were used for funerary purposes down to the end of the sequence. In this study, we also found that bodies were not only deposited but also buried in these shell-midden matrices (e.g. Alero Low, Isla Benjamín 1). Other shelters, in contrast, were used exclusively as funerary sites, reaffirming the synchronic and diachronic use of the chosen shelters (Reyes et al. 2015; present study). We found that the archaeological visibility of these contexts varies from good to nil. Some caves and shelters are visible from the shore due to the broad morphology of the rocky substrate which projects beyond the edge of the forest, while others are completely hidden and inaccessible, up vertical slopes, on small ledges with very little space and hidden by the forest. The sites which are visible from the shore and more accessible have been severely disturbed by human interference. In terms of settlement patterns and archaeological distribution, funerary contexts of this type are the only sites sometimes found a little way inland from the immediate shoreline. Their site location follows a different logic, since the object is to deposit bodies, so they do not need to be immediately on the shoreline but can be inland in the forest (e.g. Estero Sur ossuary). In this investigation, we documented 16 ossuary sites. Almost all present one or more datings, and in 3 of them 100% of the individuals recovered have been dated.

Another type of site included in this work are the post-contact archaeological shell-middens known as cholguero ports or *cholgueros* (Figs. 4.8, 4.17,4.22, 4.26, 4.27, 4.28, 4.29, 4.30, 4.32 and 4.33). These were formed by mollusc extraction for semi-industrial and industrial economic purposes (canning) at the end of the nineteenth century and during much of the twentieth, although other extractive activities contributed, such as fur-trapping, wood extraction, fishing, etc. (Curry 1991; Marín 2014; Martinic 2004, 2005; Mayorga 2016). These *cholgueros* are distinguished by the presence of historical rubbish (e.g. metal, glass, rope); they generally have a stone-ringed central hearth, used both for heating and for smoking the shellfish collected to preserve them. The shell substrate in these sites presents little diversity, consisting mainly of mussels and cholgas in large, uncompacted, homogeneous heaps, which can easily reach a height of 4 m. Sometimes, depending on the characteristics of the location (proximity of fresh water, protection from the wind, beach for pulling up boats), sites of this type are superimposed on pre-contact archaeological shell-middens, which are covered by the historical shell waste. This situation was documented in the area of Puerto Edén by Curry (1991) and by Legoupil and others (2007) in the Guayaneco Archipelago, south of the Gulf of Penas. Cultural materials, associated features and evidence of other activities (wooden shelters, plastic, hearths, vegetable gardens) can be seen on the surface. The plant cover and the trees also grow back quite quickly after the site is abandoned, so the visibility is often nil, contrary to what might be expected from the number of years since abandonment. In some cases these sites present interventions in the intertidal, such as fishing pens, and places for mooring or beaching boats. In the latter, the rocks of the intertidal near the site are cleared in order to haul boats up the beach without risk of damage. This usually leaves a clear path perpendicular to the shoreline. In this study, we found that the archaeological visibility of interventions in the intertidal is relative, since it also depends on whether prospection is carried out at high or low tide. In this investigation, we found 25 cholguero sites, 5 of which were dated.

The last type of site are open spaces in the forest which serve as *paths* or *portage routes*. They may be created to extract timber from the forest or to portage boats (Borrero 1997; Prieto et al. 2000), avoiding a long or difficult and dangerous sea passage (Fig. 4.31). In this work, we identified only one path of a historical nature (Península de Taitao 4), a portage route made with trunks laid crosswise to haul up boats. This route connects the head of Estero Thompson with the river that flows into Presidente Ríos Lake, in the centre of the Taitao Peninsula. Associated with this portage is a cleared area with overgrown hearths, evidence of sporadic camps when boats were being hauled along the route. This is a traditional overland route used by fishermen, and historically by sea lion and otter-hunters, to avoid the more dangerous passage by the Pacific Ocean and the Gulf of Penas (Álvarez et al. 2015; Urbina 2010). Identifying an archaeological path is very difficult due to the regeneration of forest cover and decomposition of wood. In this investigation, we only noted this site but did not date it chronologically.

5.2 Visibility of the Archaeological Record

In percentage terms, we found in this study that the visibility of pre-contact archaeological sites in the Chonos Archipelago is nil in the majority of cases (67%, n = 28)—mainly, obviously enough, due to the forest cover which in some cases is impenetrable. Visibility exists, i.e. elements can be distinguished which define an archaeological site, only in a small number of cases: low visibility (7%, n = 3), medium (19%, n = 8) and high (7%, n = 3) (Table 5.1). The increased visibility can be attributed to two causes: marine erosion which exposes and uncovers profiles, or anthropic disturbance which uncovers and alters part of the archaeological assemblage and the sediments. The visibility of post-contact historical archaeological sites like the *cholgueros*, contrary to expectation due to their more recent formation and abandoment, is also mostly nil (65%, n = 17), while low, medium and high visibility are each reported for 11.5% (n = 3). These historical sites, compared to pre-contact archaeological shell-middens, present higher percentages of high and low, but not medium, archaeological visibility; high visibility never dominates the whole set. This must be stressed as samples from cholguero sites, which have existed for less than 200 years, are an interesting reference for the transformation affecting sites in the archipelago, especially the effects of being overgrown by plants which make part of

Table 5.1 Archaeological visibility determined by the percentage cover of each site recorded: high (0–30%), medium (31–60%), low (61–90%) and nil (91–100%)

Archaeological sites	% Nil Visibility	% Low Visibility	% Medium Visibility	% High Visibility	Total %
Pre-contact sites	68	7	18	7	100
Post-contact or cholguero sites	65.5	11.5	11.5	11.5	100

the material record at the site invisible. This situation of nil archaeological visibility is an essential condition of the archaeological record in forested areas (e.g. Borrero and Muñoz 1999; Borrero and Borrazzo 2011; Fernández et al. 2011; García and Mena 2016; Mena 1995; Mena and Ocampo 1993; Reyes et al. 2006a, b).

5.3 Preservation of the Archaeological Record

The preservation of the archaeological sites in the archipelago area was assessed based on observation of three main factors that affect not only the state in which the deposited materials are recovered but also the situation of the archaeological context in general. Firstly, anthropic disturbance, either casual or intentional, has contributed largely to the differential preservation of some archaeological contexts. Thus many historical or cholguero camps, as well as modern fishermen's camps, are established on archaeological shell-middens, since they likewise look for favourable bays with fresh water close by and good shelter for their boats (e.g. Curry 1991; Legoupil et al. 2007). Their activities affect the upper layers of archaeological shell-middens. We must also suppose some degree of occasional, selective collection of noteworthy lithic material in the intertidal (e.g. formatted instruments such as lanceolate projectile points, stone axes). However, it is in the funerary sites, particularly ossuaries, that the greatest degree of disturbance and destruction is recorded. The deposition of human remains on the surface makes them very vulnerable to plundering and disturbance out of curiosity. Their deterioration is very clear in this study when we compare the minimum number of individuals in any ossuary with the anatomical portions by which they are represented. In general, crania are most frequently missing, while long bones are piled in packets. Of the 16 ossuaries recorded to date, only one was found with no evidence of anthropic disturbance (Isla Benjamín 5).

Second, the geology of the region must be taken into account, dominated as it is by the crystalline basement, and with the presence of flooded land and a thick layer of organic matter in permanent decomposition. These factors contribute to the generation of highly acid podzol soils (Holdgate 1961), which affect the composition of the anthropic deposits formed in the archipelago. Of the 52 pre-contact archaeological sites identified to date, only four present *lithic materials in organic soil* with no shell-midden matrix (GUA-010 Terraza, Gran Guaiteca 2 Terraza, Isla Solitaria and Isla Goñi 1; just under 8% of the total sample). These acid soils affect the differential preservation of the materials deposited (Borrazzo 2007; Hiscock 1985; Pearce 2008). We found the degradation of lithic instruments only in these contexts. The absence of bone remains may be explained by the inferred functionality of the sites, and can also depend on the type of intervention carried out and where samples were taken; nevertheless, organic material also is absent only from sites of this type. By contrast, in the 31 pre-contact archaeological shell-middens that we recorded (60% of the total sites), the matrix incorporates a high level of calcium carbonates (Claassen 1998;

Pearce 2008). These soils have a more alkaline pH, affecting the differential preservation of archaeological materials (Reynard 2014), especially human and faunal bone remains.

The other factor which has an impact on the preservation of archaeological sites is their location associated with paleo-beaches, marine terraces and immediate shorelines. This proximity to the shore, combined with the permanent changes in geomorphology and sea level recorded in the area, have repercussions for the preservation of the cultural context. Of the total of 52 sites recorded, only one is intentionally sited a little way from the coast, namely the funerary context of Estero Sur ossuary. The total universe of sites located on the shoreline, regardless of geomorphological changes, consists of 44 archaeological sites, of which 20 (45% of the sample) present marine erosion of the cultural deposits (seven sites in the Northern area, one on the Mainland coast, one in the Central area of the archipelago and 11 in the South-central area). Practically all are shell-middens, the exception being Isla Goñi 1. Today, some sites are on elevated terraces, preserved from marine erosion by tectonic emergence and/or isostatic rebound in the region (e.g. GUA-010 Terraza and Gran Guaiteca 2 Terraza, Nahuelquín 2 and 3); the deposits in others, in contrast, present evident deterioration due to tectonic sinking and higher sea levels, resulting in erosion and exposure of the profiles, and movement or collapse of the sediments (Harris 1991) and the cultural and bio-anthropological materials in the immediate intertidal. The most extreme case is the almost complete destruction of the Isla Goñi 1 context, in the central-south area, where sediments containing cultural material can be seen to have collapsed into the adjacent intertidal. Another site in the central area (Isla Benjamín 2) presents a different situation: the stratigraphy of this archaeological shell-midden presents clear ingress of marine sediments due to high energy events which covered the site; it was later reoccupied by canoe groups who settled on the newly formed beach (Reyes et al. 2018).

The historical archaeological sites, or *cholgueros*, also offer an opportunity for evaluating the dynamic of changes to the coastline and their effects on archaeological sites, but over shorter time scales. Their site location on the shoreline, like that of the archaeological shell-middens, enabled us to calibrate the impact of changes in sea levels in sites that have existed for less than 200 years on a temporal micro-scale (sensu Dincauze 2000). We found that of the 25 *cholgueros* recorded, 13 present erosion and exposure of their profiles caused by tidal action due to sinking of the land (e.g. Isla Level 1, Isla Izaza 1 and 2, Canal Darwin 1, Isla Stewart 1 and 3, Isla Goñi 3 and Taitao 6). This situation illustrates the dynamic tectonic behaviour of the area, the remodelling of the coast in the short term and the projection of these influences over a millennial scale.

Lithic materials in the intertidal were recorded in almost all the areas sampled throughout the archipelago, with the exception of the centre and the islands exposed to the Pacific. We must distinguish situations caused by the degradation of the site, with material which has fallen from profiles exposed by changes in coastal geomorphology (e.g. Isla Goñi 1, Taitao 5, Corriente del Yates 5, Posa Las Conchillas and Acuao 1), from situations in which the distribution of lithic material in the intertidal may result from human behaviours of discard and/or loss during subsistence activities in the immediate vicinity of the camp (sensu Bintliff 2000; Foley 1981b), as appears to be the case of Gran Guaiteca 4, Isla Acuao 2, Isla Goñi 2, Canquenes 2, Canal Vicuña 3 and Nahuelquín 1. In the latter contexts, there is no evidence of the direct involvement of erosion with the materials recovered. In the Northern area, in the Guaitecas Archipelago (GUA-010 Terraza and Conchal, Gran Guaiteca 2 Terraza and Conchal), and on the Mainland coast (Seno Gala 1) (Reyes et al. 2007), a series of test pits and excavations were dug in which the materials recovered include, in typological terms, raw materials and products of operative chains. These present similarities with the material recovered from the pits or under water, suggesting that discard and/or loss of instruments occurred throughout the whole occupation sequence (*ca.* 6000 years), as well as erosion due to sinking or uplifting which directly affected shoreline sites and the distribution of artefacts in the intertidal (Reyes et al. 2016b).

5.4 Site Location

Our results, together with others reported (Curry 1997; Ocampo and Aspillaga 1984, 1991; Porter 1993; Stern and Porter 1991; Stern and Curry 1995; Reyes et al. 2006b, 2007, 2013, 2015, 2016b, 2018), establish that the most frequent position of the archaeological sites in this archipelago is at the limit of the intertidal (in the first few metres above the high tide line), since the steep coastal geography and the dense forest—reaching down to the water's edge—restricted settlement possibilities. Nevertheless, even camps on the shoreline were already sited on land with forest cover. We can certainly not discount the possibility of sites further into the forest, but so far, apart from one ossuary, no evidence has been recorded inland in the islands; the recurrent use of sites may of course condition the visibility of the archaeological record (Binford 1980). Some subsistence and social activities must have occurred in the forest (e.g. portages and paths connecting sites, hunting, selection and felling of trees, plant gathering, rituals); however, as remarked above, the variables of accessibility, visibility and preservation have a decisive impact on the durability of the material record and the possibility of making finds in the forest. At the same time, economic activities associated with fishing, hunting and marine gathering, supported as we see by the faunal, bio-anthropological and technological record, implied pericoastal mobility and settlement strategies, supported by the use, maintenance and care of canoes (Ames 2002; Arnold 1995; Bjerck 2009; Fagan 2004); this is a key feature of access to certain areas of the mainland coast and the archipelagos (Fitzpatrick et al. 2016; Keegan et al. 2008; Pallo 2011). Thus, the vast majority of the archaeological record on the mainland coast and throughout the archipelago is found in such positions, informing us of the factors which conditioned human settlement decisions (sensu Barberena 2008:47).

We understand that this coastal distribution of camps favours subsistence activities oriented almost exclusively towards the marine environment while ensuring shelter; in this respect, the use of canoes is the decisive factor (e.g. Bjerck 2017;

Piana and Orquera 2009; Orquera and Piana 1999a, b). We find no temporal or geographical variation in this situation. The marked tendency in the Northern area, of Mid-Holocene sites in the Guaitecas Archipelago (GUA-010 Terraza and Gran Guaiteca 2 Terraza), is for deposits to be sited 5–6 masl. This distance from the shoreline is related to changes in sea levels in the region and around the world (e.g. Fairbanks 1989; Isla 1989; Kempf et al. 2017; Reyes et al. 2018; Shennan 2007). These changes directly affected the redistribution and positions of these sites in the coastal geomorphology (elevation and distance from the shoreline). The sediments in the matrix, at both sites, indicate that the camps were originally sited at beach level, very close to sea level. In contrast, another occupation from the end of the Mid-Holocene, in the South-central area, has been practically destroyed by the sinking of the land and the advance of the sea (Isla Goñi 1).

We stress that the interaction of marine transgressions, isostatic rebound and a tectonic tendency for the land to sink in earthquakes in the archipelago area (e.g. Atwater et al. 1992; Hervé and Ota 1993; Plafker and Savage 1970; SHOA 2000; Vita-Finzi 1996) determines a panorama which is difficult to project in clear geo-chronological sequences. This contrasts with the archipelagos of southern Patagonia, where associations can be observed between changes in the coastlines that have occurred throughout the Post-glacial period and the location of archaeological sites on an altitude gradient (Mc Culloch and Morello 2009; Reyes et al. 2018; San Román 2014; Santiago et al. 2007).

In this archipelago, with the current sea level, the bases of the deposits of archaeological sites (n = 38, excepting ossuaries) are between 0 and 3 m in the majority of cases (66%, n = 25), while in 26% of sites (n = 10) the bases of the deposits are under water, between –1 and 0 m relative to the present sea level. Only 8% (n = 3) of the campsites are between 4 and 6 m above current sea level, all in the northern area of the archipelago. The sites assigned to the Mid-Holocene present dissimilar elevations, whereas the archaeological shell-middens assigned to the Late Holocene are nowhere more than 2 m above present sea level.

The archaeological sites associated exclusively with funerary rites, i.e. ossuaries (N = 16), the logic for which differs from that followed in the installation of a camp, are found either on the forested shoreline or in rock shelters on very steep shores (Isla Elena 1, Isla Benjamín 4 and 5), only accessible by climbing up through the trees. Thus, the decision on a shelter in which to deposit bodies implies other factors of coastal geomorphology and the hierarchy established for the coastal landscape (sensu Borrero 2001b; Belardi 2003), in which protection of the bodies against the weather is important. These decisions do not vary over time or space across the archipelago. The "most distant" funerary site is no more than 50 m from the shore.

5.5 Coastal Remodelling as an Agent of Change

The results reported show that the general placing of the sites has a direct impact on how well they are preserved. Stratified cultural deposits that are eroded and/or under water, human remains and lithic material recovered from the intertidal, and emerged fishing pens all reflect the intense dynamics of the coastline and the difficulty of recording occupations in the space encompassed by the archipelago. We will concentrate on the geomorphological implications and the chronological variables of human occupation to understand the evolution of coastlines during the Post-glacial, and how this affected the archaeological record in the region (sensu Bailey 2004; Bailey and Flemming 2008).

We distinguish at least three different geomorphological scenarios in the areas studied in the archipelago (Reyes et al. 2016b, 2018). The first is the scarcity of marine terraces located above the present coastline. These indicate changes produced by tectonic activity and/or changing sea levels. We recorded this situation in the Northern area of the archipelago and in Guafo and Ipún Islands (although no associated camps were found in these latter). The second scenario is the erosion of archaeological shell-middens sited on beaches which were emerged at some time in the past but are now below the water surface, indicating that subsidence has occurred since they were formed. This situation was recorded in the Northern, Central, South-central and Mainland coast areas of the archipelago. The final case is the deposition of sandy sediments between archaeological occupations, due to high energy events with the formation of new sediments. We obtained this record from the Northern and Central areas of the archipelago.

In general terms, in the Northern area of the Chonos Archipelago, the sites recorded in the central isthmus of Gran Guaiteca Island were GUA-010 Terraza and Conchal, Gran Guaiteca 2 Terraza and Conchal and Isla Solitaria (Fig. 4.1). This area is characterised by the presence of broad marine abrasion platforms containing unconsolidated deposits of glacial and marine origin accompanied by outcrops of the metamorphic basement which predominates in the area (Hervé et al. 1976). On top of this platform, at a variable distance from the present coastline, are two terraces of marine origin on which the cultural deposits are sited. Only three sites were located on terraces between 3 and 6 masl in this portion of the island. Thus, the earliest occupations in the Northern area are situated in time between ~6200 and 4200 years cal. BP, and their position on elevated terraces is relative only to the present sea level. These terraces would have been available for the establishment of human occupations on the immediate shoreline in the Mid-Holocene, in a period of interaction between the marine transgression of the Mid-Holocene (*ca.* 7000 to 5000 years BP, Long 2001; Milne et al. 2005) and the vertical displacement caused by tectonic movement and isostatic rebound (Bartsch-Winkler and Schmoll 1993; Dietrich et al. 2010; Milne and Shennan 2007).

The shell-midden camps recorded on the shoreline, 2,000 years after the last recorded occupations of the Mid-Holocene (Reyes et al. 2016b), are not found at elevations of more than 3 m above current sea level, except for Isla Solitaria, which is a look-out point with a different function and logistical behaviour to those of a camp. Obviously, the higher terraces were available for occupation during later occupations, although as we have observed, the material record is scarce on these geoforms (Fig. 5.1); however, it is uncertain whether this is a result of the subsistence pattern of the canoe groups, or rather due to factors related with the visibility and

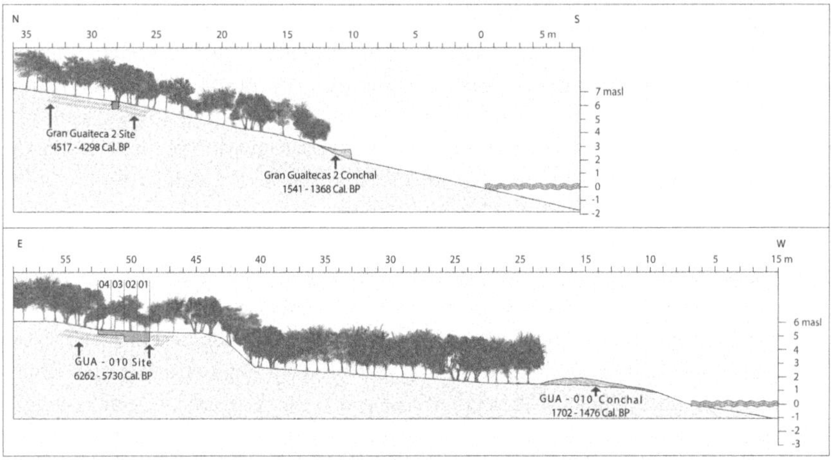

Fig. 5.1 Cross sections of Gran Guaitecas 2 Terraza, Gran Guaitecas 2 Conchal, GUA-010 Terraza and GUA-010 Conchal sites in the northern area of the Chonos Archipelago, showing the tendency for Mid-Holocene sites to be found on more elevated terraces in the northern area of the archipelago

preservation of the sites—or a combination of both. Furthermore, in the north of Gran Guaiteca Island, we recorded the Alero Low site 5 m above current sea level (a raised cave used at first as a shelter and later as a funerary site), on a rocky promontory jutting out from a steep coastline (Fig. 4.5). Towards the Mid-Holocene, during the occupation of GUA-010 Terraza and Gran Guaiteca 2 Terraza, this site would have been on the immediate shoreline exposed to the water at high tide, and thus not available for occupation. Like the sites occupied in the Late Holocene, it was occupied much later, towards ~2000 years cal. BP, when the cave had been raised above sea level and was therefore available.

In the western sector of Gran Guaiteca Island, the geomorphology and chronology associated with the marine geoforms are similar to those recorded in the central isthmus of the island. In particular, three archaeological camps are sited in this area, located on either side of the narrow channel separating Gran Guaiteca, where Gran Guaiteca 4 is located, from Marta Island, where Isla Marta 1 and 2 are sited. An eroded scarp clearly distinguishes the current coastline from a low terrace (1–2 m above present sea level) on both sides of the channel. The chronology associated with the anthropogenic deposits sited on the two low terraces provides a minimum age of ~1700 years cal BP for the origin of these geoforms (Table 4.1). Furthermore, as in the Central area, both low terraces are observed to have suffered erosion at high tide. This indicates subduction of this area, probably associated with the major earthquake of 1960 and with contemporary changes in sea levels around the world (Milne et al. 2009). The same phenomenon is observed in Gran Guaiteca 1, 3 and 4 in Gran Guaiteca Island, and Repollal Bajo 1 in Ascensión Island (Reyes et al. 2016b).

It was not possible to identify terraces located above the shoreline in the other areas of the archipelago, nor is there a well-defined abrasion platform as is observed in the Northern area. On the contrary, in the rest of the archipelago the coast tends to be sheer, which—as we shall see—has an impact on the site location of residential camps. The exceptions can be found in the offshore islands (Guafo) and those exposed to the Pacific (Ipún, Stokes, Forsyth), which rose by between 1.5 and 5 m above the current sea level in the 1960 earthquake (SHOA 2000; Plafker and Savage 1970), exposing part of the sea bed in some sectors; and in the flooded sedimentary land in the Taitao Peninsula, in the south of the archipelago (Viviani 2004).

Dated sites where there are clear erosion processes, like Seno Gala 1 on the mainland coast of the Chonos Archipelago (Reyes et al. 2007) and Posa Las Conchillas and Isla Acuao 1 in the south of Traiguén Island (Reyes et al. 2015) in the South-central area (Figs. 4.36 and 4.36), lie further east, close to important tectonic lines associated with the Liquiñe-Ofqui fault system (Hervé et al. 2015; Lara et al. 2008; Thomson 2002). The same is observed in the other records made in this study. The historical cholguero sites, dating back less than 200 years on average (e.g. Isla Izaza and Isla Level in the Central area of the archipelago and Isla Luz in the South-central area), the older pre-contact shell-midden camps (Isla Benjamín 2 and 3, Punta Cementerio, Acuao 3, Taitao 5), and Mid-Holocene occupations (Isla Goñi 1) are subject to tidal influence with clear erosion of their profiles (Figs. 4.17, 4.22, 4.23 and 4.24). The preservation and distribution of the archaeological sites reflect an important subsidence process, greater than that observed and recorded in the Northern area of the archipelago. They point to a series of vertical sinking events of the land and rises in sea level from the end of the Mid-Holocene, more than counteracting the isostatic rebound (Rivera et al. 2007). We found this difference between the two areas prospected—only 200 km apart—with an important difference in the preservation of the archaeological record. To date, no shell-midden camp dating to the Late Holocene in this archipelago has been recorded at an elevation of more than 2 m above the current sea level.

To enrich this discussion, it is interesting to consider another type of evidence, with opposite results, from the Nahuelquin 2 and 3 archaeological sites in the southwest of Traiguén Island in the South-central area. Here there are emerged fishing pens located 1–2 m above the present shoreline (Reyes et al. 2011, 2015, Fig. 4.37). These pens are located on the existing marine abrasion platform; and while there is no chronology for their abandonment, it can be seen that a floor has started to build up inside them. This evidence is not related with an as yet unconfirmed drop in sea level in recent times (Milne et al. 2009), but rather with vertical uplifting due to tectonic movement (Reyes et al. 2018).

As we noted above, not all the evidence is directly related to possible changes in sea level and/or vertical changes of the earth's crust due to plate tectonics. In Benjamín Island 2, in the Central area, it was found that the archaeological shell-midden was established on an old beach. The main stratigraphic feature of this site is the presence of a level of fine sands interrupting the deposit between ~2200 and 2400 years cal. BP with the formation of a new beach of sandy sediments; occupation was later resumed on the new beach which had formed (Fig. 4.14). We discarded the possibility that storms may have caused the incorporation of these sediments since the location is sheltered from the wind, and therefore from the high swell. If a similar stratigraphic situation is found in the future investigation of other shell-midden camps in the area, it may be possible to date a high energy event like those documented for the Pacific coast of Chiloé (Kempf et al. 2017). Isla Benjamín 2 is the first example of this that we have documented (Reyes et al. 2018), and we are on the look out for other sedimentological evidence in the area.

To summarise our conclusions about the distribution of the archaeological sites, and the site location of shell-midden camps on the shoreline and their current geomorphological situation, our work indicates that the archaeological record corresponds to eroded camps. These were formed in the Late Holocene by marine hunter-gatherers, and present cultural and bio-anthropological materials scattered in the intertidal. In the South-central area they are no older than ~3300 years cal. BP, and in the Northern area up to ~2000 years cal. BP. In the latter area, the occupations assigned to the Mid-Holocene are located on terraces elevated 5–6 m above the current sea level (Fig. 5.1). Thus in the light of the results obtained to date—and taking into account vertical movements of tectonic origin, the record of archaeological sites above or below current sea level and the age of the base deposits on coastal geoforms (sensu Erlandson and Moss 1999; Waters and Kuehn 1996)—we can predict that the material archaeological evidence of the Mid-Holocene will be found on elevated terraces in the Northern area, while in the South-central area the sites have probably disappeared or lie around present sea level (Fig. 4.24).

Finally, the cholguero sites also allow us to calibrate the impact of geomorphological changes at temporal micro-scales (sensu Dincauze), as we know or can estimate their occupation dates—less than 200 years ago. We, therefore, also know how sea levels affect recently formed sites that have not been exposed on a millennial scale to the various effects caused by the changes in sea levels and coastlines affecting pre-contact archaeological sites (Reyes et al. 2018).

5.6 Spatial Distribution of Archaeological Sites

As mentioned above, the total number of linear kilometres effectively prospected was 567, distributed through the different predefined prospection areas (Fig. 1.2) and covering most of the geographical spectrum of the archipelago. This sampling distribution allows comparisons to be made at various scales of the intensity of use of different areas (Borrero 2004), and how this varied over time.

In the Northern area, represented by the Guaitecas Archipelago, the coastline prospected totalled approximately 55 km. Sixteen archaeological sites were recorded in this area: 3 sites with lithic materials buried in organic soil, with no shell-midden matrix; 9 archaeological shell-middens; 4 shelters or caves with human bones (one a cave containing a shell-midden, with human bones in the higher levels), as well as the intertidal area with scattered lithic material. This area, according to our results, presents one site for every 3.4 km of shoreline (Table 5.2).

Table 5.2 Number of pre-contact archaeological sites recorded per linear km of coast prospected, by study area. The total effective length of shoreline prospected was 567 km

Areas covered the archipelago	Coastal prospection	Number of sites recorded	Sites per km of shoreline prospected	Estimate
Northern area	55 km	16	0.30	One site every 3.4 km
Mainland coast	70 km	4	0.06	One site every 17.5 km
Central area	120 km	17	0.06	One site every 17 km
South-central area	250 km	23	0.09	One site every 10.8 km
Offshore Islands and Pacific border area	(32 + 40 km) 72 km	0 and 2 respectively	0.02 and 0.05 respectively	One site every 36 and 20 km respectively

In the Central area, represented by the sector between Izaza, Level, Stokes and Benjamín Islands, we covered 120 km of shoreline, recording 7 archaeological sites: 2 shell-middens, 4 shelters or caves with human bones and 1 cave with shell-midden and human bones. This area, according to our results, presents one site for every 17 km of shoreline (Table 5.2).

The South-central area received the greatest coverage over several field trips, totalling 250 km of shoreline prospected, recording 23 archaeological sites: 2 shelters or caves with human bones; 2 fishing pens; 1 site with artefacts buried in organic soil, with no shell-midden matrix; 16 shell-middens and 7 areas with lithic material in the intertidal. This area, according to our results, presents one site for every 10.8 km of shoreline (Table 5.2).

In the area consisting of the offshore islands (Guafo) and islands exposed to the Pacific (Ipún, Stokes, Johnson and Forsyth), we covered 72 km of shoreline with no new archaeological record in this study. Nevertheless, we have included information on the 2 ossuaries recorded in Ipún and Sin Nombre Islands exposed to the Pacific (Reyes et al. 2013). This area contained one site for every 36 km of coastline. If we exclude the 32 km covered around the north-east coast of Guafo Island, the sole offshore island, leaving only the islands in the archipelago with exposure to the Pacific Ocean, we have one site for every 20 km of coastline.

Finally, the Mainland coast area of the archipelago was covered in different field trips; the total coverage was 70 km coastline and we recorded only 4 archaeological sites: 3 shelters or caves with human bones and 1 shell-midden with lithic material in the adjacent intertidal. This area contained one site for every 17.5 km of coastline (Table 5.2).

Based on the total number of sites recorded on the coastline prospected, we observe a series of situations more substantial than the noise generated by the variables: intensity of prospection, visibility and accessibility of each area in particular. At both macro-scale and meso-scale (Dincauze 2000), the results of the coastal prospection in the predefined areas of the Chonos Archipelago show a low density of archaeological sites, so that occupations identified can be considered as isolated points in space. The number of sites per linear km of coast prospected ranges between 0.30 and 0.02 sites per km (Table 5.2). This density is notably lower than the records reported in central-southern Patagonia, as suggested by the frequency of archaeological evidence located towards the steppes of the eastern slope of the Andes, and the far south of the continent (e.g. Barberena 2008; Borrero 2001a; Cruz and Caracotche 2006; Goñi 2010; Miotti and Salemme 2003; Orquera 1987; Orquera and Otero 2007; Zilio and Hammond 2013; Zubimendi et al. 2005, 2009). By way of comparison, including the littoral record along the coast and in similar geographical zones, the number of sites estimated for the south shore of the Beagle Channel is 2 archaeological sites per linear km (Navarino Island), while on the north shore the figure is 3 sites per linear km prospected (sensu Piana and Orquera 2009; Ocampo and Rivas 2008). Of course, the archaeological intensity and cover of prospection in this area are backed by decades of archaeological work, with a marked contrast in sampling intensity as compared to our work to date in the archipelago.

Within this general picture of low archaeological records, the behaviour of site distribution is as follows: the area with the greatest density of occupation is in the far north, in the Guaitecas Archipelago, which is also the sector where least prospection was carried out in this work. This area presents an archaeological density of 0.30 sites per linear km of coastline prospected, much higher than the other areas prospected. We may add that the chronological sequence obtained for these areas stretches from the Mid-Holocene to ~400 years cal. BP (plus the cholguero sites), but with hiati in occupation of 2,000 years. The majority of the sites are shell-middens (56%) and ossuaries (25%), while sites with lithic artefacts deposited in an organic matrix account for 19%.

The South-central area of the Chonos Archipelago presented the second-highest density of archaeological sites on the coastline prospected, with a much lower density of 0.09 sites identified per linear km. This was the area with the longest coastline prospected, with 250 km spread over different field trips. It was probably thanks to this intensity of cover that it produced a slightly higher number of sites than the other areas defined in this study. The length of coast prospected may also explain why, despite the difference in density, and like the Northern area, we could establish a chronological sequence stretching from the transition from the Mid to Late Holocene down to contact with the Spanish, inclusive, with all the types of the archaeological site represented. 70% of the archaeological sites were shell-middens, while ossuaries and fishing pens represent 9% of the total sites identified. Lithic artefacts deposited in an organic matrix account for only 4% of the sample; areas with lithic material in the intertidal, exclusively or associated with archaeological sites, are more numerous than in the other prospection areas.

The areas with the lowest density of archaeological sites are found in the Offshore Islands and the Pacific border area, where no record of any kind was found in 32 km of coastal prospection, and 0.02 sites per linear km prospected. However, if we exclude the offshore area from the rest of the archipelago, the Pacific border, the centre and the mainland coast of the archipelago present the lowest number of sites identified per km prospected, all with a similar density ranging between 0.05 and 0.06 sites per linear km of coast prospected. It is striking, and agrees with the results recorded, that these less dense areas are represented very largely by ossuaries (100% in the Pacific, 75% in the Mainland coast and 65% in the Central area), while the number of shell-midden camps in these three areas is much lower, only one on the Mainland coast (25%) and three in the Central area of the archipelago (35%) over 262 km of coast prospected. The archaeological sequence obtained in these three areas goes back no further than the Late Holocene.

Considering the distribution of archaeological sites on a regional scale, we can project the human occupation of the archipelago as isolated points in space, with some groupings observed (Figs. 4.1, 4.9, 4.12 and 4.18). As we discuss below, the distribution of sites is related with the type of occupation of the territory according to the chronological sequence on a millennial time scale, which we believe shows a hierarchisation of spaces in a continuous peopling process (Borrero 1989–90, 1999, 2001a, 2013; Borrero and Franco 1997) with differing intensities over the temporal sequence (Steele 2010). Both the location and the preservation of the sites would be strongly affected by the geomorphological characteristics of the shoreline. As we indicate, the archipelago has been modelled predominantly by the action of glacial ice and by global and regional changes in sea level. These factors produce almost vertical shores (Fig. 1.3), formed mainly of steep-sided fjords, cliffs, flooded areas and peat bogs, with sandy, rocky or mixed beaches and bays, exposed to the elements, where the thick evergreen forest cover comes down to the high tide line. There are very few marine platforms and terraces (Figs. 1.4 and 1.5) such as are observed in Gran Guaiteca Island in the north of the archipelago.

For these reasons, the archipelago should not be understood as an unmarked map with continuous shorelines where any section is suitable and available, and where the grouping or absence of sites on the coastline is directly controlled by settlement patterns under a subsistence system. "*What we know of the Chonos and Guaitecas Archipelagos suggests the exploitation of varied resources, located in what might be described as ecotone nuclei or transition nuclei distributed among the interior channels with access to oceanic and continental resources... A large enclave offering access to resources from the ocean and the interior channels on the one hand, and other nuclei established close to the mainland to give access to resources from the interior channels and the mainland...*" (Massone et al. 2016: 482). Far from agreeing with this proposal, we observe the absence of criteria on which to project a distributional archaeology of the region (e.g. Ebert 1992; Foley 1981a, b; Wandsnider 1992) and a hierarchisation of the archaeological landscape (Belardi 2003; Bintliff 2000). Accessibility to certain locations in order to establish residential camps is determined in the archipelago by multiple factors and decisions; one of these is without doubt use of the canoe and the possibility of beaching it safely close to the camp area (e.g. Bjerck 2017; Curry 1991); it was the essential technology of these groups which permitted and restricted their movements, the time they could spend at sea and the locations that they could reach (Ames 2002), and for those reasons some parts of the shoreline are unsuitable or inaccessible for setting up camp, independent of the resources available in the environment.

Following this line of argument, the environmental context serves as a frame of reference (sensu Binford 2001) which must be considered in the site location patterns and the general distribution of the archaeological record (e.g. Belardi 2003; Pallo 2011; Scheinsohn 2001). The Northern area, in the Guaitecas Archipelago, presents marine terraces and beaches on fairly flat slopes with protected coastlines, offering suitable conditions both for subsistence and for safe beaching of canoes. Geomorphological features have a direct influence on the site selected, as we see from the fact that the sites recorded in the archipelago are located pre-eminently on the shoreline. The reoccupation of archaeological sites after variable periods, as we see from the chronological sequences obtained (Tables 4.1 and 4.2) and the shell discard rates that we have inferred (Reyes et al. 2015), suggests the superposition of human occupation events (sensu Wandsnider 1992) on shorelines with locations that appear restricted—certainly not unlimited—in contexts with low occupation density, rather than the formation of nuclei of sites dictated by the acquisition of particular resources. Moreover, most of the marine resources in the archipelago are distributed throughout the length and breadth of its geography. The exceptions are mainland raw materials (e.g. rhyolithic obsidian from Chaitén Volcano) and larger terrestrial mammals (e.g. fox, pudú and huemul), which we recorded occasionally in the diet (Reyes et al. 2015, 2019a). Nevertheless, the relative abundance of these resources varies according to the geographical characteristics of this island system, such as the orientation of the channels, the proximity of mainland fjords and glacier outlets, differing water densities and tidal flows, as well as the morphology of the coastline and the sea bed, which vary from island to island and channel to channel (Viviani 2004); it is not a function of ecotones and segmented locations. Some sectors do not offer a special incentive for seasonal or provisioning visits based on a specific increase in certain resources—for example, to the Pacific border in search of birds' nesting colonies and sea lion rookeries, as may occur at the southern end of the archipelago (e.g. Borrero 2013; Chapman 1987; Horwitz and Weissel 2011). Guafo Island is an extreme example; 40 km from the nearest points of land (southern Chiloé or the Guaitecas Archipelago) and 130 km from the mainland coast across the open sea, it appears not to have been sufficiently attractive to justify the risks of the voyage. If it was visited, the archaeological signal is still invisible.

Another not insignificant factor that argues against the formation of nuclei of human activity around ecotone resources (Massone et al. 2016) is the difficulty of interpreting the type of archaeological activity which has led to the identification and recording of sites. The "nuclei" described are found where some more detailed prospection or reconnaissance has been carried out (e.g. Curry 1997; Ocampo and Aspillaga 1984, 1991; Porter 1993; Reyes et al. 2007, 2011, 2013 and judicial investigations). The absence and presence of sites must first be calibrated with the preservation, visibility, accessibility and obtrusiveness of both the coast and the archaeological sites and the materials recovered. Understanding an archaeological distribution without weighing the methodology of the work done, the types of shoreline prospected, the permanent geomorphological changes, the broad chronological scales within and between sites, the site deposition and reoccupation rates and their inferred function in relation with faunal and technological records results in a totally mistaken projection of the archaeological distribution of these groups.

On a macro-scale, the general distribution of archaeological sites in the archipelago, excepting exclusively funerary sites, tends to be heterogeneous, with irregular, low-density distributions between one site every 3.4 km and one every 20 km of shoreline (Table 5.2). Campsites are present in almost all the sectors of the archipelago sampled, except the Offshore islands and Pacific Border area. The causes of this absence of sites, and the denser grouping in some sectors, can be explained from various points of view. One is the proportionally shorter prospection route than other areas prospected; in the Northern area however, where the prospection was only a few kilometres longer, we record 16 archaeological sites. Secondly, there is our limited capacity of access to carry out sampling on exposed coasts. In this context, we recall that after the 1960 earthquake the Pacific area suffered uplifting of between 5 m (Guafo and Guamblin Islands) and 1 m (Ipún and Stokes Islands) above current sea level (Figs. 4.34 and 4.35), while the central and mainland areas sank by approximately 1.5 m (Plafker and Savage 1970, Fig. 1.4). We prospected certain forested sectors up to 20 m above present sea level, projecting the geoform before the land was uplifted. Thirdly, and consequent on the second point, these shorelines offer very little shelter for a settlement pattern in which the safety of the canoe close to the camp is vitally important (e.g. Orquera and Piana 1999b), and where difficulties of access discouraged visits or made them more infrequent or more localised in specific points (which we have not yet discovered), in a regional context where the archaeological signal of hunter-gatherer canoe groups is low. For reasons of safety, sectors more exposed to winds and ocean swell would be unattractive due to what Pallo (2011) calls the *hydrodynamic stability* of the canoe. The mainland coast likewise presents deep fjords and steep coasts not considered suitable for installing camps with a safe area for the canoe (Fig. 1.3). All these factors, at the least, lead us to understand the occupation of the archipelago in terms of spaces hierarchised around the people's mobility system, access to camps, the location of resources and site functionality, rather than imagining a completely available coastal strip, an unmarked map, in which archaeological distributions can be understood *per se*.

In concrete data, the presence of shell-midden camps on the Pacific border and the mainland coast is scarce (n = 1, Seno Gala 1 on the mainland coast). On the other hand, we recorded scattered funerary sites (n = 5, with an MNI of 55, of different sex and age, dated to between ~2000 and 300 years cal. BP, Table 4.1) on both sides of the archipelago, on steep coasts, where the location of the funerary site follows a different strategy according to its function: the location of shelters to protect bodies deposited on the surface, with subsequent reuse of these spaces. Within the funerary context and its requirements, the irregularity of the near-vertical coast and the presence of small rock shelters on steep shores would be an incentive to create temporary access in order to leave a body or bodies (e.g. Byron 1901 [1746]; Simpson 1875). This may be understood as a human signal spread throughout the archipelago, where residential camps have not yet been found or are located in other more distant but suitable sectors. Canoe transport would facilitate the hierarchisation of coastal spaces according to their functionality (hunting, logistical, camp, funerary). Human occupation extended throughout the island system, except Guafo so far, towards the Late Holocene, especially in the past 2,000 years (Reyes et al. 2019b).

5.7 Lithic Technology

The whole lithic assemblage frequently presents traces of abrasion on the edges and chipping of the thinnest borders; this is because the majority of the pieces were recovered from the intertidal, where the action of the water and sedimentary particles in suspension promote abrasion of exposed surfaces (Petraglia and Potts 1994, Figs. 4.3, 4.20, 4.24 and 4.25). These features hinder the identification of cultural traces on the pieces and the recognition of technical attributes (Borrazo 2007), as mentioned previously in our descriptions of the instruments (Reyes et al. 2007, 2015). Thus we recorded axes as pieces that were either finished or discarded in different stages of their useful life. They were manufactured from flat stones shaped by hard percussion, on one or both faces, and smoothed by flaking to form a bi-convex section and a double-bevelled active edge in the widest part of the piece. Only a few pieces show concentrated polishing of the active edge, which extended the useful life of the piece (Morello 2014) (Fig. 5.2). Another class of lithic tool recorded are anvils and hammerstones. These tools were made from flattened stones which were used for bipolar percussion, as is suggested by the traces of flaking located mainly in the centre of the faces. Net weights are also frequent; they are ovoid stones with notches

Fig. 5.2 Stone axes from the Curry Collection, Patagonian Institute, University of Magallanes. (1, 3, 5) preforms of axes with traces of percussion, (2, 4, 6, 7) stone axes with a polished surface

in their ends to take lashings, on one or both faces; some samples present a longitudinal groove instead. The manufacture of lanceolate stone projectile points (bifacial sets) involves the production of a large amount of knapping and retouching waste, found occasionally in the intertidal and in excavation (e.g. Isla Goñi 1 and GUA-010 Terraza). These pieces are generally of medium size (~8 cm) and lanceolate shape (Fig. 4.20); this type is widespread in both the northern (Gaete and Navarro 2004; Munita 2007, 2017; Porter 1993; Reyes et al. 2007, 2015; Rivas et al. 1999) and the southern channels (Morello et al. 2002; Orquera et al. 1977, 2011). Some of these pieces are made of obsidian which is thought, based on macroscopic criteria (inclusion of crystals, colour and texture) and geochemical analyses, to have come from the Chaitén Volcano source (Méndez et al. 2008/9, 2018; Reyes et al. 2006b, 2007, 2015; Stern 2018; Stern and Porter 1991; Stern and Curry 1995; Stern et al. 2002). The rocks selected for the manufacture of most of the instruments (axes, net weights, choppers) were stones gathered close to the sites and transformed into efficient tools. Only the axes and stone projectile heads present specific designs, and as such were conceived for a longer useful life. Of these categories of the artefact, only the lanceolate projectile heads required good quality raw materials (obsidian).

Whether on a macro-, meso- or micro-scale (Dincauze 2000), the lithic material recovered to date in the whole Chonos Archipelago (Porter 1994; Curry 1997; Reyes et al. 2007, 2016b and this study) must be described as scarce. The whole set consists of 1,025 lithic pieces (Table 5.3); of these, 434 pieces (42% of the set) come from excavation/test pits at 16 archaeological sites categorised as shell-midden camps (n = 12) and sites with materials in an organic soil matrix (n = 4), distributed over the northern, central, central-south and mainland coast areas of the archipelago. However, of all the excavations carried out, the lithic pieces were found in just 8 sites, with a total excavated area of 14.18 m^2. There are 10.3 m^2 excavated in eight shell-midden camps which produced no lithic evidence, from interventions ranging between 100 and 230 cm of shell-midden matrix. Surface collection, carried out over 24,000 m^2 in 12 archaeological sites with adjacent intertidal zones where materials were detected on the surface, produced 591 lithic pieces (58% of the total set); the present whereabouts of 421 of these, derived from the work of Porter (1993) in the Guaitecas Archipelago, is now unknown.

On a meso-scale, the Northern and South-central areas of the archipelago present the largest sets of lithic artefacts; not a single lithic piece was recovered in the Offshore islands and Pacific border or the Central area of the archipelago. On a micro-scale, the bay of the Cucuquén Stream, on the south coast of Gran Guaiteca Island in the Northern area, the Seno Gala 1 site on the Mainland coast and Isla Goñi 1 in the South-central area present the largest contributions in quantity and diversity of lithic pieces (Table 5.3).

The densities of artefact finds, per m^2 excavated, range from 0 lithic pieces (n = 8 shell-midden sites) up an ascending scale of 1 piece per m^2 in Taitao 5 and Corriente del Yates 5 shell-midden sites, through 1.8 pieces in the Gran Guaiteca 1 shell-midden, 2 pieces in Isla Solitaria, 4 pieces in Gran Guaiteca 2 Terraza, 9.2 pieces in GUA-010 Terraza, 16 pieces in the Seno Gala 1 shell-midden and finally 347 pieces per m^2 in Isla Goñi 1. The artefact densities per m^2 of the collection area

Table 5.3 Detail of the diversity of the artefacts and total quantity of lithic pieces recovered per site in the Chonos Archipelago, over the whole time sequence

Site	Types of Lithic Artefact														
	Axe	Hammerstone	Core	Net weight	Bifaces	Pestle	Stone projectile point	Perforators	Side-scraper	Anvil	Chopper	Scraper	Retouch flake	Knapping waste	Total
GUA-010 Terraza							12							43	55
GUA-010 and Gran Guaiteca 2 Conchal Intertidal	4	24	26	108			14	6	31		28			180	421
Gran Guaiteca 1							1							3	4
Gran Guaiteca 2 Terraza							1							3	4
Gtan Guaiteca 4				1									1		2
Isla Solitaria							1							4	5
Seno Gala						1	9					1	17	41	69
Nahuelquín 1							1								1
Posa Las Conchillas 1	6	1							1	4					12

(continued)

Table 5.3 (continued)

Site	Types of Lithic Artefact														
	Axe	Hammerstone	Core	Net weight	Bifaces	Pestle	Stone projectile point	Perforators	Side-scraper	Anvil	Chopper	Scraper	Retouch flake	Knapping waste	Total
Isla Acuao 1	1						1								2
Punta Cementerio	6														6
Acuao 2			1												1
Canal Vicuña 3				1											1
Corrientes de Yates 5										1					1
Canquenes 2			1												1
Is la Goñi 1	2		4	1	4		8		3					411	433
Is la Goñi 2	1						1							4	6
Península Taitao 5										1					1
Total	20	25	32	111	4	1	49	6	35	6	28	1	18	689	1025
Total %	2%	2.40%	3%	11%	0.40%	0.10%	4.80%	0.60%	3.40%	0.60%	2.80%	0.10%	1.80%	67%	100%

in the intertidal zone are substantially lower, ranging from 0.002 to 0.09 lithic pieces per m^2. This density should not surprise us, as it applies to very large areas, defined by exposure of the intertidal (which also varies over the year) in areas prospected ranging from 1,000 m^2 to 16,000 m^2.

Of all the lithic pieces recovered in excavations, 95% (n = 406) were obtained from the only three Mid-Holocene contexts (~6200–4400 years cal. BP), i.e. GUA-010 Terraza (n = 55), Isla Goñi 1 (n = 347) and Gran Guaiteca 2 Terraza (n = 4), with 8 m^2 excavated in total. The other 27 pieces, 5% of the total set excavated, come from 5 archaeological sites with chronological sequences assignable to the Late Holocene, with 6.38 m^2 of excavation. The differences in the numbers of pieces representing the lithic sets of the Mid and Late Holocene are also reflected in the characteristics of the deposits from which the samples are taken. The majority come from sites in which the lithic artefacts and waste are buried under a layer of acid humus, a characteristic that defines sites of the Mid-Holocene almost exclusively (except for Isla Solitaria), with dimensions between 18 and 140 m^2. The sites assigned to the Late Holocene present large quantities of shells in the sedimentary component, which define the dominant matrix of these camps. The anthropic deposit formed includes mainly discarded shells which can be considered as the sediments that alternate with artefacts (Stein 1992). These deposits alter the pH of the soil, making it more alkaline and thus allowing the preservation of bone remains; however they also substantially increase the volume of the matrix, and therefore the formal dimensions of the site. The stratigraphic depth of these deposits is between 100 and 780 cm, covering areas between 28 and 4,200 m^2. We believe that the difference in the recovery of lithic sets and the projected density is explained by cultural factors (types of site, successive reoccupations of the site and type of occupation of the geographical area). However, any interpretation must take into account the effect of the "dissolution" of the density of ecofactual and cultural materials in large volume deposits of anthropic sediments of this type.

In GUA-010 Terraza (6210–5583 years cal. BP, 6 masl, Figs. 4.1 and 4.2), with 6 m^2 excavated, whole and fragmented thick-section (~1 cm) lanceolate lithic projectile heads were recorded (n = 5), as well as preforms of such pieces (n = 7), accounting for 21.8% of the whole set recovered at the site. The rest of the evidence consists of a discrete set of knapping waste and a core without cortex, in an area that we interpreted as an excavated area where specific activities were carried out for the manufacture and/or maintenance of medium-sized lithic projectile points for hunting. The presence of some of these lithic projectile points, manufactured on rhyolithic obsidian from Chaitén Volcano (150 km further north, opposite Chiloé), supports our proposal that transport over long distances between the archipelago and the mainland coast outside the latitudes of the archipelago (Pallo and Borrero 2015) already occurred in the Mid-Holocene.

If we turn to the Gran Guaiteca 2 Terraza site (~4400 years cal., 6 masl, Figs. 4.1 and 4.4), located 200 m from GUA-010 Terraza in the same bay, the set recovered in 1 m^2 excavated included three scattered pieces of knapping waste, as well as a fragment of lanceolate lithic projectile head, also made on rhyolithic obsidian; we find a similar pattern of occupation in this site, with evidence of specific use of specialised projectile heads for hunting. Despite the representation of specific hunting activities and the total absence of any other cultural or archaeofaunal indicator (we recall that the soil is a podzol), we still need to incorporate the large set of artefacts, currently lost, which were recovered from the intertidal of the bay where these two sites are located (Porter 1993). Although the subsequent occupations in the Late Holocene, recorded as two thousand years later, may also have contributed material picked up and then discarded or lost in the intertidal, the presence under the water of lanceolate lithic projectile points on rhyolithic obsidian may be understood in part as a product of the human occupations of the bay in the Mid-Holocene. Furthermore, the existence of a diverse set of artefacts in the intertidal, consisting of net weights, axes, flakes, planes and choppers (Porter 1993), is indicative of a different set of activities, especially connected with fishing and the use of forest resources (e.g. maintenance of canoes and associated instruments), which may be related with occupations throughout the whole sequence; however, this cannot be confirmed as the contexts are in areas with permanent movement of the surface and superimposed activities (Foley 1981b; Wandsnider and Camili 1992).

The Isla Goñi 1 site (~4500 years cal. BP, 0 masl, Fig. 4.18), located in the South-central area of the archipelago 250 km south of the sites mentioned above, presents the largest lithic assemblage recovered in 1 m^2 excavated. Because the site is subject to tidal erosion, the materials recovered in the intertidal (Curry 1997; Reyes et al. 2015 and this study) come from sediments that have become detached from the shoreline in compact blocks (Figs. 4.24). We recorded a clear degradation of the site associated with geomorphological changes due to tectonic sinking (Reyes et al. 2018). To this, we may add the absence of other types of site in the area which might interact with the distribution of archaeological materials, apart from Isla Goñi 2 (1 km further south) where the preforms of lanceolate projectile heads and axes found in the intertidal can be considered an extension of the area of occupation (sensu Bintliff 2000; Foley 1981a; Binford 1980, 1982), representing an extended *locus* of activities derived from the camp at Isla Goñi 1.

Having said this, and as detailed in the previous chapter, the total lithic set not only exceeds in number any site recorded in the archipelago to date (n = 52 pre-contact archaeological sites), but also presents a greater diversity of artefacts represented: three medium-sized, thick-section lanceolate lithic projectile points, five preforms of lanceolate projectile points, four fragments of bifaces, four cores, three side-scrapers, one notched net weight and two axe preforms, as well as flakes and knapping waste with cortex on rhyolithic obsidian. The stratigraphic excavation record reveals a high frequency of lithic waste together with the cores and preforms of lanceolate lithic projectile points in various initial stages of manufacture. The almost exclusive use of rhyolithic obsidian from the Chaitén Volcano source, 330 km further north, as the principal raw material in the manufacture of instruments (stone projectile heads, bifaces and a side-scraper) is also supported by the quantity of knapping waste and by the bifacial preforms recorded (96% of the assemblage recovered). The thick-section lanceolate lithic projectile points which dominate the set of artefacts were manufactured preferentially on this obsidian, in contrast to the axe preforms and the net weight, made of raw materials obtained locally (e.g. tuff, basalt). The obsidian was transported to the site in the form of lumps with cortex (according to the evidence of the knapping waste) which was only possible by canoe; this explains the quantity and frequency of allochthonous lithic material used and discarded in a radius of action inconsistent with the *principle of declination* (Pallo and Borrero 2015) from the raw material sources. The use of canoes incorporates longer operating distances into the models of "non-terrestrial" hunters—whether or not they supplied themselves directly (Murdock 1969)—and therefore high mobility in a system of extended ranges of action (Binford 1982), or exchange systems associated with mechanisms of human organisation.

Although the Mid-Holocene sites present no faunal record, perhaps due to the podzol or to the specific sector excavated in each site, the activities inferred from the material recovered also include fishing (use of net weights); while the presence of axe preforms, and the fact that they were made on the site from local stones obtained nearby, suggests that they were used to obtain forest resources in activities which may involve cutting trunks or use of the axes as wedges for stripping the bark off trees, in inferred functions such as canoe and instrument maintenance and construction, among various other activities which involve the use of wood (e.g. firewood for hearths, to level a camping area or for hut posts) [1]. The extensive possibilities of wood use led to the proposal of a long tradition of "wood culture" (Rivas and Ocampo 2006) by the human groups who inhabited the coasts of the northern archipelago and the dense temperate forests of southern Chile.

Thus apart from the difference in the density of the lithic assemblages recovered, both in excavation and in the adjacent intertidal areas, and with an exclusive sequence in the Mid-Holocene, we find some technological continuities in the emphasis on lithic tools for hunting, with preferential manufacture of stone projectile points on raw materials obtained from the mainland coast which circulated to distances between 150 and 330 km from the source. The sites of this period mainly represent the maintenance and manufacture of these projectile heads, while artefacts for fishing and woodworking appear to be of lower priority (Table 5.3).

We may add that, in the Northern area of the archipelago (Fig. 4.1), where the Mid-Holocene sites GUA-010 Terraza and Gran Guaiteca 2 Terraza and GUA-010 Conchal and Gran Guaiteca 2 Conchal are located, Porter (1993) recorded 108 net weights. As noted above, throughout the whole sequence of occupation in this area (*ca.* 6000 years), different occupations might have contributed lithic material picked up and then discarded or lost under the sea. Nevertheless, the number of weights recovered thanks to the particularities of this intertidal—subjected to large vertical changes in geomorphology (more exposed today than in the Mid-Holocene but having sunk in the past 60 years, Reyes et al. 2016a, b)—enables us to dimension the importance of fishing and the difficulty of investigating this practice through the artifactual record in this geomorphological situation (Waters and Kuehn 1996). The possibility of using local stones with little preparation, the loss of pieces when tying them to nets or casting the nets, the absence of other evidence of manufacture such as flaking to create grooves or notches at the poles of the stone, and the presumed use of nets made of plant fibres or tendons, are all obstacles to the documentation of this practice beyond the ichthyological record in faunal columns. The record of lithic artefacts in Isla Goñi 1 makes it possible to document fishing (net weights) and use of forest resources (stone axes) since ~4500 years cal. BP at least.

To summarise, the lithic evidence recovered establishes a specific, limited set of activities carried out in the Mid-Holocene sites (GUA-010 Terraza, Gran Guaiteca 2 Terraza and Isla Goñi 1), with the production of medium-sized, thick-section lanceolate lithic projectile points the dominant category. The small proportion of other evidence may imply that the set produced was kept and not discarded in the site; or else that sampling factors specific to the area where the pieces were exposed mean that they do not necessarily reflect the whole camp. In view of the evidence, we incline to the latter alternative. The processing of lithic resources, their significant diversity, the representation of final phases of operating chains (consistent with functional relations with other camps) and the lithic categories observed suggest a strong influence of transport, and consequently high mobility of the groups who travelled the archipelago.

Of the lithic assemblages recovered from the excavations in contexts of the pre-contact Late Holocene, and which we have dated to between ~3300 and 500 years cal. BP, there are only 27 lithic pieces from the excavation of 6.38 m^2 distributed in five sites: four shell-midden camps and one site with materials in an organic soil matrix. Together with this very low frequency of finds, as remarked previously, we note the total absence of lithic material in seven other shell-midden camps and one funerary site with 10.3 m^2 excavated. These sites are GUA-010 Conchal, Gran Guaiteca 2 Conchal, Alero Low and Isla Marta 1 in the Northern area of the archipelago, Isla Benjamín 1, 2 and 4 in the Central area, and Nahuelquín 1 and Isla Prieto 1 in the South-central area.

The Late Holocene artefacts recovered in the Northern area, in Gran Guaiteca Island, are limited to a distal fragment of a thick-section lanceolate stone projectile point and four pieces of knapping waste at the Isla Solitaria site (~500 years cal. BP, 3 masl); and a fragment of a lanceolate stone projectile point preform made from a local stone and three pieces of knapping waste from Gran Guaiteca 1 (~670 years cal. BP, 1 masl). Both these archaeological contexts are situated at the end of the pre-contact archaeological sequence. The lithic material recovered and the area intervened in Isla Solitaria mark it as a site with very low occupation density, used as a look-out point (Binford 1980, 1982); the principal activities inferred are knapping of bifaces and discard of instruments. In the Gran Guaiteca 1 shell-midden camp, on the other hand, the size of the area intervened indicates successive stratigraphic reoccupations involving activities related to the manufacture and maintenance of lithic projectile points, and others connected principally with the gathering, consumption and discard of shellfish. The knapping waste recovered at both sites reveals the use of rhyolithic obsidian from the Chaitén Volcano source, 150 km to the north.

In the South-central area of the archipelago on the other hand (Fig. 4.18), two anvils made of local stones were recovered when exposed profiles were being cleared, in Corrientes del Yates 5 (~1800 and 900 years cal. BP, 0 masl) on Simpson Island, and in Península de Taitao 5 (~2500 years cal. BP, 0 masl), on the mainland peninsula. Both artefacts indicate lithic knapping with indirect percussion, in sites characterised as residential camps where multiple activities were carried out in successive reoccupations of the area, including consumption and discard of malacological fauna, hearth areas and archaeofaunal evidence of hunting birds and sea lions, as well as the burial of several individuals in the case of Corrientes del Yates 5.

Finally, in the only shell-midden identified and excavated so far on the mainland coast of the archipelago, at Seno Gala 1 (~1300 years cal. BP, –1–0 masl, Reyes et al. 2007), 16 pieces of knapping waste were recovered from 1 m^2 of excavation, and 53 lithic pieces from the adjacent intertidal. The assemblage consisted of 1 pestle/hammerstone, 9 thick-section lanceolate lithic projectile points, 1 scraper, 17 retouch flakes and 41 pieces of knapping waste; several of the projectile heads were made of obsidian from Chaitén Volcano, 120 km away. Although this intertidal artifactual context has been discussed for the contribution made by late occupation in the area and/or product of other earlier occupations, evidence of which has been eroded by geomorphological changes (Reyes et al. 2007), this site is projected to be a camp where multiple specific activities were carried on involved in the manufacture and/or maintenance of medium-sized lithic projectile points for hunting. There were successive reoccupations, with discard of the waste generated by mollusc gathering and consumption, as well as exploitation of a wide range of resources which included terrestrial mammals of the mainland (pudú), marine mammals (pinnipeds), fish and to a lesser extent birds, in a generalised hunting, fishing and gathering strategy in which resources gained by fishing predominated (Reyes et al. 2007).

The intertidal areas associated with Late Holocene shell-midden camps present a relation with the neighbouring archaeological deposits since tidal erosion leads to archaeological and bio-anthropological material and sediments falling from the exposed profile (Reyes et al. 2015, 2016b). These intertidal areas are observed in other parts of the Northern area of the archipelago, for example, Gran Guaiteca 4 (~1800 years cal. BP, 1 masl) with a retouch flake and a net weight. The exception is the south coast of Gran Guaiteca Island, which was probably a deposit resulting from the use, loss and discard of artefacts and waste over a series of occupations throughout the Mid and Late Holocene.

In the South-central area of the archipelago, we recorded material scattered in the intertidal at Nahuelquín 1 (~1355 and 400 years cal. BP, 2 masl), with a medium-sized, thick-section lanceolate stone projectile head made of rhyolithic obsidian from Chaitén Volcano, 300 km away; Canal Vicuña 3 (1 masl), with a net weight of a local stone; Punta Cementerio (0 masl), with six stone axes on local raw materials; Acuao 2 (0 masl, an extension of Punta Cementerio), and Acuao 3 and Canquenes 2 (0 masl), with one core each, on local raw materials; Posa Las Conchillas (~3300 to 1100 years cal. BP, –1 masl, Reyes et al. 2015), with six axes, a hammerstone, a side-scraper and four anvils, also made from local raw materials; and Isla Acuao 1 (~1200 years cal. BP, –1 masl), with a stone axe on local raw material and a distal fragment of a biface on rhyolithic obsidian from Chaitén Volcano, 330 km away. In all these associations, the adjacent archaeological sites present records from the Late Holocene with assemblages of artefacts derived from hunting for resources like birds and sea lions, as well as gathering and consumption of malacological fauna, and fishing; there are also hearth areas. These sites were reoccupied repeatedly and funerary activities are associated with several of them; in other words, these were residential shell-midden camps where multiple activities were carried on throughout the sequence.

As we have indicated, the artifactual density of the archaeological contexts assigned to the Late Holocene is remarkably low, even lower than the frequency in the Mid-Holocene sites recorded, with a total absence of evidence in some areas studied (Table 5.3). This low density of the lithic assemblages may have been exacerbated in part by the size and volume of the shell-midden sediments (Stein 1992). Apart from the low frequency of finds, technological continuities were detected in all the assemblages of stone artefacts defined to date for the Mid-Holocene. Medium-sized, thick-section lithic projectile points for hunting, stone axe preforms made at sites with local raw material for use in the forests, and possibly for the manufacture and/or maintenance of canoes, and net weights for fishing are the most frequent elements of the artifactual assemblages. They are present in sites with material included in organic deposits from the Mid-Holocene and also in the shell-midden matrices and adjacent intertidal areas of the recorded camps assigned to the Late Holocene.

Nevertheless, we can observe differences between these and sites assigned to the Mid-Holocene, where the stratigraphic and artifactual record is more consistent with logistical locations for the manufacture and maintenance of lithic projectile points, and with no later reoccupations. Lithic evidence from the Late Holocene, in contrast, appears in more varied contexts, associated with sites which include a range of subsistence activities, with the definition of domestic spaces (where there are hearths and discard of waste) and burial areas, in situations with the reoccupation of the same sites.

We recorded continuous use of rhyolithic obsidian from the Chaitén Volcano source as a raw material down to the end of the pre-contact period (e.g. Isla Solitaria and Gran Guaiteca 1); it was preferred for the manufacture of lanceolate lithic projectile points, which is consistent with continuity in subsistence practices (hunting) and transport over long distances. Prior to this investigation, however, there were no clear datings of contexts in which rhyolithic obsidian had been recovered, only chronological approximations. We now understand the first events in which we can record its use back to ~6200 years cal. BP (Reyes et al. 2016b), and its recurrence over time in the whole northern archipelago (Reyes et al. 2019b).

5.8 Circulation of Obsidian Pieces

Obsidian is a highly sought-after lithic raw material due to its qualities for the production of stone instruments (e.g. Andrefsky 1998); this helps to explain its transport and/or exchange over a wide spatial scale (e.g. Pallo and Borrero 2015; Méndez et al. 2018; Morello et al. 2004, 2012; Stern 2004; Stern et al. 2012; Stern 2018), regardless of its provenance. Previous works in the eastern valleys of the Andes in the Aysén Region permitted the study of these flows of allochthonous materials (Contreras et al. 2016; Méndez et al. 2006, 2018; Méndez and Reyes 2008, 2015; Reyes et al. 2009).

The principal sources detected were to the north-east and south-east, e.g. Sacanana and Telsen/Sierra Negra (~42°S) and Pampa del Asador (~47°S), between 460 and 320 km from the valleys studied (Méndez et al. 2008/9, 2018; Stern 2018). Transport not only involves the displacement of materials, but is in itself an indicator of human flows (Gamble 1999).

On a meso-scale, based on the evidence collected in the Chonos Archipelago, we observe two patterns of obsidian flow longitudinally through the Aysén Region: Andean obsidian, characterised by the Chaitén Volcano type which is found all through the northern archipelago (Munita 2007; Stern and Porter 1991; Stern and Curry 1995; Stern et al. 2002; Reyes et al. 2006b, 2007, 2015; this study); and the obsidian from inland, which moved up and down the steppe area (Méndez et al. 2008/9, 2018; Reyes et al. 2009). Our excavations in the El Toro rock shelter (~44°S), 37 km in a straight line from the coast of the Chonos Archipelago (60 km following the course of the Cisnes River), were precisely designed to clarify the proposal that contact might have existed between the two areas of marine and terrestrial interaction (Bate and Mena 2003). However, the context was found to have been formed by late incursions of a logistical nature by groups that had come from the extra-Andean steppe around 2600 to 2400 years cal. BP, with raw materials and techno-typological components of the steppe tradition (Méndez et al. 2006; Méndez and Reyes 2008; Reyes et al. 2006b, 2009).

Thus the record of rhyolithic obsidian in sites dated to the Mid and Late Holocene, its distribution in different areas of the Chonos Archipelago, and its association with conserved technologies (e.g. lanceolate lithic projectile points), combined with the absence of obsidian pieces from the interior of the mainland or with related artifactual designs (supplemented by the genetic, craniometric and subsistence evidence mentioned above), allows us to reaffirm that–in the Aysén Region—the Andes Range, with its volcanoes, montane forests and ice fields, functioned as an effective bio-geographical barrier. It conditioned human movement and excluded spheres of interaction between terrestrial and maritime hunter-gatherers.

5.9 The Faunal Record

The record of vertebrate and malacological fauna allows us to project some of the subsistence strategies used by the marine hunter-gatherer groups who occupied the archipelago (Gifford-Gonzalez 1991; Jones 2004; Lyman 2003). To do this we classified, quantified and identified the attributes or marks produced by the processing, consumption and discard of the resources obtained (e.g. Abe et al. 2002; Binford 1981, 1984; Fisher 1995; Lyman 2005). The archaeofaunal assemblage recovered from columns excavated to investigate fauna was recorded in the sites defined as shell-midden camps distributed in the Northern area of the archipelago, like GUA-010 Conchal and Gran Guaiteca 2 Conchal, on the south coast of Gran Guaiteca Island, and Alero Low and Isla Marta 1, on the north side. Evidence was also obtained from Isla Benjamín 1, 2 and 4 in the Central area of the archipelago, and Seno Gala

1 (Reyes et al. 2007) on the mainland. We also incorporated the finds of artefacts made from bones, recovered in the intertidal areas of Canal Vicuña 3 and Corrientes del Yates 6 in the South-central area.

In contrast, the faunal record for the sites assigned to the Mid-Holocene is absolutely nil; fishing and hunting can only be inferred from the artifactual assemblage recorded (principally lanceolate lithic projectile points and net weights). There are several possible explanations for this absence. One is the differential use of space, not fully represented by the area of the sites which was excavated and/or sampled. A second is the functionality attributed to the site, when no other subsistence activities are represented, such as discard and processing of fauna consumed. A third factor is the presence of podzol soils, a highly acidic substrate that affects even lithic materials and results in differential preservation of the organic material deposited (Gifford 1981; Lyman 1994). Finally, there is the major impact of the geomorphological changes which have occurred in the region: in the context of mainly coastal sites, the older sites are more affected by these events, being hidden (e.g. GUA-010 Terraza and Gran Guaitecas 2 Terraza) or destroyed (e.g. Isla Goñi 1). Organic materials in matrices transformed in this way are especially sensitive. All the sites with records of vertebrates and malacofauna present chronological sequences assigned exclusively to the Late Holocene.

Three sites on Gran Guaiteca Island, in the Northern area of the archipelago, were sampled systematically for mollusc evidence (Cárcamo 2015): GUA-010 Conchal, Gran Guaiteca 2 Conchal and Alero Low). The faunal columns in GUA-010 Conchal indicate the presence of 17 species of molluscs and one of the crustaceans; in Gran Guaiteca 2 Conchal, 10 species of mollusc and two of crustaceans were recorded. In both sites clams (*Venus antiqua*) predominate, obtained from sandy-muddy bottoms in the intertidal zone, adjacent to the shell-middens in both sites. Nine species of mollusc were recorded in Alero Low, plus one echinoderm and one crustacean; the predominant species was the snail (*Tegula atra*), obtained from the rocky substrate to which they adhere, characteristic of the coastal environment of the site.

The mollusc species represented in the fauna columns are found in a range of habitats: rocky intertidal, rocky subtidal, parasitic and mixed intertidal (sandy and/or rocky). The set, therefore, reflects diverse supply strategies; the relative importance of the species represented is consistent with the type of shoreline immediately round the site, while the variety of species recorded is related to gathering of supplies from a wider area around the camps. In other words, the variations in the fauna depend on local availability and not on the intensified gathering (sensu Binford 2001, Braje and Erlandson 2009) or the use of different methods of capture, as would appear to occur on the south coast of Chiloé (Legoupil 2005) or in Reloncaví Sound on the Chilean mainland 270 km north of this archipelago, due to inflows of new populations and cultural processes (Flores et al. 2010; Flores and Correa 2011).

Although the volume of shells discarded has not been estimated systematically in this archipelago, a first approach has been made to dimensioning the discard rates in the shell-middens formed (Reyes et al. 2015). Posa Las Conchillas, in the South-central area of the archipelago, is a shell-midden almost 8 m high and 40 m long (Fig. 4.38); it is one of the largest shell-middens recorded (Simpson 1875; Stern and Curry 1995). An ordered chronological sequence was made of this site, with nine radiocarbon datings of mollusc valves and carbon (Table 4.1). This suggests that the thick stratigraphic packets of shells have a certain stability, allowing their sedimentation rhythms to be estimated (Stein et al. 2003). The deposit rate calculated for this site was 2.3 cm per year for the formation of the mound between ~3300 and 1200 years cal. BP (Reyes et al. 2015). Although we understand that this rate was not constant, and that formation of the mound was discontinuous, it still reveals the volume of discard episodes.

This not only indicates the intensity of gathering, but also shows the restricted spaces available for human settlement in an archipelago environment defined by steep coasts where the coastal settlement pattern that we have documented made it necessary to re-use some locations. What we see as large shell accumulation results from the superimposition of residential camps where multiple activities were carried on throughout the sequence of site formation on a millennial scale.

The vertebrate fauna sample included 1,920 bone remains distributed between mammals (24.6%), birds (22.8%) and fish (52.6%) (San Román et al. 2016; Reyes et al. 2019a; Table 5.4). An exceptional case is the Isla Benjamín 2 site (shell-midden camp), where almost no bones were recovered (N = 2). In this site, a high degree of dissolution was observed in the mollusc shells. It is plausible to suggest that the very low frequency of bones is due to chemical dissolution resulting from the podzol matrix in which they were deposited. Apart from this case, the general state of conservation of the bone remains in the other sites is good.

Coypu (*Myocastor coipus*) is the most recurrent mammal species, recorded in 5 of the 8 sites where faunal columns were dug, followed by otariids (Otariidae), present in 4 of these sites, and otters (*Lontra sp.*) in 3 sites. This ranking is repeated in the frequency of NISP (Grayson 1984). Among the cervids, the greatest frequency of pudú bones (*Pudu pudu*) is found on the mainland coast, in Seno Gala 1 (Reyes et al. 2007), with datings ~1200 years cal. BP. Pudú remains were also recorded in an island context, in Isla Marta 1 in the Guaitecas Archipelago (Northern area), dated to around ~1400 years cal. BP. Huemul remains (*Hippocamelus bisulcus*) were recorded in an island context in the Central area of the archipelago, at Isla Benjamín 1, with ~1500 years cal. BP (Table 4.1). These finds probably represent carcasses, or parts of carcasses, transported from the mainland coast—or even, in the case of pudú, from Chiloé Island north of the Gulf of Corcovado. A similar situation must be supposed in the case of the canid remains recorded on Gran Guaiteca Island, at GUA-010 Conchal and Alero Low, where this taxon is also allochthonous.

Among birds, *Tachyeres spp.* (ducks) are the most recurrent taxa, recorded in 5 of the 8 sites controlled, followed by *Phalacrocorax spp.* (cormorants) and *Procellaria spp.* (petrels), present in 4 sites; in the NISP frequency, the ranking is dominated by *Phalacrocorax spp.,* followed by *Procellaria spp.* (San Roman et al. 2016).

Fish are represented in 7 of the 8 sites sampled; the commonest species are blenny (*Eleginops maclovinus*) and jurel (*Thrachurus murphyi*), although the most abundant remains are of snoek (*Thyrsites atun),* very frequent in one of the Pacific border sites, Isla Marta 1, in the Northern area of the archipelago (Torres 2016). Jurel (*Thrachurus*

Table 5.4 NISP and frequency of identified taxa per archaeological site in the Chono archipelago. SG-1: Seno Gala 1 (on the Mainland coast), IB 4: Isla Benjamín 4, IB 2: Isla Benjamín 2, IB 1: Isla Benjamín 1, (Central area of the archipelago) GUA-010 C: GUA-010 Conchal, GG2C: Gran Guaiteca 2 Conchal, A. Low: Alero Low, IM1: Isla Marta 1 (Northern area of the archipelago), CV3: Canal Vicuña 3 y CY6: Corrientes del Yates 6 (South-central area of the archipelago)

	Species/Sites	SG 1	IB 4	IB2	IB 1	GUA010C	GG2C	AL	IM1	CV3	CY6	Total
Mammals	Cetacea							1				1
	Otaridae	45			2			11	18			76
	Lontra sp.		.		3			3	6			12
	Myocastor coypus		26		20	1		32	16			95
	Hippocamelus bisulcus				3							3
	Pudu pudu	17							1			18
	Artiodactyla				10				1			11
	Canidae					5		1				6
	Rattus norvegicus		3									3
	Rodentia		65					3	1			69
	Mammalia		10	1	2	40		12	112		1	178
	Total (N taxa)	62 (2)	104 (2)	1 (1)	40 (4)	46 (2)		63 (6)	155 (5)		1 (1)	472
Birds	*Pelecanoides* sp.		61									61
	Phalacrocorax sp.	6			2			22	51			81
	Procellaria sp.		67	1	1			3				72
	Diomedeidae							2				2
	Procellariiforme							1	1			2
	Spheniscus sp.							3	5			8
	Tachyeres sp.		1		4	7		16	7			35

(continued)

Table 5.4 (continued)

	Species/Sites	SG 1	IB 4	IB2	IB 1	GUA010C	GG2C	AL	IM1	CV3	CY6	Total
	Anas sp.								3			3
	Chloephaga sp.								3			3
	Passeriforme							2				2
	Indeterminate		56		12	2		62	36	1		169
	Total (N taxa)	6 (1)	185 (3)	1 (1)	19 (3)	9 (1)		111 (6)	106 (6)	1 (1)		438
Fish	*Auchenionchus* sp.								1			1
	Mustelus mento						1					1
	Austrolycus sp.							2				2
	Cottoperca gobio							2				2
	Eleginops maclovinus		1		9	1		128	72			211
	Paralichthys microps							3				3
	Genypterus sp.							4				4
	Cheilodactylus variegatus							7	72			79
	Sebastes oculatus				1			4	37			42
	Thyrsites atun				15	25	1	14	310			365
	Thrachurus murphyi	267	1		5		1	20	6			300
	Total (N taxa)	267 (1)	2 (2)		30 (4)	26 (2)	3 (3)	184 (9)	498 (6)			1010
NISP Total		335	291	2	89	81	3	358	759	1	1	

Source Taken from Reyes et al. (2019a)

murphyi) is the dominant fish species overall, and of the taxa identified at Seno Gala 1 (Reyes et al. 2007) on the Mainland coast.

Analysis of taxonomic variety and the geographical location of the sites revealed a tendency to the greater taxonomic richness in the Northern area of the archipelago, in Isla Marta 1 and Alero Low on Gran Guaiteca Island, both contexts exposed to the Pacific (Torres 2016). This tendency is consistent with the ecological characteristics of the region, principally in the presence of sea birds and fish (Lozano 2006; Pequeño and Riedmann 2002; Vuilleumier 1985).

Turning to seasonality, sites in both the Central part of the archipelago (Isla Benjamín 1 and 4) and the Northern area (Isla Marta 1 and Alero Low) present indicators attributable to summer occupations based on the presence of juvenile specimens of sea lions and birds (San Román et al. 2016).

The principal anthropogenic marks recorded on bones are cutting marks (San Román et al. 2016). These were identified in mammals and birds from Alero Low (n = 11, 6.3% NISP) and Isla Marta1 (n = 7, 2.7% NISP). There were also cases with evidence of thermal alteration (hearths) in Alero Low (n = 2, coracoid and humerus of *Tachyeres sp.*) and Isla Benjamín 1 (n = 1, the humerus of *Phalacrocorax sp.*).

We were also able to identify two instruments made of birds' long bones (n = 2), punches or awls made from cormorant tarsometatarsus (*Phalacrocorax sp.*), recovered from excavation: one from the shell-midden at Seno Gala 1 and the other recovered from the intertidal adjacent to Canal Vicuña 3. At Corrientes del Yates 6, we recovered a pendant adornment made from a polished mammal rib. Despite the scarcity of finds, we know that bone artefacts were widespread in the contexts of marine hunter-gatherers in western Patagonia (Christensen 2016).

From the characteristics of the faunal assemblages obtained in this study (Table 5.4), we can offer some observations on the subsistence strategies deployed by the marine hunter populations of the Chonos Archipelago during the Late Holocene. First, the exploitation of mammals shows no emphasis or specialisation on a particular taxon; the contribution of small mammals like coypu (*Myocastor coipus*) to their diet is important, while the inclusion of canids, while insignificant, indicates exploitation of a broad spectrum of fauna (San Román et al. 2016). Contrary to what may be thought, otariids are not the main resource to be exploited in these contexts and it is slightly more abundant in only one site, Seno Gala 1. Secondly, the exploitation of birds is mostly concentrated on taxa which nest in colonies, such as cormorants (*Phalacrocorax spp.*) and petrels (*Procellaria spp.*). Thirdly, the variety of fish species captured (n = 11) suggests the use of coastal fishing strategies for blenny (*Eleginops maclovinus*) and jurel (*Thrachurus murphy*i), and boats to fish for snoek (*Thyrsites atun*) in the neritic-pelagic habitats of the species recorded (Torres 2016). The high frequency of net weights in the intertidal of GUA-010 Conchal and Gran Guaitecas 2 Conchal (Porter 1993) is indicative of how this resource was obtained, the variety of species recovered and the number present in the archaeofaunal record. Fourth, isotope analysis is consistent with a predominantly marine diet (Reyes et al. 2015, 2019a), in which terrestrial mammals are scarcely represented. Fish tend to be the most important element in consumption, accounting for 56.2% of the total sample of vertebrate fauna recovered. This needs to be evaluated in future in order to determine whether bias factors exist which affect trends (e.g. sampling of few columns, absence of extensive excavations, the greater presence of skeletons only in the Late Holocene), or whether this is in fact a late process in which the capture of some resources—especially fish—was intensified, as in the area of the Beagle Channel (Zangrando 2009a, b). Fifth, as in the representation of the malacological fauna, the variation in the taxonomic richness in the different areas of the archipelago is explained as a reflection of the taxonomic richness available locally, and not as a product of intensified capture of certain resources or the use of different and/or specialised capture techniques. Sixth, the presence of canids in Gran Guaiteca Island (GUA-010 Conchal and Alero Low), of pudú in Isla Marta 1 and of huemul in Isla Benjamín 1 cannot be explained by the local availability of these species. The possible explanations include transport of animal carcasses, or parts of them, on wide mobility and/or exchange circuits to and from the mainland coast and the northern end of the archipelago, as is confirmed by the wide circulation of rhyolithic obsidian from the Chaitén Volcano source.

5.10 Human and Faunal Isotopes

Another important set of data for the subsistence of the hunter-gatherers of this archipelago is the information obtained from the bone collagen isotope records from the human remains recovered, as well as those obtained from archaeological and modern fauna which, together with the information from the sampling columns of the shell-midden camps, indicates dependence on and specialisation in the capture of resources from the littoral and maritime media (Coltrain et al. 2016; Erlandson 2001; Perlman 1980; Yesner 1980; Orquera and Piana 2005; Orquera et al. 2006; Richard and Hedges 1999; Tables 5.5, 5.6 and 5.7).

Integrated $\delta^{13}C$ and $\delta^{15}N$ analysis was performed on 38 individuals of different sexes and ages, obtained from archaeological contexts distributed over the whole archipelago: Pacific border islands, Mainland coast, and the Northern, Central and South-central areas (Reyes et al. 2019a). The contexts included burials in open-air shell-middens, shell-middens in shelters, and caves used exclusively for funerary purposes. All the individuals were dated directly, allowing us to situate both the funerary events and the individual context of subsistence in time (Table 5.5).

We had access to 11 samples of archaeological fauna, 5 of which were obtained from previous studies in neighbouring areas (Barberena et al. 2011; Reyes et al. 2007) and 6 which we added corresponding to 4 different taxa of terrestrial (pudú and huemul) and estuarine mammals (marine otter and coypu), all with radiocarbon dating (Table 5.6). Finally, we brought together a series of studies on modern isotope ecology carried out in areas adjacent to the archipelago, including malacological fauna, crustaceans, marine mammals, fish and birds (Forero et al. 2004; Huckstadt et al. 2007; Mayrs et al. 2011; Muñoz et al. 2013) which we found represented in the archaeological record of the different sites investigated, establishing reference values for the resources consumed. To this end, we equated the isotope values obtained from

Table 5.5 Stable isotope values, dated by ^{14}C and calibrated for Marine Reservoir Effect (Merino et al. 2019) and associated information of human remains samples

Site	Treatment lab	Sample	Age ^{14}C (years BP)	Calibrated range (cal BP) 2-sigma	Collagen yield (%)	C (%)	N (%)	C:N	$\delta^{13}C$ (‰)	$\delta^{15}N$ (‰)	Skeletal element	Sex	Age
Gran Guaiteca 2 Conchal	UGAMS 8294 [a]	REP-01	830 ± 30	624–491	12.6	36.5	12.4	3.4	−13.9	15.3	Rib	I	Adu
Alero Low	UGAMS 26,283[d]	ALOW 1–1	820 ± 25	531–362	19.2	42.6	16.0	3.1	−11.3	17.1	2nd right metatarsal,	I	Adu
Alero Low	UGAMS 26,284[d]	ALOW 1–2	990 ± 25	650–510	11.6	22.9	8.2	3.2	−11.1	18.5	2nd right metatarsal,	I	Adu
Repollal Caverna	UGAMS 8295 [d]	REP-02	730 ± 25	482–300	12.1	42.8	15.0	3.3	−11.7	18.2	Tibia	I	Adu
Puquitín	UGAMS 8293[a]	PUQUIT-01	1050 ± 30	666–508	11.2	43.2	15.0	3.3	−10.1	18.5	Rib	I	Adu
Isla Sin Nombre	UGAMS 8290[a]	ISN-01	870 ± 30	546–335	12.7	43.6	14.7	3.4	−10.4	17.3	Rib	I	Adu
Isla Ipún 1	UGAMS 10,450[c]	IPUN 1	870 ± 25	625–490	12.3	45.1	17.2	3.0	−12.8	16.8	Metacarpal	F	Ado
Isla Ipún 1	UGAMS 10,451[c]	IPUN 2	1590 ± 25	1223–962	8.1	45.4	16.3	3.2	−9.3	16.9	Ulna	M	Adu
Estero Sur	UGAMS 10,818[d]	ES1	1980 ± 25	1677–1412	–	45.0	16.0	3.2	−11.5	17.0	Hip	M	Adu
Estero Sur	UGAMS 10,819[d]	ES2	2030 ± 25	1736–1507	–	17.0	6.0	3.3	−11.9	17.4	Vertebra	I	Juv

(continued)

Table 5.5 (continued)

Site	Treatment lab	Sample	Age ^{14}C (years BP)	Calibrated range (cal BP) 2-sigma	Collagen yield (%)	C (%)	N (%)	C:N	$\delta^{13}C$ (‰)	$\delta^{15}N$ (‰)	Skeletal element	Sex	Age
Estero Sur	UGAMS 10,820[d]	ES7	2030 ± 25	1736–1507	–	43.7	15.3	3.3	−11.9	16.3	Cranium	M	Ado
Estero Sur	UGAMS 10,821[d]	ES9	2030 ± 25	1734–1501	–	45.3	16.1	3.2	−11.8	16.7	Cranium	F	Adu
Estero Sur	UGAMS 10,822[d]	ES11	1960 ± 25	1683–1428	–	40.1	14.0	3.3	−12.4	17.4	Cranium	M	Adu
Osario Melimoyu	UGAMS 26,279[d]	MEL 1–1	1010 ± 25	667–528	15.4	44.1	16.5	3.1	−11.65	18.0	Cranium	I	Adu
Osario Melimoyu	UGAMS 26,280[d]	MEL 1–3	920 ± 25	633–498	15.0	43.3	15.8	3.2	−12.1	17.7	Cranium	M	Adu
Isla Yalac 1	UGAMS 26,281[d]	YAL1-1	710 ± 25	479–301	17.1	42.8	15.9	3.1	−12.2	17.9	Cranium	M	Adu
Isla Yalac 1	UGAMS 26,282[d]	YAL1-6	780 ± 25	514–330	21.4	41.1	15.3	3.1	−11.9	18.0	Cranium	F	Adu
Isla Harris	UGAMS 8289[a]	HAR-01	650 ± 25	450–277	14.0	4.2	13.9	3.3	−12.5	16.9	Rib	I	Adu
Seno Canalad	UGAMS 8288[a]	CAN-01	740 ± 30	493–308	14.1	45.5	15.9	3.3	−12	18.5	Rib	I	Adu
Isla Benjamín 1	UGAMS 8286[a]	BEN-01	700 ± 30	463–239	12.9	44.2	15.3	3.3	−10.6	18.7	Rib	I	Adu
Isla Benjamín 1	UGAMS 8287[a]	BEN-02	770 ± 30	482–288	13.2	39.4	13.5	3.3	−10.5	17.7	Rib	I	Adu

(continued)

Table 5.5 (continued)

Site	Treatment lab	Sample	Age ^{14}C (years BP)	Calibrated range (cal BP) 2-sigma	Collagen yield (%)	C (%)	N (%)	C:N	$\delta^{13}C$ (‰)	$\delta^{15}N$ (‰)	Skeletal element	Sex	Age
Isla Benjamín 4	UGAMS 21,292[d]	IBEN4	820 ± 25	516–323	–	43.2	15.5	3.2	−10.9	16.9	Rib	I	Adu
Isla Benjamín 5	UGAMS 21,291[d]	IBEN5-2	890 ± 25	573–433	–	41.9	14.9	3.2	−11.1	17.5	Rib	F	Adu
Isla Benjamín 5	UGAMS 21,451[d]	IBEN5-3	770 ± 25	489–302	–	39.3	14.4	3.1	−11.0	18.8	Tooth	M	Adu
Isla Benjamín 5	UGAMS 21,452[d]	IBEN5-4	850 ± 25	542–361	–	43.3	16.2	3.1	−10.9	20.3	Rib	I	Inf
Isla Benjamín 5	UGAMS 21,453[d]	IBEN5-5	790 ± 25	508–231	–	40.6	14.6	3.2	−11.4	18.0	Tooth	I	Adu
Isla Elena 1	UGAMS 8119[a]	ELE 1–1	1880 ± 25	1517–1298	6.7	40.3	14.0	3.3	−10.6	16.5	Clavicle	F	Adu
Isla Elena 1	UGAMS 8120[a]	ELE 1–2	1880 ± 25	1517–1298	7.2	44.8	16.7	3.1	−10.6	16.7	Clavicle	F	Adu
Isla Elena 1	UGAMS 8121[a]	ELE 1–3	1750 ± 30	1393–1188	6.1	38.2	14.2	3.1	−11.2	15.8	Clavicle	I	Ado
Isla Elena 1	UGAMS 8122[a]	ELE 1–4	1820 ± 30	1475–1244	6.6	41.1	14.5	3.3	−10.4	19.5	Clavicle	I	Inf
Nahuelquín 1	UGAMS 04,949[b]	NAH 1–1	430 ± 25	505–330	12.1	–	–	–	−21	11.3	Scapula	F	Adu
Nahuelquín 1	UGAMS 13,127[d]	NAH1-2	1030 ± 25	683–537	–	14.2	4.6	3.6	−11.8	15.3	Rib	M	Adu

(continued)

Table 5.5 (continued)

Site	Treatment lab	Sample	Age ^{14}C (years BP)	Calibrated range (cal BP) 2-sigma	Collagen yield (%)	C (%)	N (%)	C:N	$\delta^{13}C$ (‰)	$\delta^{15}N$ (‰)	Skeletal element	Sex	Age
Isla Victoria 2	UGAMS 10,452[c]	VIC 2–1	1820 ± 25	1473–1257	7.1	43.3	16.6	3.0	−10.6	16.1	Cranium	M	Adu
Isla Victoria 2	UGAMS 10,453[c]	VIC 2–2	2330 ± 25	2073–1820	5.8	44.5	16.8	3.0	−11.2	16.3	Cranium	F	Adu
Isla Acuao 1	UGAMS 8117[a]	ACUAO-1	1630 ± 25	1537–1413	8.0	36.1	13.0	3.2	−10.8	16.5	Parietal	I	Adu
Corrientes del Yates 5	UGAMS 13,128[d]	CY5-1	1420 ± 20	1088–880	–	38.1	13.3	3.3	−11.4	16.5	Rib	M	Adu
Puqueldón	UGAMS 8291[a]	PUQUEL-01	210 ± 25	297/ -1	15.6	37.6	12.8	3.4	−20.1	8.8	Rib	I	Adu
Puqueldón	UGAMS 8292*	PUQUEL-02	210 ± 30	299/-1	16.3	45.0	15.2	3.4	−19.7	9.3	Rib	I	Adu

UGAMS: Center for Applied Isotope Studies, University of Georgia;[a]Reyes et al. (2013); [b]Reyes et al. (2011); [c]Reyes et al. (2015). [d]Reyes et al. (2019a); I: Indeterminate; F: Female; M: Male. Adu: Adult (>20 years); Ado: Adolescent (<20–> 13 years); Juv: Juvenile (<13–> 7 years); Inf: Infant (<3 years)

Table 5.6 Stable isotopes in fauna samples obtained from archaeological sites, Chonos Archipelago and adjacent mainland coast

Site	Species	Sample	Lab. Code	$\delta^{13}C$ (‰)	$\delta^{15}N$ (‰)	Age ^{14}C (years BP)	Calibrated range (cal BP) 2-sigma[a]	Diet, habitat
alero low	*M. coypus*	Radius	UGAMS 26,285	−22.5	4.5	1940 ± 25	1906−1747[c]	Herbivore, aquatic and riverine environments and wetlands
alero low	*M. coypus*	Radius	UGAMS 26,286	−19.3	10.3	2050 ± 25	2012−1901[c]	
I. Benjamín 1	*M. coypus*	Metapodial	UGAMS 26,288	−16.7	10.1	1820 ± 25	1748−1607[c]	
I. Benjamín 1	*L. felina*	Vertebra	UGAMS 26,289	−9.8	22.0	Modern	Modern[c]	Carnivore/piscivore, shoreline and riverine
I. Benjamín 1	*H. bisulcus*	Metapodial	UGAMS 26,287	−22.6	1.7	1590 ± 25	1518−1371[c]	Herbivore, forest
Los Ñadis	*H. bisulcus*	Femur	UGAMS 6676	−21.1	1.3	Modern	Modern*	
El Toro	*H. bisulcus*	Tooth	UGAMS 6671	−20.8	0.7	-	2760−2360*	
El Toro	*H. bisulcus*	Diaphysis	UGAMS 6672	−20.6	-0.6	-	2760−2360*	
Isla Marta 1	*P pudu*	Metapodial	UGAMS 26,290	−22.29	2.07	1520 ± 25	1406−1310[c]	Herbivore, forest
El Toro	*P. pudu*	Diaphysis	UGAMS 6670	−21.6	1.8	-	2760−2360*	
Gala Sound 1	*P. pudu*	Tooth	BETA 230,515	−23.6	-	1340 ± 40	1299−1093[b]	

*Barberena et al. (2011) (chronological range of the stratigraphic unit a); [b]Reyes et al. (2007); [c]Reyes et al. (2019a)

Table 5.7 Comparative table of stable isotopes of present and archaeological fauna. The collagen samples of archaeological fauna were projected (sensu Fernandes 2015) in order to compare the results of the archaeofaunal assemblage with the isotope results obtained from present fauna for reference purposes

Area	Approx. Latitude	Species	Material	Chronology	δ13C coll ‰	Standard Dev	δ15N coll ‰	Standard Dev	N	References
Talcahuano coast	37°	*Otaria flavescens* (common sea lion)	skin/fur	present	−12.4	0.7	20.9	0.8	27	Huckstadt et al. (2007)
Comau Fjord, Mainland Chiloé	42°30′	*Aulacomya ater* (Cholga)	organic tissue	present	−17.7	0.9	10.6	0.9	8	Mayr et al. (2011)
Comau Fjord, Mainland Chiloé	42°30′	*Crepidula sp.* (Snail)	organic tissue	present	−16.3	0.7	9.0	1.5	8	Mayr et al. (2011)
Comau Fjord, Mainland Chiloé	42°30′	*Mytilus chilensis* (Chilean mussel)	organic tissue	present	−18.0	0.9	10.4	0.8	139	Mayr et al. (2011)
Comau Fjord, Mainland Chiloé	42°30′	*Tegula sp.* (Snail)	organic tissue	present	−15.4	1.0	11.3	2.5	4	Mayr et al. (2011)
Comau Fjord, Mainland Chiloé	42°30′	*Helicolenus lengerichi* (Chancharro)	organic tissue	present	−15.2	0.4	17.0	0.7	6	Mayr et al. (2011)
Comau Fjord, Mainland Chiloé	42°30′	*Sebastes capensis* (Vieja/Cabrilla)	organic tissue	present	−15.7		16.7		1	Mayr et al. (2011)
Chiloé Archipelago	42°-43°	*Trachurus murphyi* (Jurel)	organic tissue	present	−15.1	0.4	17 .3	0.4	3	Muñoz et al. (2013)

(continued)

Table 5.7 (continued)

Area	Approx. Latitude	Species	Material	Chronology	δ13C coll ‰	Standard Dev	δ15N coll ‰	Standard Dev	N	References
Chiloé Archipelago	42°-43°	*Helicolenus lengerichi* (Chancharro)	organic tissue	present	−14.5	0.3	17.3	1.0	5	Muñoz et al. (2013)
Chiloé Archipelago	42°-43°	*Genypterus sp.* (Cusk eel)	organic tissue	present	−14.2	0.3	17.7	0.9	11	Muñoz et al. (2013)
Chiloé Archipelago	42°-43°	*Thyrsites atun* (Snoek)	organic tissue	present	−15.1	0.3	17.8	0.9	7	Muñoz et al. (2013)
Chiloé Archipelago	42°-43°	*Paralabrax humeralis* (Common cabrilla)	organic tissue	present	−14.2	0.9	16.2	0.5	9	Muñoz et al. (2013)
Chiloé Archipelago	42°-43°	*Eleginops maclovinus* (Blenny)	organic tissue	present	−13.7	1.2	15.6	1.0	12	Muñoz et al. (2013)
Chiloé Archipelago	42°-43°	*Homalaspis plana* (Jaiva crab)	organic tissue	present	−13.3	0.6	14.7	0.7	2	Muñoz et al. (2013)
Chiloé Archipelago	42°-43°	*Cancer setosus* (Jaiva peluda)	organic tissue	present	−12.6		15.5		1	Muñoz et al. (2013)
Atlantic Coast of Patagonia	40°-46°	*Phalacrocorax magellanicus* (Lile Cormorant)	organic tissue	present	−16.3	0.1	19.3	0.1	48	Forero et al. (2004)
Atlantic Coast of Patagonia	40°-46°	*Phalacrocorax atriceps atriceps* (Imperial shag)	organic tissue	present	−16.3	0.1	20.0	0.1	15	Forero et al. (2004)

(continued)

Table 5.7 (continued)

Area	Approx. Latitude	Species	Material	Chronology	δ13C coll ‰	Standard Dev	δ15N coll ‰	Standard Dev	N	References
Arch. Chonos/Aysén	44°40′S	*H. bisulcus* (Huemul)	Collagen	Archaeological	−22.8	0.2	3.0	0.4	4	Barberena et al. (2011)/this study
Arch. Chonos	43°50-44°42′S	*Pudu pudu* (Pudú)	Collagen	Archaeological	−24.6	1.4	3.9	0.1	3	Reyes et al. (2007)/this study
Arch. Chonos	43°50-44°40′S	*M coypus* (Coypu)	Collagen	Archaeological	−21.5	2.9	10.3	3.3	3	This study
Arch. Chonos	44°40′S	*Lontra felina* (Marine otter)	Collagen	Archaeological	−10.9		24.0		1	This study

the bone collagen of the archaeofaunal samples to muscle (sensu Fernándes 2015) to homologate the regional faunal universe projected for human consumption (Table 5.7), and so progress towards the construction of a regional isotope ecology (Reyes et al. 2019a).

Of the 38 samples studied, all presented C:N ratios within the normal range (Ambrose 1990; De Niro 1985), indicating that the isotope signals are primary, with no post-deposit alterations; the mean value was 3.2 and the extremes 3.0 and 3.6. Within this group, 36 individuals present chronologies that are pre-contact or around the date of contact according to the calibration ranges applied with new Marine Reservoir Effect values (Merino et al. 2019), i.e. ~2000 to 350 years cal. BP. To these, we added two individuals from a shell-midden on Lemuy Island in the Chiloé Archipelago (~200 years cal. BP), used previously in comparative analyses (Reyes et al. 2013, Tables 5.6 and 4.1). It is as well to clarify that various calibration and correction formulas exist for establishing chronological ranges when dating human samples (and any type of datable organic material). It must, therefore, be remembered that marine hunter-gatherers present the Reservoir Effect, i.e. the datings obtained from these individuals are "aged" by their consumption of a pre-eminently marine diet. This is important for cultural interpretations, especially at the temporal boundary of European contact (Reyes et al. 2019a, b).

The 35 individuals from moments which are pre-contact or around the date of contact present high $\delta^{15}N$ values, between 15.3‰ and 20.3‰, and $\delta^{13}C_{col}$ values between −9.3 and −13.9‰. The only infants sampled, in the Central and South-central areas, tend to present higher $\delta^{15}N$ and $\delta^{13}C_{col}$ values probably due to the effects of breastfeeding, i.e. one trophic level higher than their mothers (Fogel et al. 1989), so the sample tends to vary slightly within the archaeological set. These were eliminated in the subsequent analysis of paleo diet. Nevertheless, this sample forms a fairly homogeneous group, with typical values for marine hunter-gatherers, similar to those recorded in the far south of Patagonia, in the area of the Magellan Strait, the Beagle Channel and the southern archipelago (Alfonso et al. 2015; Barberena 2002; Borrero et al. 2001; Kochi 2017; Panarello et al. 2006; Tafuri et al. 2017; Tessone et al. 2003; Yesner et al. 2003; Zangrando et al. 2004). Samples with post-contact dates from Lemuy Island (Chiloé) and Traiguén Island (South-central area of the archipelago) present very different values for $\delta^{13}C_{col}$, between −21 and –19.7‰, and $\delta^{15}N$, between 11.3‰ and 8.8‰ (Reyes et al. 2015).

Comparison between the $\delta^{13}C_{col}$ and $\delta^{15}N$ values of the two sets produces significant differences (Fig. 5.3). In paleodietary terms, the post-contact individuals show a significant reduction in the consumption of marine resources, which are no longer their main source of food (Reyes et al. 2019a). The incorporation of terrestrial resources into the diet, for example, domestic animals and/or plants like potatoes with a photosynthetic C_3 pathway which is recognised in historical populations, may have formed a more substantial contribution. These changes can be explained in the framework of the profound cultural alterations suffered by local people in the northern archipelagos from the sixteenth century, after European contact (García 1889 [1766]; Torrejón et al. 2013; Urbina 1988, 2017; Reyes et al. 2019a; Urbina et al. 2020).

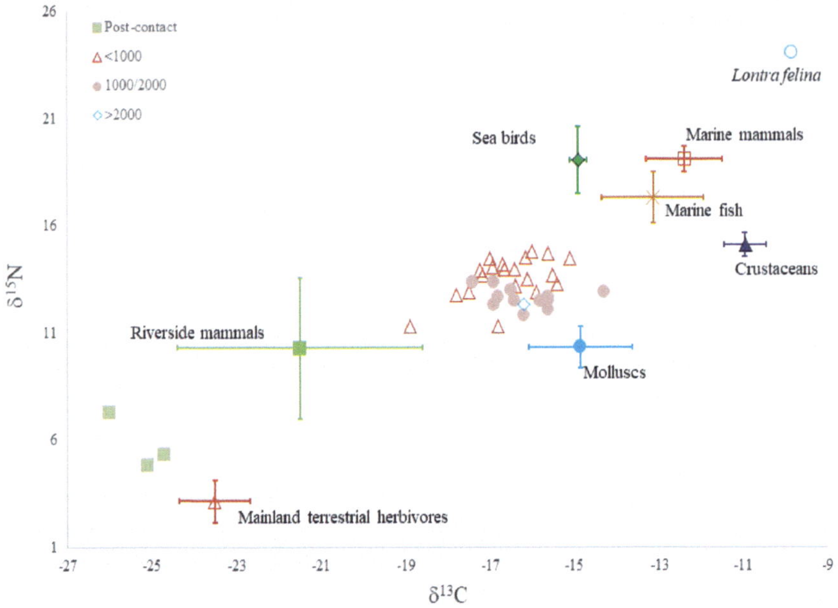

Fig. 5.3 Representation of estimated diets of the pre-contact and post-contact individuals of the Chonos Archipelago and Chiloé: averaged $\delta^{13}C_{col}$ and $\delta^{15}N$ values from marine and native terrestrial fauna, by chronology. *Source* Taken from Reyes et al. (2019a).

Even though the pre-contact group is very homogeneous, we evaluated whether any difference existed by the sex or geographical distribution of the samples, or by the funerary practices recorded, type of site or chronology (Reyes et al. 2019a): (a) No significant differences are observed between the two groups by geographical distribution; (b) If the samples are divided by funerary practices (deposition in shelters or burial in shell-middens), no significant differences are observed (Mann–Whitney $p > 0.05$), apart from those representing the numbers of individuals by type of context sampled and number of individuals recovered; (c) No significant differences are observed by the sex of the individuals (Mann–Whitney $p > 0.05$) or their age (Mann–Whitney $p > 0.05$); (d) Finally, when other arbitrary groups are formed by time in blocks of millennia, a slight differentiation is observed in the $\delta^{15}N$, but not in the $\delta^{13}C$ values—the pre-contact individuals of the past 1000 years record slightly higher nitrogen values, 1‰ higher than the time block 1000–2000 years cal. BP. The difference is statistically significant (Mann–Whitney $p < 0.05$). The two isotope markers in individuals from these two time blocks differ significantly from those in the post-contact group (Fig. 5.3).

The differences observed within the pre-contact set, with slightly higher nitrogen values in the past 1000 years cal. BP as compared to earlier samples, might be explained ecologically by changes in the N cycle in the marine environment over time (Saporiti et al. 2014). Another possibility is that the emphasis in resource acquisition shifted towards certain resources with higher $\delta^{15}N$ values, e.g. fish, sea birds and marine and riverside mammals, perhaps associated with the introduction of new techniques for obtaining these resources; however, the frequency of the latter in the zooarchaeological record in the archipelago is lower than the frequency of fish, which dominate the assemblages recovered (Reyes et al. 2007; San Román 2016). The common sea lion (*Otaria flavescens*), present in the archaeofaunal samples and hunting of which can be inferred from the lithic technology (thick-section lanceolate lithic projectile points), is not observed to be the predominant resource in the diet consumed. Terrestrial fauna (Canidae, *Hippocamelus bisulcus* and *Pudu pudu*) are present in some island sites, allowing us to establish parameters for measuring ranges of prey transport, mobility and/or exchange between the mainland coast where they were obtained and the islands. Consumption of these species by marine hunter-gatherers is only opportunistic, as has been recorded in other contexts in the Patagonian Archipelago (Fernández et al. 2016; Legoupil et al. 2007; San Román et al. 2002), and has no major impact on the total diet.

To sum up, analysis of the isotope values in humans of the Chonos Archipelago, in the context of a regional isotope ecology, supports our previous interpretations based on a smaller body of data (Reyes et al. 2013, 2015). The data are consistent with adaptation as hunter-gatherer-fishers based on a pattern of littoral and marine subsistence. Secondly, the profound cultural changes which occurred after European contact had profound implications for the subsistence and feeding strategies of these populations (Reyes et al. 2019a, Fig. 5.3).

5.11 Comparison with the Human Diet of the Patagonian Archipelago: The Supra-Regional Scale

When the data set of individuals from the Chonos Archipelago is compared with the $\delta^{13}C$ and $\delta^{15}N$ values published for other canoe populations of the far south of the Patagonian Archipelago (Table 5.8), we observe that the pre-contact samples form quite a homogeneous group for the whole archipelago (Alfonso et al. 2015; Barberena 2002; Borrero et al. 2001; Borrero and Barberena 2006; Guichón et al. 2001; Kochi 2017; Kochi et al. 2018; Panarello et al. 2006; Tafuri et al. 2017; Tessone et al. 2003, Tykot et al. 2009; Yesner et al. 2003); they differ moreover from the records for the south-central area of Chile, apart from similarities with the contexts of Santa María and Mocha, islands off the coasts of the Bío-Bío and Araucanía Regions, respectively (Campbell et al. 2020). The values clearly reflect the diet patterns of marine hunter-gatherer-fishers. However, pre-contact individuals of the central-southern and southern areas of the Patagonian Archipelago present more strongly negative $\delta^{13}C$ values than the Chonos Archipelago group (Fig. 5.4). We explain this statistically significant difference (Mann–Whitney < 0.05) by the lower emphasis on hunting terrestrial herbivores in the Chonos Archipelago as compared to the faunal components recorded further south (Reyes et al. 2019a, b). As we have documented in the archaeofaunal analysis, terrestrial resources did not make a large

Table 5.8 Descriptive statistics about the human groups that inhabited the western Patagonian channels

Region/Chronology	Southern archipelago/Pre-contact		Southern archipelago/Post-contact		Central-Southern archipelago/Pre-contact		Chonos archipelago/Pre-contact		Chonos and Chiloé Archipelago/Post-contact	
	$\delta^{13}C$	$\delta^{15}N$	$\delta^{13}C$	$\delta^{15}N$	$\delta^{13}C$	$\delta^{15}N$	$\delta^{13}C$	$\delta^{15}N$	$\delta^{13}C$	$\delta^{15}N$
N	14	14	28	28	5	5	33	33	3	3
Mean	−12.9	17.2	−12.2	18.4	−12.8	17.6	−11.4	17.2	−20.3	9.8
Standard deviation	2.2	2.5	0.8	1.5	1.4	1.6	0.9	0.9	0.7	1.3
Minimum	−18.5	10.6	−15.0	12.3	−14.1	15.8	−13.9	15.3	−21.0	8.8
Maximum	−11.1	19.2	−11.2	20.4	−10.6	20.0	−9.3	18.8	−19.7	11.3

Source Taken from Reyes et al. (2019a)

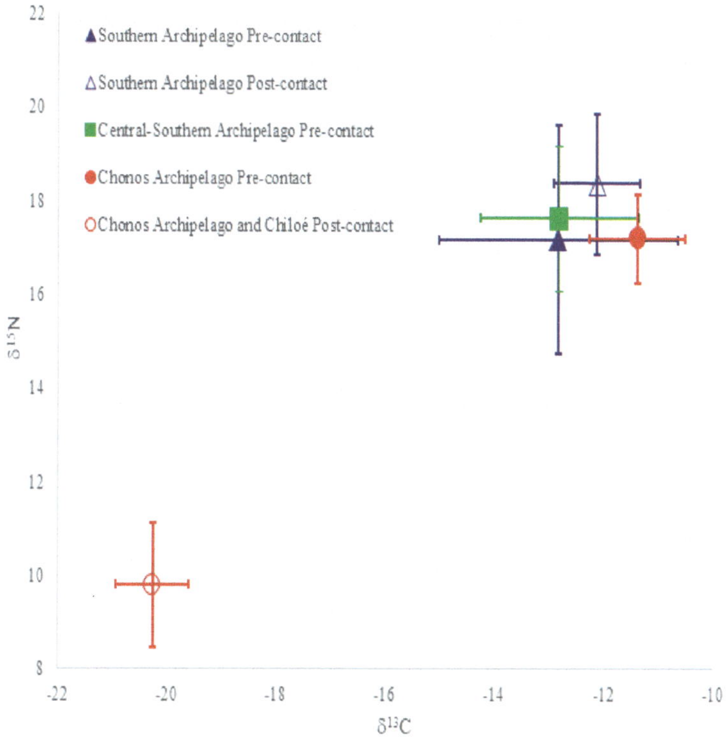

Fig. 5.4 Comparison of the averaged $\delta^{13}C$ and $\delta^{15}N$ isotope values of different pre-contact and post-contact marine hunter-gatherer groups of the Patagonian Archipelago. *Source* Taken from Reyes et al. (2019a)

contribution to diet in the Chonos Archipelago, such as is indicated in the bone assemblages with a marked presence of guanaco in the Beagle Channel (Orquera and Piana 1999a, b) and huemul in the central-southern archipelago (Emperaire 1963; Fernández et al. 2016). These data mark differences in the spectrum of the "marine diets" of the different hunter-gatherer-fisher groups throughout the western channels of Patagonia, probably conditioned by the ecological diversity present in this extensive territory as well as by diverse cultural trajectories in terms of subsistence patterns and the consumption of faunal resources.

Comparison with post-contact individuals produces highly significant differences. While the diets consumed in the southern archipelago after European contact continued to be mainly marine (Tafuri et al. 2017), the diets of individuals in the Chonos Archipelago and Chiloé underwent a profound change, probably induced by the dismantling of their culture and the transformation of their modes of subsistence to include horticultural practices and consumption of terrestrial fauna (farmyard fowls, domestic livestock), minimising the role of marine resources which had been predominant prior to contact (Fig. 5.4).

Another indirect indicator of specialisation in the consumption of marine resources and the undeniable impact of geographical barriers in this region is clearly shown by the comparison of the $\delta^{15}N$ and $\delta^{13}C_{col}$ values of the marine hunter-gatherers of the Chonos Archipelago with the isotope values presented by the terrestrial hunter-gatherer groups present to the east of the archipelago, on the other side of the Andes Range where it gives way to the eastern steppe (Méndez et al. 2014). This situation reminds us of the continuous barriers (sensu Barberena 2008), which mark the ecological, environmental and cultural division between the hunter-gatherer groups on either side of the Andean watershed in the Aysén region (Méndez et al. 2006, 2008/9; Méndez and Reyes 2008, 2015; Reyes et al. 2006b, 2009). Thus the samples obtained from steppe hunter-gatherers from the Early to Late Holocene (Méndez et al. 2014; Reyes et al. 2012) present $\delta^{15}N$ values between 9.4‰ and 16‰, and $\delta^{13}C_{col}$ values between −24.5 and −18.5‰, while the isotope signals generated in the archipelago environment, commented above, range between 15.3‰ and 20.3‰ for $\delta^{15}N$ and between –13.9 and –9.3 for $\delta^{13}C_{col}$. This comparison only goes to confirm our affirmations based on different lines of evidence about the marked bio-geographical barriers which separate the Chonos Archipelago from the extra-Andean steppe in this latitude, where the circulation of raw materials, the technologies used, the diet and subsistence, and the genetic and craniometric distances project two completely separate, unconnected cultural traditions on either side of the Andes (Kuzminsky et al. 2018; Méndez et al. 2006, 2008/9, 2018; Méndez and Reyes 2008, 2015; Moraga et al. 2009; Reyes et al. 2006b, 2009, 2012, 2015, 2016a, 2019b). A remarkable difference is observed between the diets consumed by hunter-gatherers of the steppe and by marine hunter-gatherers, consisting as they do of mutually exclusive resources when the distance between them from east to west is less than 140 km.

5.12 Chronology of the Archaeological Record, Use of Space and Function of Sites

We carried out a systematic programme of radiocarbon datings ($n = 86$) to understand on different scales of analysis (sensu Dincauze 2000) the temporal dimension of occupation of the archipelago territory by marine hunter-gatherer groups (Reyes et al. 2019b, Fig. 5.5). We evaluated the temporal distribution of these occupations and the peopling dynamic using chronological information to explain human activity in the landscape. The samples were obtained from all the types of sites described for the area (Reyes et al. 2015), except for fishing pens and historical paths that could not be dated, covering the past 6200 years cal. BP. These include strata buried below organic soils with lithic material, rock shelters, individual burials and ossuaries, and open-air shell-midden deposits, some of which were formed by historical mollusc gathering (*cholgueros*). We evaluated the intensity of the chronological human signal

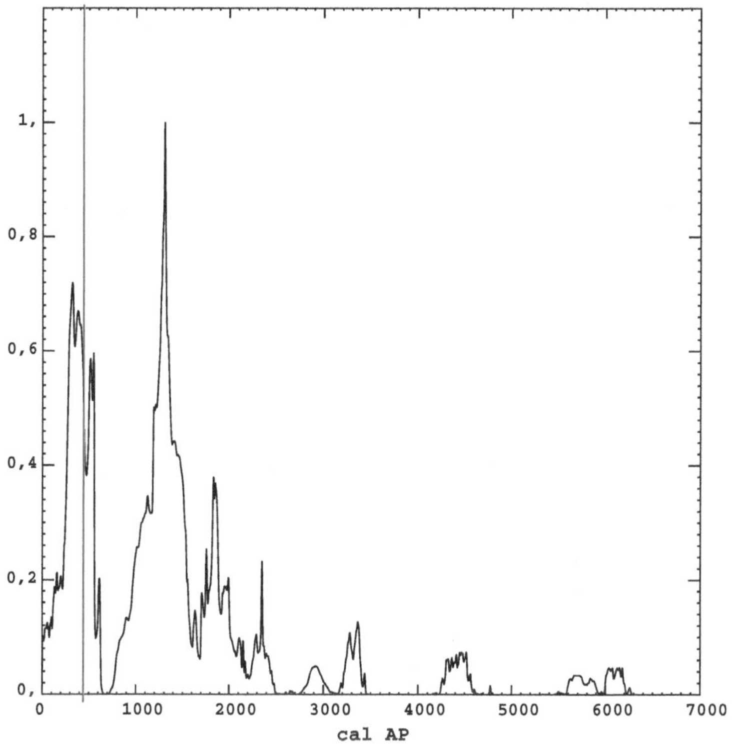

Fig. 5.5 Chronological distribution of the occupation of the Chonos Archipelago. The red line indicates the date of local contact with Europeans in the Taitao Peninsula according to the written records (1553 AD)

from the Mid-Holocene to recent moments, discussing the possible explanations for (dis)continuities in the use of the archipelago space.

The set of dates obtained is based on ^{14}C dating by AMS from 42 sites out of a total of 78 identified (carbonised organic material N = 18, mollusc valves N = 23, human bone remains N = 37, faunal bone remains N = 7 and one sample of worked tree bark from a funerary context, Table 4.1). We dated the start of occupations in 24 of the 32 pre-contact shell-midden camp sites discovered (75% of the total pre-contact camps). They were selected by weighting various elements like the importance of the contextual information available on the archaeological characteristics in order to document the first occupations on coastal geoforms, to assess the various geomorphological changes subsequent to human settlement and to understand the spatial distribution of the human occupation and its recurrence in time. Thus we have datings for the start of the deposits in the majority of the archaeological sites, and also another set of datings carried out to project occupation and abandonment sequences, the deposition and discard rates of the anthropic sediments formed, and changes in the use of the space and in the function of some of these sites, as we shall see below. Of the 24 camps with dated bases, 18 sites present two or more datings.

We also dated almost all the funerary contexts characterised as ossuaries (except the Osamentas Repollal site in the Northern area of the archipelago). In some of these sites we dated 100% of the individuals identified (n = 2), in order to understand the recurrence of visits to specific places and the continuity of certain funerary cultural practices and the change in function of some of these contexts. Thirty-six individuals were dated directly with this object (Table 5.5). Of the 26 historical archaeological sites (cholguero camps), five were dated (Table 4.2), also at the start of the deposit, to determine whether they were pre or post-contact and the impact of the geomorphological changes which occurred after their initial occupation on a time scale of less than 200 years.

The oldest occupation recorded in the Chonos Archipelago was at GUA-010 Terraza, on Gran Guaiteca Island in the Northern area of the archipelago (Fig. 4.1), dated to nearly 6200 years cal. BP in the Mid-Holocene. Two statistically different ages were recorded, representing different chronological events, in brief moments at the start of the cultural sequence in the area. This is the only site with such an early dating (Porter 1993; Reyes et al. 2016b); specific activities can be inferred for the maintenance of medium-sized, thick-section lanceolate lithic projectile points, used for hunting, and we note the use of rhyolithic obsidian from the Chaitén Volcano source, 150 km away to the north-east on the mainland coast opposite Chiloé Island. The coastal settlement pattern provides evidence of the profound geomorphological changes that have occurred since the first human signal was established.

Subsequently, more than 1000 years after the first occupations of the area, we detect two other occupations, dated between ~4300 and 4500 years cal. BP, in both the Northern area (close to GUA-010 Terraza in the bay of the Cucuquén Stream) and the South-central area (Fig. 4.18). These are the sites identified as Gran Guaiteca 2 Terraza, on Gran Guaiteca island (Reyes et al. 2016b), and Isla Goñi 1 (Reyes 2017; Reyes et al. 2019b) on Goñi Island, framing distribution over a much wider range of latitude in the fifth millennium, with two sites located at opposite ends of the archipelago. These sites present similar characteristics to one another and to GUA-010 Terraza. They also represent areas of activities for the maintenance of hunting instruments, such as medium-sized thick-section lanceolate lithic projectile points, and the use of rhyolithic obsidian, this time with a greater range of dispersion from the source, around 350 km from Isla Goñi 1. Furthermore, the artifactual record of this latter site shows evidence of the use of other instruments like net weights and stone axes, technologies used for fishing and woodworking. The use of the canoe is made more evident by the record of lumps of obsidian with cortex, transported to the site by means of this indispensable technology.

The sites recorded for the Mid-Holocene (n = 3) do not present subsequent occupation sequences, as we observe in the stratigraphic packets of organic sediments that define the horizontal distribution of lithic materials and carbon specks. All these sites, in coastal locations, present differential distribution consistent with permanent geomorphological changes to coastlines. Thus we find uplifting in the Northern area, as a result of which the shoreline is moved further away from sites (GUA-010 Terraza and Gran Guaiteca 2 Terraza), and sinking and destruction of sites in the South-central area of the archipelago (Isla Goñi 1). There is also an absence of archaeofaunal and malacological evidence. The lithic materials were recovered in dark, organic matrices of a very acid podzol.

The areas of the sites where we carried out interventions represent sectors where specific activities were carried on for the manufacture and maintenance of specialised instruments; however, we cannot rule out the possibility that areas used for other activities may also be hidden and/or destroyed. These sites are to be understood as the material representation of exploratory logistical ranges of high mobility and low intensity; signs of their lithic technology have been preserved, with high artifactual specificity, low diversity, little formatting (except for axes and projectile points) and with a supply of allochthonous raw materials (rhyolithic obsidian). The latter is used exclusively for the manufacture and maintenance of hunting instruments, and the current evidence indicates that the sites were not revisited and/or reoccupied after use.

Other areas with gathering activities, or the occupation of other types of space (e.g. rock shelters), were not detected. The shelters tested and recorded in neighbouring areas to these sites produced no evidence of occupation and use during this period. The archaeological signal, the stratigraphy defined and the low artefact density suggest the low intensity of occupation in a context of exploration, logistics and high mobility (Borrero 1989–90, 1999, 2001a). This was made possible by the use of the canoe as an indispensable aid to hunting and fishing and the transport of people and possessions (Ames 2002; Bjerck 2017; Martial et al. 2007; Orquera and Piana 1999b), such as tool kits, prey and lumps of rhyolithic obsidian as a source of raw material (reserved particularly for lanceolate lithic projectile points).

This situation contrasts with older evidence reported for the northern end of the Patagonian channels where, unlike our records, the archaeological sites of the Mid-Holocene located in Reloncaví Sound (e.g. Piedra Azul, Punta Metri) and on Chiloé Island (Puente Quilo, Chepu, Yaldad 2; Álvarez et al. 2008a, b; Flores et al. 2010; Flores and Correa 2011; Gaete et al. 2004; Gaete and Navarro 2004; Galarce 2005a, b, c; Legoupil 2005; Munita 2017; Ocampo and Rivas 2004; Rivas and Ocampo 2005, 2006; Rivas et al. 1999) record subsequent occupation sequences in the same site. These present the same artifactual technology, presence of rhyolithic obsidian and somewhat older chronologies, and fairly well-marked continuities and discontinuities, as the case may be, ending with Late Ceramic occupations.

This reflects not only the greater relative stability of the coastline—in what we know to be a highly seismic zone—allowing site locations to be maintained over time (and therefore making it easier to project where to search), but also the fact that the locations remain available for use by human groups over the whole chronological sequence. The situation in the Chonos Archipelago would illustrate a more exploratory context of occupation (sensu Borrero 1989–90, 1999, 2001a), more ephemeral and discontinuous over time, in which the places visited were not reused for long periods (of the order of millennia) by groups which probably came from the Chiloé Archipelago and neighbouring areas, using the same technology (e.g.

medium-sized, thick-section lanceolate points, stone axes, net weights) and familiar with the same raw materials, such as the rhyolithic obsidian from the Chaitén Volcano source.

Thus from the last occupations situated towards the end of the Mid-Holocene, a millennium passed before the first occupations characterised as shell-middens start to appear around ~3400 years cal. BP in the southern part of the archipelago. This would be a phase of colonisation (sensu Borrero 1989–90, 1999, 2001a), an initial consolidation of human groups in certain sectors, with repeated use of locations and increased archaeological visibility. We now start to find the inclusion of malacological material in the sediments, with shells forming an important part of this anthropic deposition (Stein 1992). Various activities related to the formation of camps (hearths, discard of waste, knapping waste, burials) are recorded in these shell-middens, with successive reoccupations in the same place giving rise to this diversity of features, all activities carried on in residential camps (sensu Binford 1980). We start to identify this evidence in the South-central area of the archipelago, in Posa Las Conchillas on Traiguén Island and Canal Darwin 2 on Luz Island, where again we see a pulse of occupation but only over a small range of latitude (Fig. 4.18).

Reoccupations of the same locations, which give rise to and define the shell-midden sites of the Late Holocene, are dependent on two conditions. Firstly, the particular place selected on the shoreline is important in an environment dominated by steep coasts, fjords, flooded land and peat bogs, with forest cover coming right down to the water's edge; canoes are an essential technological component for the residential and logistical mobility of these human groups (Ames 2002; Curry 1991; Pallo 2011), and their protection and maintenance require good, sheltered bays and accessible beaches. In view of the current geography of the area (Binford 2001), the reuse of some coastal spaces implies a hierarchisation of preferred sites (e.g. Borrero 2001b; Belardi 2003), linked in this case to local barriers and suitable sites for beaching/mooring/protecting the canoe. We, therefore, understand that geomorphological factors exist which limit site location to certain locations. The use of canoes defines not only the high mobility of these groups, but also the fact that access was restricted to certain sectors of the coast; it also defines their different subsistence and technology strategies (e.g. Bjerck 2017; Piana and Orquera 2009; Orquera and Piana 1999a, b). This pattern of coastal settlement, recurrent over time, may be related with Binford's concept of "tethered nomadism", indicating extreme redundance in the reuse of sites over time, in this case, shell-midden camps; this ties the type of settlement pattern to certain specific locations, while others are very little used due to their remoteness and the costs of obtaining limited but crucial resources (Binford 1980). It has further been claimed that the first occupation of a site may be prompted by various reasons, but subsequently, once human activity is established there, this becomes of itself a spatial attraction which promotes later reoccupations of the space in different moments and conditions (Piana and Orquera 2010:270).

Secondly, the reuse of camps goes hand in hand with an increase in the chronological signal of human occupation events, as we have shown (Fig. 5.5). This speaks of a larger number of visits and occupations, increasing the intensity of the archaeological signal (relative and proportional in the context—in this archipelago—of low general representation of archaeological sites), particularly in the past ~2700 years cal. BP. This in turn translates into an increase in the quantity of sites and reoccupations, and the wider distribution range recorded in all the areas of the archipelago—also associated with greater diversity, both contextual and with respect to the qualitative inferences of the functionality of sites and associated activities. Thus anthropic deposits, on a millennial scale, increase in area and depth during the Late Holocene.

Nevertheless, it is only after ~2700 years cal. BP in the Central and South-central areas of the archipelago, and towards ~2000 years cal. BP in the Northern area, that archaeological occupations start to appear with greater intensity around shell-midden camps, extending the dispersion of the human signal semi-continuously to the whole archipelago. We find a peak at ~1300 years cal. BP, a well-defined trough towards ~700 years cal. BP and a new peak towards ~300 years cal. BP. In the Northern area of the archipelago, the long hiatus of 2,000 years with no cultural evidence after the occupation of Gran Guaiteca 2 Terraza (~4400 years cal. BP), ends with the start of a series of sustained occupations around 2000 cal. BP, distributed in various sectors of Gran Guaiteca Island and Marta Island (n = 5: Alero Low, Isla Marta, GUA-010 Conchal and Gran Guaiteca 2 and 4). Dispersion of an important quantity of lithic material on the surface of the intertidal may have occurred during this interval, in view of its spatial association with dated contexts (Reyes et al. 2007, 2016b). An analogous situation may be suggested for the fishing pens recorded (Reyes et al. 2015).

After ~2000 years cal. BP, the site location pattern continues to be exclusively on the shoreline, with reoccupations in the same place, contributing to the formation of large shell-midden mounds. Other locations are also occupied, associated with the presence of rock shelters, used either for funerary purposes or domestic activities such as sheltered camps. This kind of use is not recorded for the occupations identified in the Mid-Holocene, although the locations were available. Geomorphological changes are also observed in these later occupations. Throughout the Chonos Archipelago, we found that the sites located on the immediate shoreline have a tendency to sink; this can result in tidal erosion of exposed profiles and the presence of lithic and bio-anthropological material scattered in adjacent intertidal areas.

In the archaeological sites of the Late Holocene, we recorded a concentration of inferred activities for residential camps in addition to the identification of sites of specific activities (e.g. Isla Solitaria 1 as the only logistical site serving as a look-out point). The factors which explain this situation are related with the differential visibility of the material and ecofactual record (shell-midden versus materials in organic sediments), as well as a different organisation of space in which specific activities carried on outside the immediate area of the site, such as hunting, fishing or gathering raw materials, would not leave much evidence in the associated sites. Furthermore, some activities may be carried out in the canoe or actually in the water.

The activities represented in this latter period, inferred from the scarce technological components and from the fauna columns dug, have to do with the hunting of marine mammals (common sea lion), estuarine terrestrial fauna (coypu and otter), birds (cormorants, ducks and petrels) and to a lesser extent mainland terrestrial fauna (huemul, pudú, fox). Fishing also made an important contribution, as is shown by the diversity and quantity of the ichthyological record as compared to other faunal assemblages, and by the presence of net weights in the intertidal, forming an important part of the set of lithic artefacts. Furthermore, the isotope signal obtained from the human bone remains (Table 5.5, Figs. 5.3 and 5.4) indicates a predominantly marine diet. The gathering and discard of malacological fauna, despite the area and volume of the shell-middens, is to be understood more as localised gathering close to the shoreline of the site, in contexts of reoccupations with the limitations affecting a coastal location, where the deposition rate on a millennial scale continues to be low (Reyes et al. 2015).

Above we commented that the artifactual density in the Late Holocene shell-middens is almost non-existent. First Bird (1988) and then Legoupil (2005) remarked on the shell-middens on Chiloé Island and their poverty of cultural materials, inversely proportional to the size of these deposits. They explained this by taphonomic factors or by a differential settlement pattern in which shell-middens could serve as selective waste dumps away from living areas (Legoupil 2005). In this respect, they differ widely from the broad range of fauna and cultural materials recorded in canoe group contexts in the far south (e.g. Christensen 2016; Legoupil 1997; Legoupil et al. 2011; Orquera and Piana 1999a, b; Orquera et al. 2011; San Román 2010; Zangrando and Tivoli 2015). We also note that within this increase in the human signal in the Chonos Archipelago during the Late Holocene, the number of sites per linear km of shoreline prospected is very low, while this signal is understood as being of "higher relative intensity" of occupation of the archipelago but in a general context of low occupation. This low human occupation could explain in part the scarce quantity of artifactual materials deposited and recovered.

We started to record the bio-anthropological and funerary evidence (Table 5.9), as part of this contextual variability in the Late Holocene, from ~2000 years cal. BP, initially in the South-central area (Isla Victoria 2 ossuary site) and on the Mainland coast (Estero Sur). We then projected this funerary pattern throughout the archipelago, starting from this moment (Fig. 5.6) which coincides with the increase in the number of camps, spread over a wider latitudinal and longitudinal range of distribution. However, we observe a greater emphasis on the record during the past millennium and moments subsequent to European contact. For decades, the human remains recovered have come indifferently from ossuaries which were used in a single segment of time, or over a short period, or else reused and revisited as a funerary space over the whole sequence, with intervals of centuries, marking long continuity of this practice in which individual depositions generated diachronic collective assemblages (Reyes et al. 2013).

One particularity of the use of rock shelters for funerary purposes is that at the end of the occupation sequence of the shell-midden camps formed under rock shelters, as at Alero Low and Isla Benjamín 1 (Figs. 4.5, 4.13 and 4.14) in the Northern and Central areas of the archipelago, we recorded changes in the function of the sites. Between 650 and 300 years cal. BP, individuals were buried and/or deposited in these shelters (MNI = 14 and 11 individuals respectively), just as in the exclusively ossuary contexts; as a result the sites ceased to be used as residential camps, given the saturation of the space delimited by the shelter and/or the transformation of the

Table 5.9 Summary of the archaeological sites presenting mortuary contexts in the Chonos Archipelago, the number of individuals recorded (MNI) and the radiocarbon datings carried out on both the individuals and deposits in their site of provenance

Sites with human remains	Type of site	MNI individuals	Dated ^{14}C skeletons	Dated ^{14}C site	Chronology[a] (cal. BP)	Total Datings
GUA-010 Conchal	Shell-midden camp	1	–	1	1.840–1554	**1**
Gran Guaiteca 2 Conchal	Shell-midden camp	3	1	1	1.585–491	**2**
Isla Marta 1	Shell-midden camp	2	–	2	1733–1310	**2**
Alero Low	Shell-midden shelter camp	14	2	3	2.012–362	**5**
Repollal Caverna	Ossuary	17	1	–	482–300	**1**
Puquitín 01	Ossuary	19	1	–	666–508	**1**
Osamentas Repollal	Shell-midden shelter camp	2	–	–	–	–
Isla Sin Nombre	Ossuary	13	1	–	546–335	**1**
Isla Ipún 1	Ossuary	2	2	–	1223–490	**2**
Estero Sur	Ossuary	16	5	–	1736–1412	**5**
Osario Melimoyu	Ossuary	11	2	–	667–498	**2**
Isla Yalac 1	Ossuary	13	2	–	514–330	**2**
Isla Harris	Ossuary	34	1	–	450–277	**1**
Seno Canalad	Ossuary	18	1	–	493–308	**1**
Isla Benjamín 1	Shell-midden shelter camp	11	2	3	1.748–239	**5**
Isla Benjamín 4	Ossuary	8	1	–	516–323	**1**
Isla Benjamín 5	Ossuary	4	4	1	573–231	**5**
Isla Elena 1	Ossuary	4	4	–	1517–1188	**4**
Nahuelquín 1	Shell-midden camp	3	2	1	1.461–330	**3**
Isla Victoria 2	Ossuary	3	2	–	2073–1257	**2**
Isla Acuao 1	Shell-midden camp	1	1	–	1280–1167	**1**
Corrientes del Yates 5	Shell-midden camp	4	1	1	1.890–880	**2**

(continued)

Table 5.9 (continued)

Sites with human remains	Type of site	MNI individuals	Dated ^{14}C skeletons	Dated ^{14}C site	Chronology[a] (cal. BP)	Total Datings
Total sites n = 22	Total ossuaries recorded n = 13 Total camps with burials recorded n = 9	Ossuaries n = 162 ind Camps n = 41 ind Total = 203 ind	36	13	Only Late Holocene	49

[a]This chronology is based on the calibrated extremes of the maximum and minimum datings obtained

Fig. 5.6 Recreation of funerary rite in a cave by marine hunter-gatherers in the Chonos Archipelago

place into a funerary area. Using a space that had previously been used for domestic purposes, or a protected rocky space, would not imply a major change in the funerary pattern practiced; however, the change in the use of the site from a residential camp to an ossuary at the end of its sequence of occupation would be significant. No ossuary in a shelter, even the oldest, like Estero Sur (~1700 years cal. BP), was ever used thereafter as a residential camp or for any specific activity other than funerary practices.

This late change in the function of these sites has led some investigators assume, in broad general terms, that this was a late funerary pattern from the end of the archaeological sequence of the ethnohistorical groups known as Chono (e.g. Aspillaga et al. 2006; Cooper 1917, 1946; Munita et al. 2011; Ocampo and Aspillaga 1984). Today, thanks to direct radiocarbon dating, we know that this practice was installed from an early date (Reyes et al. 2013, 2015, 2019b).

The burials recovered from shell-midden camps so far are no older than ~1200 years cal. BP at most (Isla Acuao 1 in the South-central area of the archipelago, Table 4.1. This diverse situation should not surprise us; an ossuary is easier to find and record—or indeed plunder—and contains a larger number of individuals. Burials in shell-middens, on the other hand, unless exposed in profiles, are invisible, requiring excavation with slower recovery and containing fewer individuals. In our work, we recovered human remains (MNI = 14) from our interventions in new shell-midden camps. We also knew that human remains could be observed in the profiles of various shell-middens of the archipelago, or scattered in the adjacent intertidal by tidal action (Curry 1997; Simpson 1875). The human remains recovered to date in the shell matrices do not reflect any social or age distinctions, nor any great complexity in their mortuary treatment (sensu Binford 1971, Carr 1995). We recorded no obvious trappings or offerings (the intentional presence of shells beside the body in these anthropic sediments is not easily resolved); however, the bodies are placed in flexed or hyperflexed lateral decubitus in the same pattern recorded in other areas of the northern archipelago (e.g. Díaz and Garretón 1972/73; Munita et al. 2011).

The two modes of mortuary practice described for the Chonos Archipelago, i.e. individual or multiple burials in shell-middens and individual or collective depositions in rock shelters, are related with, and show similarities to, the funerary contexts described for the northern archipelago, in the area of Chiloé, Reloncaví Sound (Munita et al. 2011) and in the rest of the western channels of Patagonia (Byron 1901 [1746]; Emperaire 1963; Legoupil and Sellier 2004; Legoupil and Prieto 1991; Prieto and Cárdenas 2001; San Román and Morello 2001). However, they do not reflect the same artifactual diversity (trappings/offerings), especially compared with the diverse mortuary contexts and treatments recorded in the southern end of the Patagonian Archipelagos in the Beagle Channel (Álvarez. M. et al. 2008; Aspillaga et al. 1999; Orquera and Piana 1999b; Piana et al. 2006; Tessone 2014).

With respect to the space dedicated to funerary use in open-air residential camps, we observe that all the human remains were recovered on the periphery of the sites, at the edges of the shell-middens (e.g. Gran Guaiteca 2 Conchal, GUA-010 Conchal, Isla Marta 1, Nahuelquín 1, Acuao 1, Corrientes del Yates 5), in the Northern and South-central areas of the archipelago. This intra-site distribution may be understood as a mortuary pattern in which the body (or bodies, since multiple burials have been recorded) is buried in the domestic space of the camp, but in marginal sectors, where it will not affect other central activities of the occupants; or perhaps when the site is abandoned, supposing that the human remains will not be altered by future occupants. The latter was not always the case, since subsequent occupations, centuries later, altered the older deposits with new camp activities as they spread or were moved to a different space in the site. Future, more extensive excavations in various shell-middens may confirm this diversity in the use of space and segregation of funerary activities within the camp.

For the moment, the chronology of some ossuary contexts in the Chonos Archipelago is older than the first shell-midden burials recorded in the archipelago. The human remains recovered to date (MNI = 203) present no emphasis on age or sex in specific geographical locations, i.e. there is no association that can be made with provisioning trips for specific resources (e.g. if only men were identified in ossuaries towards the Pacific border, or a greater presence of women and children in the interior channels). On the contrary, all the individuals recorded in the ossuaries throughout the archipelago reflect a diversity of age and sex, just like the buried individuals who were recovered from our excavations and searches in shell-middens; no social distinctions or special treatments (sensu Binford 1971) were found in the human remains investigated. From this evidence, we propose that family units were distributed in residential camps throughout the island system without excluding any of their members.

As mentioned above, after a marked trough in the archaeological signal around ~700 years cal. BP, a second important peak is recorded in the northern area of the archipelago around ~300 years cal. BP, which coincides exactly with European contact, when—in an apparent contradiction—we observe an increase in occupation events (Fig. 5.5). The trends proposed for this chronological distribution in moments just before and after European contact are still subject to evaluation while adjustments are made in the local reservoir effect, especially in human beings with a marine diet (Reyes et al. 2019a, b).

The reasons for the marked pre-contact trough (~700 years cal. BP) are still unknown. Taphonomic variables (e.g. destruction of sites due to geomorphological changes, plundering of funerary sites) and archaeological search design are not sufficient to explain completely this dynamic in the record (Reyes et al. 2015, 2018). A useful argument can be identified in the diverse evidence of traumas, enthesopathies and infectious pathologies recorded in a series of individuals recovered from the Guaitecas Archipelago and the area close to Magdalena Island (N = 96, University of Chile Collection). It is suggested, from the health evidence obtained, that these canoe groups were going through an adaptive crisis, with a declining population due to the expansion of farming populations from the north, before local contact with Europeans (Aspillaga et al. 2006). Unfortunately, the samples taken were all assumed to be later than their contextual funerary associations, which—as we have argued—present older datings. Wider chronological ranges should, therefore, be expected for the individuals analysed in the study cited. Regardless of this, it is a plausible explanation that coincides with the decline observed in the pre-contact chronological record.

Turning to the increase in the chronological signal towards ~400–300 years cal. BP, we remarked in Chapter 2, Background, that the human groups which inhabited this archipelago, known generically as Chono (Cooper 1917, 1946), suffered a rapid process of cultural transformation in post-contact historical moments, with exile and depopulation of the archipelago in barely 200 years, being forcibly displaced first as slave labour (under the Encomienda system) and subsequently by the formation of Jesuit missions. They also suffered diseases, confrontations that turned into massacres, and interbreeding with groups of Huilliche origin from the Chiloé Archipelago on the one hand and southern canoe groups when they started to avoid more frequented routes in favour of other parts of the southern archipelago on the other (Emperaire 1963; Martinic 2005; Moreno 2011; Urbina 1988, 2016). All this led to the rapid cultural dismantling of the canoe groups who inhabited and visited the archipelago, and no further signal of any kind is recorded from the end of the eighteenth century (e.g. Cárdenas et al. 1991; Cooper 1946; Darwin 1845; de Vea 1886 [1675–1675]; FitzRoy 1939; García 1889 [1766]; Ibar 1959; Latcham 1928, 1930; Machado 1889 [1769]; Moraleda 1888 [1786]; Simpson [1871, 1875]; Urbina 1983, 1988, 2014, 2016).

Thus the post-contact increase in chronological events could be due to some canoe groups adapting to, and/or being obliged to adopt, new conditions of interethnic and commercial relations during this debacle which stimulated and reoriented economic and geopolitical visits to this area during the process of cultural dismantling (Urbina et al. 2020). This greater presence may have affected a slight increase in the post-contact radiocarbon signal (peak: ~400–300 years cal. BP). Later, after no more cultural or bio-anthropological evidence or chronological signal, we observe the appearance of a new type of archaeological record, the post-contact shell-middens or *cholgueros,* in a new occupation sequence of the area in historical, post-independence moments.

As early as the nineteenth century, changing commercial relations involved new types of trades in the archipelago. Both the archaeological evidence documented in the record of the cholguero campsites and the datings obtained for the start of the formation of five of these sites reconfirm this new way of occupying the archipelago—after the cultural dismantling of the canoe groups in the area—on a recurrent seasonal basis for commercial and productive purposes. A clear intensification is observed in the capture of specific marine resources, concentrating mainly on the large-scale gathering of commercial shellfish species (especially mytilids). There was also an increase in occupations by fishermen, fur-trappers and lumberjacks, among other extractive activities carried on mainly by a floating population originating largely from the Chiloé Archipelago (Curry 1991; Legoupil et al. 2007; Martinic 2005). Although these historical and sub-present occupations were dated, they are not represented in the radiocarbon curve because they are modern or out of calibration range. However, the profusion and magnitude of sites of this type show an intensity of use unprecedented in this archipelago.

Felling of forests, burning of whole islands, decimation of native fauna, intrusion of exogenous fauna and over-exploitation of natural resources have set the general tone of how modern society has marked its presence and left its imprint in this archipelago.

To sum up the situation on a macro-scale, the record of pre-contact archaeological sites and the chronological sequence of the human occupation events that we have documented (Table 4.1, Fig. 5.5) project greater intensity of use and occupation of the archipelago since ~2700 years cal. BP, and especially around ~1300 years cal. BP. This is related to a greater number and diversity of sites identified (look-out points, open-air and sheltered shell-midden camps, ossuaries, burials in shell-middens) than in the Mid-Holocene, where we recorded hiati on a millennial scale. This is consistent with the ^{14}C data sets collected for Patagonia, Pampa and other regions, which show a tendency towards increased occupations during the Late Holocene, in some parts during the past 1000 years, with a strong, general "late signature" for hunter-gatherer populations in the region (Barberena et al. 2015; Berón 2015; Campbell and Quiroz 2015; García Gureieb et al. 2015; Goñi 2010; Goñi et al. 2004; Martinez et al. 2015; Perez et al. 2016; Zubimendi et al. 2015).

Note

[1] In this context we would add a note in parenthesis. A small bivalve mollusc exists in the sea, *Teredo navalis*, the best-known species of the family *Terenidae*; it is of xylophagous habits, in other words, it eats wood. This mollusc, known as *broma* in Spanish, is described in the chronicles of sailors and travellers (e.g. Moraleda 1888 [1786]). It attacks the wooden hulls of sea-going vessels, slowly boring into the wood and causing severe damage and total loss if timely maintenance is not carried out (replacing damaged planks). On an archaeological field trip in the Bernardo O'Higgins National Park near Puerto Edén (Magallanes Region), in the central channels of Patagonia, Francisco Arroyo (personal communication 2012), who was born in the Guayaneco Archipelago and was one of the last five speakers of the Kawésqar language, talked to us about boat maintenance. He remarked that the canoes that they made of coihue (*Nothofagus dombeyi*) lasted about one year due to *broma*. This single mention is a graphic indication that canoe maintenance was no minor task, since it required not only the right wood but also a set of technical elements, foresight in visiting certain locations (good places to stay, islands with suitable trees) and technical details associated with use and maintenance, perhaps on a seasonal basis (for more details see the summarised information about the canoes of the Yámana, Orquera and Piana 1999b).

References

Abe Y, Marean W, Nilssen P, Assefa Z, Stone E (2002) The analysis of cutmarks on archaeofauna: a review and critique of quantification procedures, and a new image-analysis GIS approach. Am Antiq 67(4):643–663

Alfonso M, Giles B, Misarti N, San Román M, Morello F (2015) Antiquity and geographic distribution of cranial modification among the prehistoric groups of Fuego- Patagonia. Am J Phys Anthropol 158:607–623

Álvarez M, Vázquez M, Piana EL (2008) Prácticas mortuorias entre los cazadores-recolectores del canal Beagle: el caso de Shamakush Entierro. Magallania 38(2):7–124

Álvarez M, Briz I, Balbo A, Madella M (2010) Shell Middens as archives of past environments, human dispersal and specialized resource management. Quatern Int 239:1–7

Álvarez R, Munita D, Fredes J, Mera R (2008) Corrales de pesca en Chiloé. Valdivia, Imprenta América

Álvarez R, Munita D, Mera R, Borlando I, Ther F, Núñez D, Hidalgo C, Hayward P (2019) rebounding from extractivism: The history and re-assertion of traditional weir-fishing practices in the Interior Sea of Chiloé. Shima 13(2):155–173

Álvarez R, Navarro M, Saavedra G, Donoso C (2015) Referencias exploratorias sobre el lago Presidente Ríos, para sortear el Istmo de Ofqui, Península de Taitao, Regióm de Aysén. Magallania 43(1):91–101

Ambrose S (1990) Preparation and characterization of bone and tooth collagen for isotopic analysis. J Archaeol Sci 17:431–451

Ames K (2002) Going by boat. In Fitzhugh B, Habu J (eds.), Beyond foraging and collecting, pp 19–52. Springer

Andrefsky W (1998) Lithics. Macroscopic approaches to analysis. Cambridge University Press, Cambridge

Arnold J (1995) Transportation innovation and social complexity among maritime hunter-gatherer societies. Am Anthropol 97(4):733–747

Aspillaga E, Castro M, Ocampo C (1990) Paleopathology and life-style: the Chono and Fuegian examples (Abstract). Am J Phy Anthropol, Supplement

Aspillaga E, Castro M, Rodríguez M, Ocampo C (2006) Paleopatología y estilo de vida: El ejemplo de los Chonos. Magallania (Punta Arenas) 34(1):77–85

Atwater B, Núñez H, Vita-Finzi C (1992) Net Late Holocene emergence despite earthquake-induced submergence, south-central Chile. Quatern Int 15–16:77–85

Bailey G (2004) World prehistory from the margins: the role of coastlines in human evolution. J Int Stud History Archaeol 1:39–50

Bailey G, Flemming N (2008) Archaeology of the continental shelf: marine resources, submerged landscapes and underwater archaeology. Quatern Sci Rev 27(23):2153–2165

Barberena R (2002) Los límites del mar: isótopos estables en Patagonia meridional. Sociedad Argentina de Antropología. Buenos Aires

Barberena R (2008) Arqueología y biogeografía humana en Patagonia Meridional. Sociedad Argentina de Antropología. Buenos Aires

Barberena R, Méndez C, Mena F, Reyes O (2011) Endangered species, archaeology, and stable isotopes: Huemul (Hippocamelus bisulcus) isotopic ecology in Central-Western Patagonia (South America). J Archaeol Sci 38(9):2313–2323

Bartsch-Winkler S, Schmoll H (1993) Evidence for late holocene relative sea-level fall from reconnaissance stratigraphical studies in an area of earthquake- subsided intertidal deposits, Isla Chiloé, Southern Chile. In: Tectonic Controls and Signatures in Sedimentary Successions, pp 91–108. Blackwell Publishing Ltd

Bate F, Mena F (2003) Alero El Toro: un campamento indígena en el bosque siempreverde cercano al litoral en Aisén. En Actas del XVI Congreso Nacional de Arqueología Chilena. Tomé. (pp. 675–680). Museo de Historia Natural de Concepción, Tomé

Belardi J (2003) Paisajes Arqueológicos: Un estudio comparativo de diferentes ambientes patagónicos. Tesis doctoral inédita Facultad de Filosofía y Letras, Universidad Nacional de Buenos Aires. Buenos Aires

Berón M (2015) Chronological distribution and disturbance factors to evaluate population dynamics in Western Pampas, Argentina. Quatern Int 356:74–88

Binford L (1971) Mortuary practices: their study and their potential. In: Brown JA(ed.), Approaches to the social dimensions of mortuary practices, 25, pp 6–29. Memoirs of the society for American archaeology, Washington, DC

Binford L (1980) Willow smoke and dogs' tails: hunter-gatherer settlement systems and archaeological site formation. Am Antiq 45(1):4–20

Binford L (1981) Bones: ancient men and modern Myths. Academic Press, New York

Binford L (1982) The archaeology of place. J Anthropol Archaeol 1:5–31

Binford L (1984) Faunal remains from Klasies River mouth. Academic Press, Orlando

Binford L (2001) Constructing frames of reference. An Analytical method for archaeological theory building using ethnographic and environmental data sets. University osf California Press, California

Bintliff J (2000) The concepts of 'site' and 'offsite' archaeology in surface artefact survey. Non-destructive techniques applied to landscape archaeology 4

Bjerck H (2009) Colonizing seascapes: comparative perspectives on the development of maritime relations in Scandinavia and Patagonia. Arctic Anthropol 46(1–2):118–131

Bjerck H (2017) Settlements and seafaring: Reflections on the integration of boats and settlements among marine foragers in Early Mesolithic Norway and the Yámana of Tierra del Fuego. J Island Coastal Archaeol 12(2):276–299

Borrazzo K (2007) Aporte de la tafonomía lítica al estudio de distribuciones artefactuales en ambientes lacustres: el caso del sistema lacustre al sur del Lago Argentino. Comechingonia Virtual 1(3):132–153

Borrero L (1989–90) Evolución cultural divergente en la Patagonia austral. Anales del Instituto de la Patagonia Ser. Cs. Hs. 19:133–139

Borrero L (1997) The origins of ethnographic subsistence patterns in Fuego-Patagonia. In: McEwan C, Borrero LA, Prieto A (eds.), Patagonia. Natural history, prehistory and ethnography at the uttermost end of the earth, pp 60–81. British Musem Press, London

Borrero L (1999) The prehistoric exploration and colonization of Fuego-Patagonia. J World Prehistory 13(3):321–355

Borrero L (2001a) El poblamiento de la Patagonia. Toldos, milodones y volcanes. Emecé editores, Buenos Aires

Borrero L (2001b) Regional taphonomy: the scales of application to the archaeological record. In: Buitenhuis H, Prummel W (eds.), Animals and man in the past, pp 17–20. ARC-Publicatie 41. Groningen. Netherlands

Borrero L (2004) The archaeozoology of andean "Dead Ends" in patagonia: living near the continental ice cap. In: Mondini M, Muñoz S, Wickler S (eds.), Colonisation, migration, and marginal areas. A zooarchaeological approach, pp 55–61. Oxbow Books, Oxford

Borrero L (2013) Paisajes desconocidos, geografía cultural y tafonomía total. Anuario de Arqueología, Rosario 5:17–30

Borrero L, Borrazo K (eds.) (2011) Geografía cultural del sudoeste de Patagonia continental. In: Bosques, montañas y cazadores. Investigaciones arqueológicas en Patagonia meridional, pp 7–36. Consejo Nacional de Investigaciones Científicas y Técnicas (CONICET) e Instituto Multidisciplinario de Historia y Ciencias (IMHICIHU), Buenos Aires

Borrero L, Guichon R, Tykot R, Kelly J, Prieto A, Cárdenas P (2001) Dieta a partir de isótopos estables en restos óseos humanos de Patagonia Austral. Estado actual y perspectivas. Anales del Instituto de la Patagonia 29:119–127

Borrero L, Franco N (1997) Early patagonian hunter-gatherer: subsistence and technology. J Anthropol Res 57:219–239

Borrero L, Muñoz S (1999) Tafonomía en el bosque patagónico. Implicaciones para el estudio de su explotación y uso por poblaciones humanas de cazadores-recolectores. Soplando en el viento, pp 43–56. Actas de las III Jornadas de Arqueología de la Patagonia. Universidad Nacional del Comahue/INAPL

Braje T, Erlandson J (2009) Mollusks and mass harvesting in the Middle Holocene. California Archaeol 1(2):269–289

Byron J (1901 [1746]) Rclato del Honorable John Byron que contiene una esposicion de las grandes penurias sufridas por él i sus compañeros en la costa de la Patagonia desde el año 1740 hasta su arribo a Inglaterra en 1746.... Imprenta Cervantes. Santiago

Caldwell M, Lepofsky D, Combes G, Washington M, Welch J, Harper J (2012) A bird's eye view of Northern Coast Salish intertidal resource management features, Southern British Columbia, Canada. J Island Coast Archaeol 7(2):219–233

Campbell R, Quiroz D (2015) Chronological database for Southern Chile (35° 30'–42° S),∼ 33000 BP to present: Human implications and archaeological biases. Quatern Int 356:39–53

Campbell R, Santana F, Munita D, Mera R, Massone M, Andrade P, Sánchez M, Márquez T (2020) Diet in Southern Chile (36°-42°S). A synthesis from the isotopic data. Quat Int. En prensa https://doi.org/10.1016/j.quaint.2020.01.015
Cárdenas R, Montiel D, Hall K (1991) Los Chonos y los Veliche de Chiloé. Editorial Olympo. Chile
Carr C (1995) Mortuary practices: their social, philosophical-religious, circumstantial, and physical determinants. J Archaeol Method Theory 2(2):105–200
Chapman A (1987) La Isla de los Estados en la prehistoria. Primeros datos arqueológicos, EUDEBA, Buenos Aires
Christensen M (2016) La industria ósea de los cazadores-recolectores: el caso de los nómades marinos de Patagonia y Tierra del Fuego. Colección Poblamiento Humano de Fuego-Patagonia. Ediciones Universidad de Magallanes, Punta Arenas, Chile
Claassen C (1998) Shells. Cambridge Manuals in Archaeology. Cambridge University Press
Coltrain JB, Tackney J, O'Rourke DH (2016) Thule whaling at point barrow, Alaska: the Nuvuk Cemetery stable Isotope and radiocarbon record. J Archaeol Sci Reports 9:681–694
Cooper J (1917) Analytical and critical bibliography of the tribes of Tierra del Fuego and adjacent territory. Smithsonian Institution, Washington
Cooper J (1946) The Chono. Handbook of South American Indians. The marginal tribes. In: Steward J (ed.), Bureau of American ethnology bulletin, vol 1, p 143. Smithsonian Institution. Washington, DC
Contreras C, Méndez C, Reyes O (2016) Gestión de recursos líticos y organización espacial de cazadores recolectores en la estepa de los valles de Cisnes y Simpson, Región de Aysén Chile. Magallania 44(1):167–185
Cruz I, Caracotche, M (2006) Arqueología de la costa patagónica, perspectivas para su conservación. Arqueología de la costa patagónica: perspectivas para su conservación. Edited by Cruz I, Caracotche M. Universidad Nacional de la Patagonia Austral–Subsecretaría de Cultura de la Provincia de Santa Cruz, Río Gallegos
Curry P (1991) Distribución de sitios e implicaciones para la movilidad de los canoeros en el canal Messier. Anales del Instituto de la Patagonia 20:145–154
Curry P (1997) Manuscrito. Archaeological exploration of the central Chonos Archipelago. Informe Arqueológico Consejo Monumentos Nacionales, Santiago
Darwin Ch (1845) The voyage of the beagle (Journal of researches) into the natural history and geology of the countries visited during the voyage of H.M.S. Beagle round the world, under the Command of Capt. Fitz Roy, R.N. 2ª edición, Londres: John Murray. Cap. 13:273–290
De Niro MJ (1985) Post-mortem preservation and alteration of in-vivo bone collagen isotope ratios in relation to paleodietary reconstruction. Nature 317:806–809
De Vea A (1886) Espedicion de Antonio de Vea (1675–1676). Anuario Hidrográfico de la Marina de Chile. Año XI. Imprenta Nacional, Santiago
Díaz C, Garretón M (1972/73) El poblamiento prehispánico del área insular septentrional chilena. In: Actas del VI Congreso de Arqueología chilena, pp 559–584
Dietrich R, Ivins E, Casassa G, Lange H, Wendt J, Fritsche M (2010) Rapid crustal uplift in Patagonia due to enhanced ice loss. Earth and Planet Sci Lett, 22–29
Dincauze D (2000) Environmental archaeology. Principles and practice. Cambridge University Press
Ebert J (1992) Distributional archaeology. University of Utah Press
Emperaire J (1963) Los nómades del mar. Ediciones de la Universidad de Chile, Santiago
Erlandson J (2001) The archaeology of aquatic adaptations: paradigms for a new millennium. J Archaeol Res 9(4):287–350
Erlandson J, Moss M (1999) The systematic use of radiocarbon dating archaeological survey in coastal and other erosional environments. Am Antiq 64(3):431–443
Fairbanks R (1989) A 17,000 year glacio-eustatic sea level record: influence of glacial melting rates on the Younger Dryas event and deep ocean circulation. Nature 342:637–642
Fagan B (2004) The house of the sea: an essay on the antiquity of planked canoes in southern California. Am Anti, 7–16

Favier Dubois C, Borella F (2007) Consideraciones acerca de los procesos de formación de concheros en la costa Norte del golfo San Matías (Río Negro, Argentina). Cazadores-recolectores del Cono sur 2:151–165

Favier Dubois C, Borella F, Steffan P, Ortega F (2015) Aportes al estudio de la contemporaneidad en las acumulaciones de valvas de origen antrópico en ambientes litorales. Cuadernos del Instituto Nacional de Antropología y Pensamiento Latinoamericano–Series Especiales, 2(4)

Fernandes R (2015) A simple (r) model to predict the source of dietary carbon in individual consumers. Archaeometry. https://doi.org/10.1111/arcm.12193

Fernández P, Carballido Calatayud M, Bellelli C, Podestá M, Scheinsohn V (2011) Marcas en la piedra, huellas en la tierra. El poblamiento del Bosque del suroeste de Río Negro-noroeste de Chubut. In: Valverde S, Maragliano G, Impemba M (eds.), Procesos históricos, transformaciones sociales y construcciones de fronteras. Aproximaciones a las relaciones interétnicas. Estudios sobre Norpatagonia, Argentina y Labrador, Canadá, pp 195–221. Facultad de Filosofía y Letras, Universidad de Buenos Aires, Buenos Aires

Fernández P, Cruz I, Belardi J, de Nigris M, Muñoz S (2016) La explotación del Huemul (*Hippocamelus bisulcus*, Molina 1782) en la Patagonia a lo largo del Holoceno. Magallania 44(1) 187–209, Punta Arenas

Fisher J (1995) Bone surface modifications in zooarchaeology. J Archaeol Method and Theory 2(1):7–68

Fitzpatrick S, Thompson V, Poteate A, Napolitano M, Erlandson J (2016) Marginalization of the margins: the importance of smaller islands in human prehistory. J Island Coast Archaeol 11(2):155–170

Fitz-Roy R (1839) Voyages of the adventure and beagle. Volume II. Proceedings of The Second Expedition, 1831–1836. Henry Colburn (ed. London.)

Flores C, Broitman J, Rivas P (2010) Changes in the subsistence strategy of prehistoric intertidal gathering: the pre-ceramic and ceramic coastal huntergatherers of Reloncaví Sound, Chile. Comparative Perspectives on the Archaeology of Coastal South America. Co-published: Center for Comparative Archaeology, University of Pittsburgh, Pontificia Universidad Católica del Perú, Ministerio de Cultura del Ecuador

Flores C, Correa I (2011) Explotación de ambientes costeros, intensificación de recursos y transformaciones culturales en la transición holoceno medio a holoceno tardío en la costa de los canales patagónico septentrionales, Chile. *Actas XVII Congreso Nacional de Arqueología Argentina.* Tomo I:321–326

Foley R (1981a) A model of regional archaeological structure. In Proceedings of the Prehistoric Society, vol 47, pp 1–17. Cambridge University Press

Foley R (1981b) Off-site archaeology: an alternative approach for the short-sited. Pattern of the past: studies in honour of David Clarke, 157–183

Forero M, Bortolotti G, Hobson K, Donazar J, Bertelloti M, Blanco G (2004) High trophic overlap within the seabird community of Argentinean Patagonia: a multiscale approach. J Anim Ecol 73(4):789–801

Gaete N, Navarro X, Constantinescu F, Mera R, Selles D, Solari M, Vargas L, Oliva D, Duran L (2004) Una mirada al modo de vida canoero del mar interior desde Piedra Azul. Chungara. Tomo I:333–346

Gaete N, Navarro X (2004) Estrategias de vida de canoeros cazadores pescadores recolectores del seno de Reloncaví. Entre el bosque siempreverde y el mar interior. Región de Los Lagos, Chile. In: Contra viento y marea. Arqueología de Patagonia, pp 217–235. INAPL-Sociedad Argentina de Antropología, Buenos Aires

Galarce P (2005a) Manuscrito. Tecnología lítica de grupos cazadores-recolectores-pescadores holocénicos en el Seno de Reloncaví (Puerto Montt, X Región). Informe de análisis lítico entregado a Ciprés Consultores

Galarce P (2005b) Manuscrito. Análisis de los instrumentos líticos de los sitios PM005, PM009 y PM012, Seno de Reloncaví (Puerto Montt, X Región). Proyecto Mejoramiento Ruta 7 (Carretera

Austral) entre las localidades de Quillaipe—Caleta La Arena. Ciprés Consultores Ltda. Santiago de Chile
Galarce P (2005c) Manuscrito. Análisis de los instrumentos líticos de los sitios Puente Quilo y Chepu 005, costa de Chiloé. X Región. Informe de análisis lítico entregado a Ciprés Consultores
Gamble C (1999) The Palaeolithic societies of Europe. Cambridge University Press, Cambridge
García C, Mena F (2016) ¿La frontera del oeste?: Prospecciones arqueológicas en el bosque montano del extremo occidental del valle medio del río Ibáñez (Andes patagónicos, Chile). Int Antropol 17(1):49–61
García J (1889) [1766] "Diario del viaje y la navegación hechos por el padre José Joseph García de la Compañía de Jesús, desde su misión de Cailin, en Chiloé, hacia el sur, en los años 1766–1767. Imprenta Nacional Santiago, Chile, Anuario hidrográfico de la Marina de Chile, XIV
García Guraieb S, Goni R, Tessone A (2015) Paleodemography of Late Holocene hunter-gatherers from Patagonia (Santa Cruz, Argentina): an approach using multiple archaeological and bioarchaeological indicators. Quatern Int 356:147–158
Gifford D (1981) Taphonomy and paleocology: a critical review of Archaelogy's sister disciplines. In: Schiffer M (ed.), Advances in archaeological method and theory, 4, 365–438. Academic Press, New York
Gifford-Gonzalez D (1991) Bones are not enough: analogues, knowledge, and interpretive strategies in zooarchaeology. J Anthropol Archaeol 10(3):215–254
Goñi R (2010) Cambio climático y poblamiento humano durante el Holoceno tardío en Patagonia Meridional. Una perspectiva arqueológica. Tesis de Doctorado, Facultad de Filosofía y Letras, Universidad de Buenos Aires. Manuscrito
Goñi R, Belardi J, Espinosa S, Savanti F (2004) Más vale tarde que nunca: cronología de las ocupaciones cazadoras- recolectoras en la cuenca del lago Cardiel (Santa Cruz, Argentina). In: Civalero T, Fernández P, Guraieb G (eds.), Contra Viento y Marea. Arqueología de la Patagonia, pp 237–248. Buenos Aires
Grayson D (1984) Quantitative zooarchaeology: topics in the analysis of archaeological faunas. Academic Press, Orlando, Florida
Guichón R, Borrero L, Prieto L, Cárdenas P, Tykot R (2001) Nuevas determinaciones de isótopos estables para Tierra del Fuego. Revista Argentina de Antropología Biológica, 3
Hammond H (2015) Manuscrito. Sitios concheros en la costa norte de Santa Cruz: su estructura arqueológica y variabilidad espacial en cazadores recolectores patagónicos. Tesis Doctoral, Facultad de Ciencias Naturales y Museo, Argentina
Harris E (1991) Principios de estratigrafía arqueológica. Editorial Crítica, Barcelona
Hervé F, Calderón M, Fanning CP, Quezada R (2015) Holocene marine deposits at Huinay: evidence of paleoseismic activity on the Liquiñe-Ofqui Fault Zone. 14 Congreso Geológico Chileno. Sociedad Geológica de Chile, La Serena, Chile, pp 375–376
Hervé F, Ota Y (1993) Fast Holocene uplift rates at the Andes of Chiloé, southern Chile. Andean Geology 20(1): 15–23
Hervé F, Thiele R, Parada M (1976) El basamento metamórfico del Archipiélago de las Guaitecas, Aysén. Chile. In: Primer Congreso Geológico Chileno. Sociedad Geológica de Chile, Santiago, Chile
Hiscock P (1985) The need for a taphonomic perspective in stone artefact analysis. Queensland Archaeol Res 2:82–95
Holdgate M (1961) Vegetation and soils in the South Chilean Islands. J Ecol 49(3):559–580. https://doi.org/10.2307/2257223
Horwitz V, Weissel M (2011) Arqueología de Isla de los Estados: la frontera de la abundancia. Los cazadores-recolectores del extremo oriental fueguino. Arqueología de Península Mitre e Isla de los Estados. Sociedad Argentina de Antropología, Buenos Aires, 27
Hückstädt L, Rojas C, Antezana T (2007) Stable isotope analysis reveals pelagic foraging by the Southern sea lion in central Chile. J Exp Mar Biol Ecol 347(1):123–133
Ibar J (1959) Ensayo sobre los indios Chonos e interpretación de sus toponimias. Anales de la Universidad de Chile, Santiago, Chile

Isla F (1989) Holocene sea-level fluctuation in the southern hemisphere. Quatern Sci Rev 8:359–368
Jones E (2004) Dietary evenness, prey choice, and human–environment interactions. J Archaeol Sci 31(3):307–317
Keegan W, Fitzpatrick S, Sealey K, Le Febvre M, Sinelli P (2008) The role of small islands in marine subsistence strategies: Case studies from the Caribbean. Human Ecology 36(5):635–654
Kempf P, Moernaut J, Van Daele M, Vandoorne W, Pino M, Urrutia R, De Batist M (2017) Coastal lake sediments reveal 5500 years of tsunami history in south central Chile. Quatern Sci Rev 161:99–116
Kochi S (2017) Paleodietas en cazadores-recolectores del Canal Beagle durante el Holoceno tardío. Intersec Antropol 18:329–339
Kochi S, Pérez S, Tessone A, Ugan A, Tafuri M, Nye J, Tivoli A, Zangrando F (2018) $\delta^{13}C$ and $\delta^{15}N$ Variations in terrestrial and marine foodwebs of beagle channel in the holocene. Implications for human paleodietary reconstructions. J Archaeol Sci Reports 18:696–707
Kuzminsky S, Reyes O, Arriaza B, Mendez C, Standen V, San Roman M, Muñoz I, Duran A (2018) Investigating cranial morphological variation of early human skeletal remains from Chile: A 3D geometric morphometric approach. Am J Phys Anthropol. https://doi.org/10.1002/ajpa.23344
Lara L, Cembrano J, Lavenu A (2008) Quaternary vertical displacement along the Liquine-Ofqui fault zone: differential uplift and coeval volcanism in the Southern Andes. Int Geol Rev 50(11):975–993
Latcham R (1928) Prehistoria de Chile. Sociedad Impresora y Litográfica Universo, Santiago
Latcham R (1930) La dalca de Chiloé y los canales patagónicos. Boletín del Museo Nacional de Chile, Tomo XIII, 63–72. Santiago
Legoupil D (1997) Bahía Colorada (île d'Englefield): Les Premiers Chasseurs de Mammifères Marines de Patagonie Australe. Éditions Recherche sur les Civilisations, Paris
Legoupil D (2005) Recolectores de mariscos tempranos en el sureste de la isla de Chiloé. Magallania 33(1):51–61
Legoupil D, Béarez P, Lefèvre C, San Román M, Torres J (2011) Estrategias de subsistencia de cazadores recolectores de Isla Dawson (Estrecho de Magallames) durante la segunda mitad del Holoceno: primeras aproximaciones. Magallania (Punta Arenas) 39(2):153–164
Legoupil D, Bearez P, Lepetz S, San Román M, Salas K (2007) De la pesca a la caza: evolución económica del sitio Stuven 1, al sur del Golfo de Penas. In: Morello F, Martinic M, Prieto A, Bahamonde G (eds.), Arqueología de Fuego-Patagonia. Levantando piedras, desenterrando huesos… y develando arcanos, pp 279–294. Punta Arenas: Ediciones CEQUA
Legoupil D, Prieto A (1991) Sepultura de niños canoeros en un abrigo pintado en Ultima Esperanza, Chile. Anales del Instituto de la Patagonia 20:133–138. Punta Arenas, Chile
Legoupil D, Sellier P (2004) La sepultura de la cueva Ayayema (isla Madre de Dios, archipiélagos occidentales de Patagonia). Magallania 32:115–124
Long A (2001) Mid-Holocene sea-level change and coastal evolution. Prog Phys Geogr 25(3):399–408
Lozano, P (2006) Descripción y caracterización biogeográfica de las grandes unidades ambientales de la Patagonia chilena. Manuscrito
Lyman L (1994) Vertebrate taphonomy. Cambridge University Press, Cambridge
Lyman L (2003) The influence of time averaging and space averaging on the application of foraging theory in zooarchaeology. J Archaeol Sci 30(5):595–610
Lyman L (2005) Analyzing cut marks: lessons from artiodactyl remains in the northwestern United States. J Archaeol Sci 32(12):1722–1732
Machado A (1889 [1769]) Viajes del piloto Don francisco Machado a los archipiélagos occidentales de Patagonia. Anuario hidrográfico de la Marina 14
Marín A (2014) Palena, Quitralco y Melimoyu: Intentos fallidos de colonización en el litoral de la Región de Aysén (1889–1983). Rev Austral Cienc Soc 27:137–156
Martial L, Hyades P, Deniker J (2007) Etnografía de los indios Yaghan en la misión científica del Cabo de Hornos, 1882–1883. Ediciones de la Universidad de Magallanes. Punta Arenas

Martínez G, Prates L, Flensborg G, Stoessel L, Alcaráz P, Bayala P (2015) Radiocarbon trends in the Pampean region (Argentina). Biases and demographic patterns during the final Late Pleistocene and Holocene. Quatern Int 356:89–110

Martinic M (2004) Archipiélago Patagónico. La última frontera. Ediciones de la Universidad de Magallanes, Punta Arenas, Chile

Martinic M (2005) De la Trapananda al Aysén. Ediciones Pehuén, Santiago

Massone M, Morello F, Borrero L, Legoupil D, Mena F, Prieto A, Ocampo C, Rivas P, San Román M, Martin F, Méndez C, Reyes O, Munita D (2016) Cazadores-Recolectores en la Patagonia chilena desde 11.000 años a.p. a la colonización occidental. In: Falabella F, Uribe M, Sanhueza L, Aldunate C, Hidalgo J (eds.), Prehistoria en Chile. Desde sus primeros habitantes hasta los Incas. Editorial Universitaria. Santiago, Chile

Mayorga M (2016) Antecedentes históricos referidos a la caza de Lobos Marinos y su interacción con el medio geográfico y humano en el extremo Austral Americano: El caso del lobero escocés William Low. Magallania (Punta Arenas) 44(2):37–64

Mayrs C, Försterra G, Häussermann V, Wunderlich A, Grau J, Zieringer M, Altenbach A (2011) Stable isotope variability in a Chilean fjord food web: implications for N-and C-cycles. Mar Ecol Prog Ser 428:89–104

McCulloch R, Morello F (2009) Evidencia glacial y paleoecologica de ambientes Tardiglaciales y del Holoceno temprano. Implicaciones para el poblamiento temprano de Tierra del Fuego. In: Salemme M, Santiago F, Alvarez M, Piana E, Vásquez M, Mansur E (eds.), Arqueología de la Patagonia. Una mirada desde el último confín, pp 119–136. Utopias, Ushuaia

Mena F (1995) Reflexiones desde la antropología: El ser humano y su larga relación con los bosques. Ambiente y Desarrollo, Marzo, pp 63–69. Chile

Mena F, Ocampo C (1993) Distribución, localización y caracterización de sitios arqueológicos en el Río Ibáñez (XI Región). Boletín del Museo Regional de la Araucanía, Temuco 4(1):33–58

Méndez C, Velásquez H, Reyes O, Trejo V (2006) Tras los moradores del bosque Conjuntos arqueológicos de Alero El Toro (Valle del río Cisnes, XI Región de Aisén). Revista Werkén 8:101–115. Santiago

Méndez C, Reyes O (2008) Late Holocene human occupation of Patagonian forests: a case study at Cisnes River basin (44° S, Chile). Antiquity 82:560–570

Méndez C, Reyes O (2015) Archaeology near the southern ice-end. Current advances in human interdisciplinary research in central western Patagonia. SAA Archaeol Record 15(3):21–26

Méndez C, Stern C, Reyes O (2008/9) Transporte de obsidianas a lo largo de los Andes de Patagonia Central (Aisén, Chile). Cazadores-Recolectores del Cono Sur. Revista de Arqueología 3:51–68. Mar del Plata

Méndez C, Barberena R, Reyes O, Nuevo A (2014) Isotopic ecology and human diets in the forest-steppe ecotone, Aisén Region, central-western Patagonia, Chile. Int J Osteoarchaeol 24:187–201

Méndez C, Stern C, Nuevo A, Reyes O, Mena F (2018) Spatial and temporal distributions of exotic and local obsidians in Central Western Patagonia, southernmost South America. Quatern Int 468:155–168

Merino V, De Pol-Holz R, Southon J, Latorre C, Collado S (2019) Marine radiocarbon reservoir age along the chilean continental margin. Radiocarbon 61(1):195–210

Milne G, Long J, Bassett S (2005) Modelling holocene relative sea level observations from the Caribbean and South America. Quarter Sci Review 24(10–11):1183–1202

Milne G, Gehrel W, Hughes C, Tamisiea M (2009) Identifying the causes of sea-level change. Nat Geosci 2(7):471–478

Milne G, Shennan I (2007) Sea level studies. Isostasy. In: Elias S (ed), Encyclopedia of quaternary science, pp 3043–3051. Oxford, Elsevier

Miotti L, Salemme M (2003) When Patagonia was colonized: people mobility at high latitudes during Pleistocene/Holocene transition. Quatern Int 109:95–111

Moraga M, Aspillaga E, Mena F (2009) Search for founder mitochondrial lineages in Holocene human remains in Patagonia. Am J Phy Anthropol Suppl. 48:193–193

Moraleda J (1888 [1786]) Esploraciones jeograficas e hidrográficas... Imprenta Nacional, Santiago, Chile
Morello F (2014) Manuscrito. Colección patricia curry: avances sobre la tecnología lítica de los canoeros del Archipiélago de los Chonos central. Informe Proyecto Fondecyt 1130151. Santiago
Morello F, San Román M, Prieto A (2002) Puntas de proyectil lanceoladas en Patagonia meridional y Tierra del Fuego. Anales del Instituto de la Patagonia, Serie Ciencias Humanas 30:155–166
Morello F, San Román M, Prieto A (2004) Obsidiana verde en Fuego-Patagonia: distribución y estrategias tecnológicas. In: Contra Viento y Marea. Arqueología de Patagonia, pp 149–166. Sociedad Argentina de Antropología. Buenos Aires
Morello F, Borrero L, Massone M, Stern C, Garcia-Herbst A, McCulloch R, Martinez I (2012) Hunter-gatherers, biogeographic barriers and the development of human settlement in Tierra del Fuego. Antiquity 86(331):71–87
Moreno R (2011) El archipiélago de Chiloé y los jesuitas: El espacio geográfico para una misión en los siglos XVII y XVIII. Magallania, Punta Arenas 39–2:47–55
Munita D (2007) Materias primas líticas en sitios costeros del extremo sur septentrional de chile. Dispersión y aprovisionamiento. Boletín de la Sociedad Chilena de Arqueología 39:103–112
Munita D (2017) Ocupaciones arqueológicas en el borde costero del seno de Reloncaví, el caso de Bahía Ilque. Facultad de Ciencias Sociales, Universidad de Chile, Memoria para optar al título de Arqueóloga
Munita D, Álvarez R, Ocampo C (2005) Corrales de piedra, pesca pasiva en la costa interior de Chiloé. Boletín de la Sociedad Chilena de Arqueología 37:61–74
Munita D, Mera R, Arregui I, Manneschi M (2011) Funebria de grupos canoeros durante el holoceno tardío en la región de los lagos. El conchal de Yaco Alto–1, Calbuco, Chile. Revista cazadores recolectores del cono sur. Argentina
Munita D, Mera R, Alvarez R (2016) Una historia de seis mil años. In: Chiloé. Editor Carlos Aldunate. Colección Santander Museo Chileno de Arte Precolombino, pp 58–87
Muñoz L, Pavez G, Quiñones R, Oliva D, Santos M, Sepúlveda M (2013) Diet plasticity of the South American sea lion in Chile: stable isotope evidence. Revista de Biología Marina y Oceanografía 48(3):613–622
Ocampo C, Aspillaga E (1984) Breves notas sobre una prospección arqueológica en los archipiélagos de las Guaitecas y de los Chonos. Revista Chilena de Antropología 4:155
Ocampo C, Aspillaga E (1991) Problemas del registro arqueológico de los sitios del archipiélago de los Chono y las Guaitecas. Actas del XII Congreso Nacional de Arqueología Chilena 1:17–18
Ocampo C, Rivas P (2004) Poblamiento temprano de los extremos geográficos de los canales patagónicos: Chiloé e isla Navarino 1 . Chungara Volumen Especial, tomo 1:317–331
Ocampo C, Rivas P (2008) Manuscrito. El poblamiento marítimo de los canales occidentales de la Patagonia: desde Chiloé hasta Cabo de Hornos, pp 57–69
Orquera L, Gómez-Otero J (2007) Los cazadores-recolectores de las costas de Pampa, Patagonia y Tierra del Fuego. Relaciones. Sociedad Argentina de Antropología 32:75–100
Orquera L, Piana E (1991) La formación de los montículos arqueológicos de la región del Canal Beagle. Runa 19:59–82
Orquera L, Piana E (1992) Un paso hacia la resolución del palimpsesto. *Análisis espacial en la arqueología patagónica*, editado por Luis Alberto Borrero y José Luis Lanata. Ayllu, Buenos Aires, pp 21–52
Orquera L, Piana E (1999a) Arqueología de la Región del Beagle. Sociedad Argentina de Antropología. Buenos Aires
Orquera L, Piana E (1999b) La vida material y social de los Yámana. Eudeba-IFIC, Buenos Aires
Orquera L, Piana E (2000) Composición de conchales de la costa del canal Beagle (Tierra del Fuego. República Argentina)-Primera Parte, Relaciones de la Sociedad Argentina de Antropología 25
Orquera L, Piana E (2005) La adaptación al litoral sudamericano sudoccidental: qué es y quiénes, cuándo y dónde se adaptaron. Relaciones-Sociedad Argentina de Antropología 30:11–32

Orquera L, Piana E, Alvarez M, Fiore D, Vásquez M, Zangrando F, Tessone A, Tivoli A (2006) El Proyecto arqueológico Canal Beagle. In: Cruz I, Caracotche S (eds.), Arqueología de la costa patagónica: perspectivas para su conservación, pp 267–290. UNPA. Río gallegos, Argentina

Orquera L, Sala A, Piana E, Tapia A (1977) Lancha Packewaia. Editorial Huemul, Buenos Aires

Orquera L, Legoupil D, Piana E (2011) Littoral adaptation at the southern end of South America. Quatern Int 239:61–69

Pallo C (2011) Condicionamientos de la dinámica ambiental en las decisiones humanas sobre asentamiento y circulación a lo largo del estrecho de Magallanes durante el Holoceno tardío. Magallania 39(2):177–192

Pallo C, Borrero L (2015) ¿Intercambio o Movilidad?: Una Evaluación sobre el Uso de Escalas de Análisis Espaciales y Curvas de Declinación en Patagonia Centro-Meridional (Argentina). Latin American Antiquity 26(3):287–303

Panarello H, Zangrando F, Tessone A, Kozameh L, Testa N (2006) Análisis comparativo de paleodietas humanas entre la región del canal Beagle y península Mitre: Perspectivas desde los isótopos estables. Magallania 34(2):37–46

Pearce T (2008) When a snail dies in the forest, how long will the shell persist? Effect of dissolution and micro-bioerosion. Am Malacol Bull 26(1–2):111–117

Pequeño G, Riedemann A (2002) Los peces bentónicos de los canales orientales de Aysén (XI Región, Chile), con nuevos avances sobre su distribución geográfica. Resultados del Crucero CIMAR 7:149–154

Pérez I, González P, Bernal V (2016) Past population dynamics in Northwest Patagonia: an estimation using molecular and radiocarbon data. J Archaeol Sci 65:154–160

Perlman S (1980) An optimum diet model, coastal variability, and hunter-gatherer behavior. Adv Archaeological Method Theory 3:257–310

Petraglia M, Potts R (1994) Water flow and the formation of early Pleistocene artifact sites in Olduvai Gorge, Tanzania. J Anthropol Archaeol 13(3):228–254

Piana E, Orquera L (2009) The southern top of the world: the first peopling of Patagonia and Tierra del Fuego and the cultural endurance of the Fuegian Sea-Nomads. Arctic Anthropol 46(1–2):103–117

Piana E, Orquera L (2010) Shell midden formation at the Beagle Channel (Tierra del Fuego, Argentine). In: Calado D, Baldía M, Boulanger M (eds.), Monumental questions: prehistoric megaliths, mounds, and enclosures. BAR Int Series 2122:263–271

Piana E, Tessone A, Zangrando F (2006) Contextos mortuorios en la región del canal Beagle: del hallazgo fortuito a la búsqueda sistemática. Magallania 34(1):103–117

Plafker G, Savage JC (1970) Mechanism of the Chilean Earthquakes of May 21 and 22, 1960. Geol Soc Am Bull 81(4):1001–1030

Porter C (1993) GUA-010, un sitio costero erosionado en una zona sísmica activa. Actas del XII Congreso Nacional de Arqueología Chilena. Boletín del Museo Regional de la Araucanía 4:81–88

Porter S, Stuiver M, Heusser C (1984) Holocene sea-level changes along the strait of magellan and beagle channel, southernmost South America. Quatern Res 22(1):59–67

Prieto A, Cárdenas R (2001) Ajuar de cueros en entierros de canoeros en el archipiélago de Tierra del Fuego. Anales del Instituto de la Patagonia 29:183–188. Punta Arenas, Chile

Prieto A, Chevallay D, Ovando D (2000) Los pasos de indios en Patagonia Austral. Desde el país de los gigantes I. Universidad Nacional de la Patagonia Austral, Río Gallegos, pp 87–94

Reyes O, Méndez C, Velásquez H, Trejo V (2006a) Distribuciones espaciales y contextos arqueológicos de cazadores recolectores esteparios en alto río Cisnes: (XI Región de Aisén). Magallania. (Punta Arenas) 34(2):75–90

Reyes O, Méndez C, Trejo V, Velásquez H (2006b) De las estepas a los canales septentrionales: uso humano de distintos ambientes. Bosques del valle del río Cisnes (Holoceno tardío, XI Región de Aisén). Actas del XVII Congreso Nacional de Arqueología Chilena [2010].Tomo 2: 1487–1495.Valdivia

Reyes O, Méndez C, San Román M, Cárdenas P, Velásquez H, Trejo V, Morello F, Stern C (2007) Seno Gala 1; Nuevos Resultados En La Arqueología De Los Canales Septentrionales (44° S, XI Región de Aisén, Chile). Magallania 35(2):105–119

Reyes O, Méndez C, Maldonado A, Velásquez H, Trejo V, Cárdenas M, Abarzúa A (2009) Uso del espacio de cazadores recolectores y paleoambiente Holoceno en el valle del río Cisnes, Región de Aisén Chile. Magallania 37(1):91–107

Reyes O, Trejo V (2010) Manuscrito. Informe antropológico físico de restos óseos recuperados en la localidad de Repollal, isla Ascensión, Archipiélago de las Guaitecas, XI Región de Aisén. Proyecto FONDECYT 1130151. Santiago

Reyes O, San Román M, Moraga M (2011) Archipiélago de los Chonos: Nuevos registros arqueológicos y bioantropológicos en los canales septentrionales. Isla Traiguén, XI Región de Aisén. Magallania 39(2):295–303

Reyes O, Méndez C, Mena F, Moraga M (2012) The bioanthropological evidence of a ca. 10,000 CALYBP ten-individual group in central Patagonia. Southbound: Late Pleistocene Peopling of Latin America, 167–171

Reyes O, Moraga M, Aspillaga E (2013) El registro bioantropológico y las evidencias de ocupación en el Archipiélago de los Chonos (XI Región de Aisén, Chile). Avances en la arqueología de los canales septentrionales del extremo sur. In: Tendencias Teórico-Metodológicas y Casos de Estudio en la Arqueología de la Patagonia, pp 227–232. Argentina

Reyes O, Moraga M, Méndez M, Cherkinsky A (2015) Maritime hunter-gatherers in the chonos archipelago (43°50'–46°50' S), Western Patagonian Channels. J Island Coast Archaeol 10(2):207–231

Reyes O, Kuzminsky S, Méndez C, San Román M (2016a) Examining biological variation among the maritime hunter-gatherers of the Chonos Archipelago, Western Patagonian Channels, Chile. Libro Resúmenes 81 The Annual Meeting of the Society for American Archaeology. Orlando

Reyes O, San Román M, Morello F (2016b) Search for maritime hunter-gatherer archaeological record in the shifting shorelines of the South Pacific Coast (Chonos and Guaitecas Archipelago, Chile). In: Marine Ventures: Archaeological Perspectives on Human-Sea Relations, pp 141–155. Equinox Publishing, United Kingdom. https://doi.org/10.1558/equinox.24563

Reyes O, Méndez C, San Román M, Francoise J (2018) Earthquakes and coastal archaeology: assessing shoreline shifts on the southernmost pacific coast (Chonos Archipelago 43°50′–46°50′ S, Chile, South America). Quatern Int 463:161–175

Reyes O, Tessonne A, San Román M, Méndez C (2019a) Dieta e isótopos estables de cazadores recolectores marinos en los canales occidentales de Patagonia Chile. Latin Am Anti 30(3):550–568

Reyes O, Méndez C, San Román M (2019b) Cronología de la ocupación humana en los canales septentrionales de la Patagonia occidentalChile. Intersec Antropol 20(2):195–210

Reynard J (2014) Trampling in coastal sites: an experimental study on the effects of shell on bone in coastal sediment. Quatern Int 330:156–170

Richards M, Hedges R (1999) Stable isotope evidence for similarities in the types of marine foods used by late mesolithic humans at sites along the Atlantic coast of Europe. J Archaeol Sci 26:717–722

Rivas P, Ocampo C (2005) El antiguo curanto chilote. In: Chile País Oceánico. Edited by Badal G, Montanari M, Santa Cruz C, Altamirano C: 70–72.

Rivas P, Ocampo C (2006) La adaptación humana al bosque en la isla de Chiloé. Estrategias adaptativas en el litoral septentrional de los canales patagónicos. Actas del XVII Congreso Nacional de Arqueología Chilena (2010). T2:1449–1460. Valdivia

Rivas P, Ocampo C, Aspillaga E (1999) Poblamiento Temprano de los Canales Patagónicos: El Núcleo Ecotonal Septentrional. Anales del Instituto de la Patagonia 27:221–230

Rivera A, Benham T, Casassa G, Bamber J, Dowdeswell J (2007) Ice elevation and areal changes of glaciers from the Northern Patagonia Icefield Chile. Glo Planet Change 59(1–4):126–137

Roksandic M, de Souza S, Eggers S, Burchell M, Klokler D (2014) The cultural dynamics of shell-matrix sites. UNM Press

San Román M (2010) La explotación de recursos faunísticos en el sitio Punta Santa Ana 1: estrategias de subsistencia de grupos de cazadores marinos tempranos de Patagonia meridional. Magallania 38(1):183–198

San Román M (2014) Sea-level changes and coastal peopling in southernmost Pacific South America: Marine hunters from Patagonia. In: Encyclopedia of global archaeology, pp 6515–6525. Springer, New York

San Román M, Morello F (2001) Canal Maule: Nuevos antecedentes sobre prácticas funerarias en el Archipiélago Fueguino. Anales del Instituto de la Patagonia 29:149–161. Punta Arenas, Chile

San Román M, Morello F, Prieto A (2002) Nuevos antecedentes sobre la explotación de recursos faunísticos en el Mar de Otway y canales adyacentes. Anales del Instituto de la Patagonia Serie Ciencias. Humanas 30:147–154

San Román M, Reyes O, Méndez C, Cárcamo J, Torres J (2016b) Western Patagonia subsistence strategies: Zooarchaeological studies of marine hunter-fisher-gatherers of the Chonos archipelago, Chile. Libro Resúmenes 81 The Annual Meeting of the Society for American Archaeology. Orlando

Santiago F, Bujalesky G, Salemme M (2007) Prospección arqueológica en la cuenca del río Chico. Tierra del Fuego, Argentina. In Arqueología de Fuego-Patagonia. Levantando piedras, desenterrando huesos... y develando arcanos (pp 357–363). CEQUA, Punta Arenas, Chile

Saporiti F, Bala L, Gómez-Otero J, Crespo E, Piana E, Aguilar A, Cardona L (2014) Paleoindian pinniped exploitation in South America was driven by oceanic productivity. Quatern Int 352:85–91

Scheinsohn V (2001) Odisea del espacio: Paisajes y distribuciones artefactuales en arqueología. Relaciones de la Sociedad Argentina de Antropología, 26

Shennan I (2007) Sea level studies. Overview. In: Elias S (eds.), Encyclopedia of quaternary science, pp 2967–2974. Oxford, Elsevier

SHOA (2000) El maremoto del 22 de mayo de 1960 en las costas de chile. Ediciones Servicio Hidrográfico y Oceanográfico de la Armada de Chile, Valparaíso

Simpon E (1871) Memorias Científicas i Literarias. Anales de la Universidad, Agosto

Simpon E (1875) Esploraciones hechas por la Corbeta Chacabuco en 1870 .... Anuario Hidrográfico de la Marina, Tomo I

Steele J (2010) Radiocarbon dates as data: quantitative strategies for estimating colonization front speeds and event densities. J Archaeol Sci 37:2017–2030

Stein J (1992) Sediment analysis of the British Camp shell midden. In: Stein J (ed.), Deciphering a shell midden, pp 135–162. Academic Press, San Diego

Stein J, Deo J, Phillips L (2003) Big sites-short time: accumulation rates in archaeological sites. J Archaeol Sci 30(3):297–316

Stern C (2004). Obsidian in Southern Patagonia: review of the current information. Contra viento y marea. Arqueología de Patagonia, 167–176

Stern C (2018) Obsidian sources and distribution in Patagonia, southernmost South America. Quatern Int 468:190–205

Stern C, Porter C (1991) Obsidiana en yacimientos arqueológicos de Chiloé y las islas Guaitecas. Anales del Instituto de la Patagonia. Serie Ciencias Humanas 20:205–209

Stern C, Curry P (1995) Obsidiana del sitio Posa Las Conchillas, Isla Traiguén (45°30′S), Archipiélago de los Chonos. *Anales del Instituto de la Patagonia*. Serie Ciencias Humanas. 23:119–124

Stern C, Navarro X, Muñoz J (2002) Obsidiana gris translúcida del volcán Chaitén en los sitios arqueológicos de Quilo (Isla Grande de Chiloé) y Chanchán (X Región) Chile, y obsidiana del Mioceno en Chiloé. Anales del Instituto de la Patagonia, Serie Ciencias Humanas 30:167–174

Stern C, Caracotche S, Cruz I, Charlin J (2012) Obsidiana gris porfírica calco-alcalina del volcán Chaitén en sitios arqueológicos al sur del río Santa Cruz, Patagonia meridional. Magallania 40(1):137–144

Tafuri M, Zangrando F, Tessone A, Kochi S, Cecchi J, Di Vincenzo F, Profico A, Manzi G (2017) Dietary resilience among hunter-gatherers of Tierra del Fuego: Isotopic evidence in a diachronic perspective. PloS one 12(4):e0175594

Tessone A (2014) Conductas mortuorias en el Canal Beagle. Una visión desde el registro etnohistórico y etnográfico. In: Oría J, Tivoli AM (eds,), Cazadores de mar y tierra. Estudios recientes en arqueología fueguina. Museo del Fin del Mundo, pp 149–167

Tessone A, Zangrando F, Valencio S, Panarello H (2003) Isótopos estables del carbono en restos óseos humanos de la región del canal Beagle, Isla Grande de Tierra del Fuego. Revista Argentina de Antropología Biológica. 2:33–43

Thomson S (2002) Late Cenozoic geomorphic and tectonic evolution of the Patagonian Andes between latitudes 42°S and 46°S: an appraisal based on fission-track results from the transpressional intra-arc Liquiñe-Ofqui fault zone. Geol Soc Am Bull 114(9):1159–1173

Torrejón F, Bizama G, Araneda A, Aguayo M, Bertrand S, Urrutia R (2013) Descifrando la historia ambiental de los archipiélagos de Aysén, Chile. El influjo colonial y la explotación económica mercantil republicana (siglos XVI-XIX). Magallania 41(1):29–52

Torres J (2016) Manuscrito. Informe análisis ictioarqueológico de seis sitios del archipiélago de las Guaitecas y los Chonos, XI Región de Aisén. FONDECYT 1130151. Santiago

Tykot R, Falabella F, Planella M, Aspillaga E, Sanhueza L, Becker C (2009) Stable isotopes and archaeology in central Chile: methodological insights and interpretive problems for dietary reconstruction. Int J Osteoarchaeol 19:156–170

Urbina R (1983) La periferia meridional indiana, Chiloé en el siglo XVIII. Ediciones Universitarias de Valparaíso y Universidad Católica de Valparaíso. Valparaíso

Urbina R (1988) Los chonos en Chiloé: Itinerario y Aculturación. *Revista Chiloé* No 9–Agosto, pp 29–42

Urbina X (2010) La navegación por los canales australes en la Patagonia occidental insular en los siglos coloniales: la ruta del istmo de Ofqui. Magallania, Punta Arenas 38–2:41–67

Urbina X (2014) Fuentes para la historia de Patagonia occidental en el periodo colonial. Primera parte: Siglos XVI y XVII. Ediciones Universitarias de Valparaíso. Pontificia Universidad Católica de Valparaíso. Chile.

Urbina X, Reyes O, Belmar C (2020) Canoeros en Chiloé: de facilitadores de las navegaciones españolas en los archipiélagos de los Chonos y de Guayaneco, a productores y comerciantes. Chungara Revista Chilena de Antropología 52(2):335–346

Urbina X (2016) Traslados de indígenas de la Patagonia occidental insular a Chiloé en los siglos XVI, XVII y XVIII. In: Jaime V (ed.), América en diásporas: esclavitudes y migraciones forzadas (siglos XVI-XIX), Santiago, Instituto de Historia, Pontificia Universidad Católica de Chile/RiL editores

Vázquez M, Zangrando F (2017) Estructuras de pesca en el canal Beagle. Magallania (Punta Arenas) 45(1):101–122

Vuilleumier F (1985) Forest birds of Patagonia: ecological geography, speciation, endemism, and faunal history. Ornithol Mono 36:255–304

Vita-Finzi C (1996) Paleoseismology in coastal Chile. J Geophy Res Solid Earth (1978–2012) 101(B3): 6109–6114

Viviani C (2004) Manuscrito. *Caracterización del litoral de la Región de Aisén.* Universidad de Magallanes, Punta Arenas

Wandsnider L (1992) The spatial dimension of time. In: Rossignol J, Wandsnider L (eds.), Space, time, and archaeological landscapes pp 257–282, New York, Plenum Press

Wandsnider L, Camilli E (1992) The character of surface archaeological deposits and its influence on survey accuracy. J Field Archaeol 19(2):169–188

Waselkov G (1987) Shellfish gathering and shell midden archaeology. Adv Archaeol Method Theory 10:93–210

Waters M, Kuehn D (1996) The geoarchaeology of place: the effect of geological processes on the preservation and interpretation of the archaeological record. Am Anti, 483–497

Zangrando F (2009a) Historia evolutiva y subsistencia de cazadores-recolectores marítimos de Tierra del Fuego. Sociedad Argentina de Antropología. Buenos Aires

Zangrando F (2009b) Is fishing intensification a direct route to hunter-gatherer complexity? a case study from the Beagle Channel region (Tierra del Fuego, southern South America). World Archaeol 41(4):589–608

Zangrando F, Tessone A, Valencio S, Panarello H, Mansur M, Salemme M (2004) Isótopos estables y dietas humanas en ambientes costeros. *Avances en Arqueometría* 2003. Universidad de Cádiz, España, pp 91–97

Zangrando F, Tivoli A (2015) Human use of birds and fish in marine settings of southern Patagonia and Tierra del Fuego in the Holocene: A first macro-regional approach. Quatern Int 373:82–95

Zilio L, Hammond H (2013) Distribución de concheros y estructuras de entierro (chenques) en la Bahía del Oso Marino, Costa Norte de Santa Cruz. Tendencias teórico metodológicas y casos de estudio en la arqueología de la Patagonia, pp 535–544

Zubimendi M (2012) La variabilidad del registro arqueomalacológico en la costa norte de Santa Cruz (Patagonia argentina): resultados exploratorios a partir de estudios estratigráficos. Intersec Antropol 13(2):359–375

Zubimendi M, Ambrústolo P, Zilio L, Castro A (2015) Continuity and discontinuity in the human use of the north coast of Santa Cruz (Patagonia Argentina) through its radiocarbon record. Quatern Int 356:127–146

Zubimendi M, Castro A, Moreno E (2005) Procesos de ocupación de la costa norte de Santa Cruz (Argentina): una síntesis. Relac Soc Argent Antropol 30:225–234

Zubimendi M, Castro A, Moreno E, Mazzitelli L (2009) Tiempo y espacio en el uso de la Costa Norte de Santa Cruz, Patagonia Argentina. In: La arqueología como profesión: los primeros 30 años. XI Congreso Nacional de Arqueología Uruguaya,755–768

Chapter 6
Conclusions and Projections

Abstract The final considerations derived from a regional archaeology developed on multiple analytical scales are presented. We analyse the characteristics that define the human occupation of the Patagonian Archipelago by groups of marine hunter-gatherers-fishermen, highlighting the results achieved and their relevance to the archaeology of Patagonia and future study possibilities. In addition, some reflections related to the development of the investigations are made.

Keywords Western patagonia · Chonos archipelago · Maritime hunter-gatherers · Maritime adaptation · Biogeographic barrier · Coastal survey · Archipelagic settlements · Coastal archaeology · Coastal adaptation · Coastal erosion

In this work, we summarise our investigations carried out in the northern portion of western Patagonia. We propose and discuss our characterisation of the human occupation of the Chonos Archipelago, setting it in a regional context and evaluating how these human groups related to their environment. We also consider the geomorphological variables which have such an important influence on the ability to find and record sites in this archipelago environment, and the cultural contexts which may be found.

In our prospection of 567 km of shoreline, we reported 52 pre-contact sites and 25 post-contact sites known as *cholgueros*; seven types of archaeological sites were distinguished in the archipelago. In this study, we determined that the archaeological record generated by the interaction between natural processes and the type of occupation of this space by marine hunter-gatherers was mostly invisible throughout the time sequence on the coastlines in the predefined study areas (Northern, Central, South-central, Mainland Coast, and Offshore Islands and Pacific Border). This is because the forest cover recolonises spaces occupied by people within a short time after abandonment, as we found in the historical and sub-present sites. This situation has a direct impact on our ability to find cultural contexts.

We evaluate the processes by which the archaeological record was formed, reporting the effects of the rise in sea level, isostatic rebound and local tectonics in the different study areas defined within the archipelago; specifically, we consider

O. Reyes, *The Settlement of the Chonos Archipelago, Western Patagonia, Chile*, The Latin American Studies Book Series,
https://doi.org/10.1007/978-3-030-54326-6_6

the fact that archaeological sites formed by marine hunter-gatherers are found pre-eminently along the coastline throughout the occupation sequence initiated in the Mid Holocene. The occupations of this period suffered the effects, on a millennial scale, of marine transgressions, earthquakes and glacio-eustatic changes, showing marked geomorphological transformations in the sites which reflect uplifting or sinking of the land, concealing or destroying evidence. In Late Holocene sites, we recorded the destruction of profiles and a general tendency to sinking and erosion. Adjacent to several of these shell-midden camps, we recorded intertidal areas with scattered lithic and bio-anthropological material, produced by the loss or discard of cultural materials, or by erosion of the exposed faces of archaeological sites.

Likewise, the historical sites known as *cholgueros* allowed us to calibrate and dimension the magnitude of the geomorphological changes to the coast, i.e. fluctuations in sea level and their effects on the anthropic sediments formed on the shoreline, on a temporal microscale. The constant, strong tectonic movements in the region not only are the principal cause of the destruction of archaeological sites, but also make it difficult to model palaeo-coastlines in order to guide searches for human occupations based on coastal geoforms, as can be done in other areas of the Fuegian and Patagonian Archipelagos.

Based on evidence of the first human occupation detected in the northern area (GUA-010 Terraza), which has been dated to around 6200 years cal. BP, and the various palaeo-environmental records for the region, it can be argued that environmental conditions at the time of the first occupation of the archipelago were analogous to current conditions, and had been for at least a thousand years previously (~8000 to 7000 years cal. BP). This situation and the Late Pleistocene record of Monte Verde on the mainland just north of the Patagonian Archipelago suggest the possibility of the existence of older occupations in the Chonos Archipelago. Nevertheless, environmental changes from the end of the Pleistocene also contributed to the topographical formation of the landscape of the archipelago by glacial action, modelling the physiognomy of the coast as fronts of ice advanced and retreated and sea levels rose and fell. This certainly had a direct impact on the possibility of finds and the preservation of archaeological sites earlier than those recorded from around the 6th millennium BP. Although this possibility cannot be discarded, the search design and the cover achieved over these years do not support a longer chronological sequence for human occupation.

We have also maintained for over a decade that in the Aysén Region (43°38′ to 49°16′S), the Andes Range, with its forested foothills, volcanoes, cul-de-sac valleys and ice fields, formed a biogeographical barrier which made interaction between two totally different cultural spheres—the terrestrial and marine hunter-gatherer cultural traditions—difficult or even impossible. Our understanding is based initially on the material records (technologies, circulation of raw materials, archaeofauna), site analysis, chronologies and palaeo-environmental information recovered in the valleys of the eastern Andes, and on the early information obtained from the mainland coast of the Chonos Archipelago (Seno Gala 1). It was initially proposed that contact might have existed between terrestrial and marine hunter-gatherer groups at the Alero El Toro site, located in evergreen forest in the upper Cisnes River, ~37 km from the mainland coast of the archipelago; however, this could not be proved under any archaeological parameter (raw materials, technology, archaeofauna). Studies carried out further north, in Comarca Andina in Argentina (42°S), also support the basic role of the Andes in segregating the human groups on either side of the mountains and restricting their mobility. Subsequent works on the distribution of raw materials allochthonous to the region, supported by information on the isotope signal of the human remains of terrestrial hunters of the Early and Late Holocene, and preliminary mitochondrial DNA information, have continued to discuss the subsistence patterns and mobility circuits of hunter groups in the exclusively terrestrial environment of the eastern slope of the Andes. Nevertheless, despite this archaeological clarity derived principally from investigations carried out in the mainland part of Aysén, comparisons with the archipelago world are based on specific earlier works about the distribution of raw materials and associated lithic technologies, without any systematic prior investigation which would allow this human occupation to be characterised and compared in detail with the records obtained from the mainland.

Approaches to the mainland coast for which archaeological documentation exists in the form of residential camps, choice of funerary sites and obtaining of specific resources (terrestrial fauna and obsidian) present no evidence of contacts or permeable situations in which interaction could occur between human groups with different cultural trajectories. If such sites exist, they may be hidden in a forested stretch of shoreline, or further inland. The fact remains that despite planned, systematic searching, this type of evidence has not yet been found. Furthermore, the $\delta^{15}N_{col}$ and $\delta^{13}C_{col}$ isotope values obtained in human remains from the archipelago and the steppe area indicate marine and terrestrial diets, respectively, segregating and defining the areas in which these groups subsisted. Craniometric evidence (mentioned in Chapter 2, Background) is only just starting to be explored, but it already suggests that the marine hunter-gatherers of the Chonos Archipelago share a greater biological affinity with coastal groups of the Pacific coast than with the mainland terrestrial hunters who resided less than 150 km further east. The conclusion of all this is that the peopling and occupation of the archipelago was carried out exclusively by marine hunter-gatherer groups with specialised subsistence strategies, independently of extra-Andean cultural processes, throughout the whole archaeological sequence of the region.

We also evaluated population flows through the material record recovered in the Chonos Archipelago. This space has its own defining characteristics, with active, continuous biogeographical barriers which separated it absolutely from the cultural processes of the extra-Andean steppe. We understand the occupation of the territory as proposed by Borrero (1989–90, 1999, 2001), within a context of continuous peopling that commenced with a long stage of exploration and reconnaissance in the Mid-Holocene by hunter-gatherers who were accomplished sailors. They had their own conserved technology and they knew about and transported raw materials from distant sources exclusively in the littoral environment, like the rhyolithic obsidian from the Chaitén Volcano source.

Consistent with their specialised sea-travel technology, the flows and movements of the human groups in the archipelago occurred over great distances. Movements did not occur as linear displacements on a flat space, but would have been structured in accordance with the use of socially constructed routes, articulated by organisational systems in which we distinguish the disposition and function of the camps and the technological instruments either documented or inferred. The funerary deposition of human remains representing all categories of age and sex in shell-middens and caves, in all the areas of the archipelago where we took samples, is related to the presence of family units with residential mobility in the area; no exclusion of individuals from certain geographical locations is observed. From the lithic technology that has been preserved (e.g. lanceolate lithic projectile points and stone axes), the shared use of known sources of raw materials, the similar coastal settlement pattern and the characteristics of the fauna remains found at the sites, together with the development of similar cultural elements like the funerary pattern in shell-middens and ossuaries, we understand these population flows as multidirectional movements occurring throughout the coastal strip and the islands of the northern archipelago. They would have originated in the Chiloé Archipelago, the Reloncaví Sound and the adjacent mainland coast, and present pulses of increasing frequency in the Chonos Archipelago as time passed.

This may be especially significant in a scenario of initial exploration in which a low demographic presence must be assumed, and the extensive ranges of mobility would allow large social networks which would maintain social reproduction and information flows. Some regional continuities can, therefore, be expected. In future, it will be important to gain a better understanding of the occupation processes which occurred in the Chiloé Archipelago, Reloncaví Sound and adjacent areas in order to articulate them with the activity recorded in the cultural sequence of the Chonos Archipelago; however, there is no doubt that the archaeological evidence recovered to date is related to that area.

To define the characteristics of the technological assemblages used by the hunter-gatherer-fishers of the Chonos Archipelago, we recorded the lithic material available in collections and our own work. The lithic assemblage reported to date consists of a total of 1,025 pieces collected from almost the whole of the study area, apart from the Pacific border; these artefacts were recovered from contexts with chronologies between 6200 and 300 years cal. BP. There are some formatted instruments, mainly lanceolate lithic projectile points, stone axes and net weights; other items were recorded to a lesser degree, including knapping waste, anvils, side-scrapers, cores and choppers.

The predominant items in the assemblages from the Mid-Holocene are medium-sized thick-section lanceolate lithic projectile points, used for hunting; many are made on rhyolithic obsidian, transported in lumps by canoe from the Chaitén Volcano source on the mainland coast, 150 km from the northern edge of the archipelago and 360 km from the southern edge. The processing of lithic resources, the representation of the final phases of operating chains (consistent with functional relations with other camps) and the lithic categories observed suggest a strong influence of transport, and consequently high mobility of the groups who travelled the archipelago. Lanceolate designs are widely distributed throughout western Patagonia, from the Reloncaví Sound to the Beagle Channel, and even on the south coast of mainland Chile. The excavations and archaeological sequences carried out show that the manufacture of these tool-heads lasted throughout the sequence, so too did the use of net weights and stone axes, which are recorded in shell-midden camps and in adjacent intertidal zones. No new instruments or forms were found for use in any of the activities implied by their use. Production of stone axes would be related to the use of forest resources and with wood-working—these uses would include the manufacture and maintenance of canoes. Net weights are obviously related to fishing, an activity which is much better documented in the Late Holocene by the faunal record and the $\delta^{15}N_{col}$ and $\delta^{13}C_{col}$ isotope values of the individuals sampled.

We have been able to date the use and transport of rhyolithic obsidian; although it is only infrequently recorded (except in Mid-Holocene sites), it is distributed throughout the whole archaeological sequence in the Northern, Central, South-central and Mainland coast areas, providing evidence of high mobility and/or exchange networks among these canoe groups.

The lines of evidence associated with the faunal component by which we can evaluate the specialised subsistence of hunter-gatherers in the Chonos Archipelago are the locations of the sites recorded, the results of the isotope values measured in a sample of 38 individuals, and the lithic technology recovered. Based on the information compiled, we maintain that the subsistence strategies deployed by the human groups were oriented towards the consumption of a combination of resources from the marine and pericoastal environments. Although the archaeological contexts of the Mid-Holocene present no associated fauna, from the lithic instruments recorded (thick-section lanceolate lithic projectile points and net weights), we can assume hunting and fishing activities; the strongest evidence for these comes from the archaeological assemblages of the Late Holocene. We observe that mammal exploitation presents no emphasis or specialisation in a particular taxon. The contribution of small estuarine mammals to diet (*Myocastor coipus*) is observed to be important, while the presence of terrestrial mammals (*Pudu pudu, Hippocamelus bisulcus,* Canidae), recorded in both island contexts and on the mainland coast, is never more than opportunistic consumption which is not reflected in the diet consumed by the individuals, in contrast to the consumption documented in other canoe groups of the Patagonian Archipelago.

Consumption of birds on the other hand is concentrated on taxa which nest mainly in colonies (*Phalacrocorax spp., Procellaria spp., Pelecanoides spp.*). The variety of fish recorded, especially those caught in the greatest numbers, suggests the use of both coastal fishing strategies (*Eleginops maclovinus, Trachurus murphy*i) and fishing from canoes (*Thyrsites atun*). Likewise, the high frequency of net weights recorded in the intertidal of the central coast of Gran Guaiteca Island is indicative of both the technique by which these resources were obtained and the variety of species and the numbers present in the archaeofaunal record. Based on the record of net weights in dated contexts, this practice is recorded from at least 4500 years cal. BP (Isla Goñi 1) to the end of the sequence. It is quite probable, however, that fishing was practised from the first occupations around 6200 years cal. BP.

The habitats of the malacological species represented in the faunal columns are varied: rocky intertidal, rocky subtidal, parasitic and mixed intertidal (sandy and/or rocky), reflecting varied strategies of provisioning with molluscs in which the emphasis on the species represented is consistent with the type of shoreline immediately around the site. The variety of species recorded is related to provisioning from a wider area around the camp, i.e. variations in the fauna according to local availability. Analysis of taxonomic variety and the geographical location of the sites revealed a tendency to greater richness in the northern area of the archipelago, particularly Isla Marta 1 and Alero Low, both contexts exposed to the deep Pacific. This tendency is consistent with the ecological characteristics of the region, principally in the presence of sea birds and fish. As in the representation of the malacological fauna, the variations in taxonomic richness in the different areas of this archipelago are explained as a reflection of the richness of faunal resources available locally.

A partial regional isotope ecology can be projected, based on our fauna samples and others which have been published and homologued (both terrestrial and marine), together with analyses of the isotope values of 38 individuals with ages directly dated by radiocarbon to between *ca* 2000–200 years cal. BP. These samples were collected from different contexts throughout the archipelago. This projection supports our interpretation offered above of a hunter-gatherer adaptation based on a subsistence pattern understood to involve an exclusively marine diet. Post-contact human samples, on the other hand, reflect profound cultural changes among the canoe groups who occupied the northern archipelago, towards a terrestrial diet. Before European contact (C16th), the isotope values remained almost constant throughout the Late Holocene sequence, consistent with a preponderantly marine diet and with the faunal assemblages described. In the last 1000 years BP, however, and before European contact, a slight increase is observed in the nitrogen signal. It remains to be investigated whether this difference results from changes in some of the resources obtained (higher consumption, new catches) or other as yet unidentified circumstances (e.g. introduction of new capture technologies). Terrestrial mammals are scarcely represented in these values, while fish tend to be the most important vertebrate resource recorded.

Finally, we evaluate the archaeological record of the Chonos Archipelago in distributional and temporal terms. At macro- and mesoscale, we recorded a low density of archaeological sites; the occupations identified could be represented as isolated points in space, with a few groupings. Based on the information to date, the number of sites per linear km of shoreline prospected varied between one site every 3.4 km and one every 20 km, with notable differences depending on the type of area studied. The offshore area (Guafo Island) produced no prehistoric presence at all in the 32 km of shoreline prospected.

Site distribution is influenced by the type of occupation of the territory, following the chronological sequence over a broad time scale established by the discard rates of human activity; sites were hierarchised during a continuous peopling process with different intensities of occupation, and both their geographical distribution and their preservation and visibility were strongly influenced by the geomorphological characteristics of the coastline. Site distribution must therefore not be understood by direct cultural explanations based on their presence or absence *per se*. The reoccupation of archaeological sites after variable intervals, as indicated by the chronological sequences obtained and the discard rates inferred, should be understood more as the superposition of low-intensity events of human occupation in the same places rather than processes of intensification in obtaining local resources. They occur on shorelines in locations that appear to be restricted, where the use and maintenance of canoes plays a preponderant role, guiding and restricting site selection and camp installation.

Based on the 86 radiocarbon datings obtained so far, and the low artefactual and site density recorded throughout the chronological sequence, we propose that the occupation by marine hunter-gatherer groups in the Mid-Holocene was exploratory. It would have been of low intensity, logistical in nature, with high mobility and long periods of abandonment of the areas visited, of the order of millennia. Geomorphological remodelling of the coast occurred in the time lapse before the archipelago was reoccupied. The human signal, within the context of low-intensity occupation, presents an increase in the Late Holocene in what we believe was the beginning of a colonisation phase at the end of the third millennium; at this date, we start to record the first shell-midden camps, with reuse of sites and locations, towards the South-central area of the archipelago. There follows a period of stable occupation of the space from ~2000 years cal. BP, with a peak around ~1300 years cal. BP when the latitudinal and longitudinal ranges of distribution of the sites extended to take in the whole archipelago, except Guafo Island. We record a greater diversity of archaeological sites, for example, residential sites (shell-middens) and funerary sites (ossuaries), with the increased presence of human remains, site reoccupations and the use of new spaces such as caves and rock-shelters available outside the pattern of shoreline site location. Different activities can be inferred at the sites. Changes in the function and organisation of the space are also found in some cases towards the end of the sequence of pre-contact occupation, when some campsites were transformed into funerary sites. This increase in the archaeological signal in the Late Holocene, especially the last 2,000 years, also agrees with the pollen records which show an increase in the incidence of carbons (indicator of fire) in the palaeo-environmental record of the archipelago, and coincide with the general tendency recorded in Patagonia and adjacent areas to population increase and territorial dispersion into the different environments available.

Subsequently, prior to European contact, we observe a marked fall in the archaeological signal towards ~700 years cal. BP, perhaps—as some authors maintain—due to an adaptive crisis in which the canoe groups interacted with farming populations from southern mainland Chile, or else as part of the dynamic of low occupation of this island territory, where we find hiati and periods without occupations throughout the temporal sequence. It is in this context that European contact occurs, leading to cultural dismantling in barely 200 years. Paradoxically however, a final increase in the archaeological signal occurs around ~400–300 years cal. BP. This is probably associated with economic readjustment to the new cultural template imposed—under compulsion—by the Europeans, and the last efforts of a group specialising in littoral and marine subsistence to maintain their way of life and adapt to new circumstances.

6.1 Final Words

This book, of an exploratory and generalist nature, proposes an archaeological panorama to characterise the peopling and human occupation of an extensive area of the Patagonian Archipelago composed of channels, fjords, steep-sided islands and forests that come down to the sea, where even today access is difficult. This difficulty undeniably makes it a good area for evaluating the presence and characteristics of the occupation of a territory by marine hunter-gatherers about whom we had only a few notions.

In our minds, the Chonos Archipelago, shut in by permanent biogeographical barriers throughout its length, appeared as a space available only to groups specialised in the technology of sea travel and in the efficient procurement of the marine and coastal resources present in the archipelago system. This implied a particular way of occupying the space and relating with the environment by groups who arrived from the Chiloé Archipelago and the neighbouring mainland coast, where the Patagonian channels begin. The subsistence strategies of these marine hunter-gatherer-fisher groups can be catalogued as similar throughout the whole time sequence, with variants assumed and distinguished between the first groups who dared to explore this archipelago space 6000 years ago and their descendants, thousands of years later, who lived and deposited the bodies of their family members in their camps or in the rock shelters that they selected for the purpose. These later generations started to learn of other groups living further north who cultivated the land, raised animals and used ceramic pots. Contacts, exchanges of goods and people? Possibly. Sea travel is a means of access and transport, allowing groups to visit and reside in different areas, articulating social networks and maintaining them over time. Their mobility and ranges of action throughout the area were established long ago, in the Mid-Holocene. Perhaps the terra firma of the northern end of the Patagonian Archipelago served for these purposes, since groups with different modes of subsistence (e.g. marine hunter-gatherers, farming groups and those who lived on the mainland coast) could converge, while the Chonos Archipelago continued to be a space used only by those who could travel there, the canoe groups. So far, the evidence indicates a common cultural substrate (e.g. technology, circulation of raw materials, funerary practices) in the northern archipelago, and no evidence has appeared that takes us out of that means of subsistence specialised in the littoral and marine environments. In the medium term, we hope that genetic evidence, increasing archaeo-botanical information and more extensive excavations in large shell-midden camps, together with the development of a regional archaeology in the Chiloé Archipelago, will allow us to define more details that we have not yet been able to distinguish in our coarse-grained studies.

In this basic archaeological substrate, we present chronologies of occupation and describe the locations and sites identified, with the material density of artefacts and ecofacts recovered. We characterise subsistence from the archaeofaunal information, identifying a marine diet. We also calibrate the different areas sampled, weighting the different factors that affect our ability to make archaeological finds and discussing the geomorphological conditions in which we hope to recover information. The creation and development of this substrate will allow us in the future to propose an agenda with more specific questions arising from this initial substrate. We propose themes arising from the low density of the technological components recovered, and their relation with intensification in the faunal resources obtained, as well as variability in the organisation of space in the camps over the time sequence and the influence of the consumption of plant resources. Although these are not represented in the faunal record or in the isotope values recorded in human remains, the evidence for plant exploitation starts to shed light on the use of the forest medium for which we have little information to date. Much remains to be done. The search for answers, going beyond the immediate results expected, will enable us to feed new perspectives on the inexhaustible trajectory of the human being as a hunter-gatherer.

References

Borrero L (1989–90) Evolución cultural divergente en la Patagonia austral. Anales del Instituto de la Patagonia Ser. Cs. Hs. 19:133–139

Borrero L (1999) The prehistoric exploration and colonization of Fuego-Patagonia. J World Prehist 13(3):321–355

Borrero L (2001) El poblamiento de la Patagonia. Toldos, milodones y volcanes. Emecé editores, Buenos Aires

The manufacturer's authorised representative in the EU is Springer Nature Customer Service Centre GmbH, Europaplatz 3, 69115 Heidelberg, Germany. If you have any concerns regarding our products, please contact ProductSafety@springernature.com

Printed and bound by CPI Group (UK) Ltd, Croydon, CR0 4YY
30/09/2026
02225496-0007